ISNM
International Series of
Numerical Mathematics
Vol. 115

Edited by
K.-H. Hoffmann, München
H. D. Mittelmann, Tempe
J. Todd, Pasadena

Computational Optimal Control

Edited by

R. Bulirsch
D. Kraft

Springer Basel AG

Editors

Prof. Dr. R. Bulirsch
Mathematisches Institut
TH München
Postfach 20 24 20
D-80290 München
Germany

Prof. Dr. D. Kraft
Labor für Regelungs- und Steuerungstechnik
Fachbereich Maschinenbau
Fachhochschule München
Postfach 20 01 13
D-80001 München
Germany

A CIP catalogue record for this book is available from the Library of Congress, Washington D.C., USA

Deutsche Bibliothek Cataloging-in-Publication Data
Computational optimal control / ed. by R. Bulirsch ; D. Kraft.
– Basel ; Boston ; Berlin : Birkhäuser 1994
    (International series of numerical mathematics ; Vol. 115)
    ISBN 978-3-0348-9650-4   ISBN 978-3-0348-8497-6 (eBook)
    DOI 10.1007/978-3-0348-8497-6
NE: Bulirsch, Roland [Hrsg.]; GT

© 1994 Springer Basel AG
Originally published by Birkhäuser Verlag, P.O. Box 133, CH-4010 Basel, Switzerland in 1994
Softcover reprint of the hardcover 1st edition 1994
Camera-ready copy prepared by the editors
Printed on acid-free paper produced from chlorine-free pulp
Cover design: Heinz Hiltbrunner, Basel

9 8 7 6 5 4 3 2 1

Table of Contents

3 Algorithms for Optimal Control Calculations

4 Software for Optimal Control Calculations

5 Applications of Optimal Control

Preface

Resources should be used sparingly both from a point of view of economy and ecology. Thus in controlling industrial, economical and social processes, optimization is the tool of choice. In this area of applied numerical analysis, the INTERNATIONAL FEDERATION OF AUTOMATIC CONTROL (IFAC) acts as a link between research groups in universities, national research laboratories and industry. For this purpose, the technical committee *Mathematics of Control* of IFAC organizes biennial conferences with the objective of bringing together experts to exchange ideas, experiences and future developments in control applications of optimization. There should be a genuine feedback loop between mathematicians, computer scientists, engineers and software developers. This loop should include the design, application and implementation of algorithms. The contributions of industrial practitioners are especially important.

These proceedings contain selected papers from a workshop on CONTROL APPLICATIONS OF OPTIMIZATION, which took place at the Fachhochschule München in September 1992. The workshop was the ninth in a series of very successful biennial meetings, starting with the Joint Automatic Control Conference in Denver in 1978 and followed by conferences in London, Oberpfaffenhofen, San Francisco, Capri, Tbilisi and Paris. The workshop was attended by ninety researchers from four continents. This volume represents the state of the art in the field, with emphasis on progress made since the publication of the proceedings of the Capri meeting, edited by G. di Pillo under the title 'Control Applications of Optimization and Nonlinear Programming'.

This collection is comprised of five parts. First, the field is surveyed from several viewpoints. Then, theoretical aspects of optimal control and nonlinear programming are developed into implementable algorithms for optimal control calculations. The next step involves the software engineer providing an actual realization of an algorithm in the form of a piece of software. Such realizations are essential if optimal control is to be applied to various fields.

The treatment of large but structured problems is a continuing theme that links theory to practice via algorithms and software. J. T. Betts, P. E. Gill, W. Murray, M. A. Saunders and M. Steinbach contribute to this topic. Also the paper of G. Sonnevend on interior point methods for semi-infinite optimization problems belongs in this category.

A very important subject is the transfer of results in finite-dimensional nonlinear programming to infinite-dimensional optimization problems. E. W. Sachs gives examples of control applications of reduced sequential quadratic programming methods. The Russian authors (V. V. Dikoussar, A. V. Dmitruk, R. Gabasov et al.) contribute to continuation methods, singular extremals and adaptive control laws.

A rather new class of optimal control problems has differential-algebraic systems (DAE) as dynamical constraints. Recent results on high index systems of this kind are contributed by R.W.H. Sargent and his co-workers. Even higher complexity is found in the field of differential games, where two or more controllers (players) intend to influence the same goal to their respective benefit. N. D. Botkin and co-authors, B. Järmark, H. Bengtsson, and H. J. Pesch present recent applications and results.

The so-called Munich-School of Optimal Control (M. H. Breitner, R. Callies, B. Kugelmann, H. J. Pesch, and O. von Stryk and M. Schlemmer) is well-known for tackling large highly nonlinear complex problems in the field of aeronautics, biomechanics, chip design, economics, robotics, and vehicle dynamics.

Two major software efforts are discussed: ANDECS by G. Grübel and co-workers and OCCAL by R. Schöpf and P. Deuflhard. ANDECS is designed for the multi-objective treatment of closed-loop control systems. OCCAL is an implementation of an indirect method with semi-analytical and semi-numerical components. Variations on the theme of automatic differentiation have been given by A. Griewank (whose paper is not included) and R. Mehlhorn, M. Dinkelman, and G. Sachs.

Optimal Control has been greatly stimulated by men's steady desire to leave the Earth's atmosphere and travel in the direction of other planets. It's not surprising, therefore, that many applications arise from astronautics and aeronautics (e.g. K.H. Well and G. Sachs and their co-workers). Recently, the trajectory planning of robot paths has gained much attention. The reasons are manifold: higher productivity, greater accuracy, reduced energy consumption and less wear. A survey is given by F. L. Chernousko, and detailed technical calculations come from R. Callies, H. G. Bock, R. W. Longman, V. H. Schulz, M. Schlemmer, and O. von Stryk. Other very interesting applications arise in the field of biological, electrical, and mechanical systems (M. D. Ardema and H.-C. Chou, E. Arnold and H. Puta, D. Bestle and P. Eberhard, D. Claude and N. Nadjar, G. Ericsson and co-authors, K. Malinowski, W. Schenker and H. P. Geering, and Z. Schindler).

The Workshop has been generously supported by the 'Deutsche Forschungsgemeinschaft', the 'Bavarian Ministry of Economic Affairs and Transport', the 'Stifterverband für die Deutsche Wissenschaft', and several industrial companies.

The editors would like to thank K.-H. Hoffmann, H. D. Mittelmann, and J. Todd for accepting these proceedings to be published in their International Series of Numerical Mathematics (ISNM).

The editors would also like to acknowledge the work of the following colleagues who contributed invaluable efforts and ideas either as members of the International Program Committee or of the National Organizing Committee: H. G. Bock, J. F. Bonnans, A. E. Bryson, G. Feichtinger, H. P. Geering, T. Glad, R. Göhl, G. Grübel, J. Höcht, J. L. de Jong, F. M. Kirillova, D. Q. Mayne, A. Miele, H. J. Oberle, H. J. Pesch, G. di Pillo, G. Sachs, J. L. Speyer, I. Troch, K. H. Well.

Our thanks go to D. E. Knuth, L. Lamport, and T. Rokicki for putting TeX, LaTeX, and dvips into the public domain.

Finally, the continuing interest and support of Th. Hintermann of Birkhäuser Publishers is gratefully appreciated.

ROLAND BULIRSCH
DIETER KRAFT
München, Fall 1993

1 A Survey on Computational Optimal Control

International Series of Numerical Mathematics, Vol. 115, © 1994 Birkhäuser Verlag Basel

Issues in the Direct Transcription of Optimal Control Problems to Sparse Nonlinear Programs[*]

John T. Betts[†]

Introduction

Direct transcription of an optimal control problem into a sparse nonlinear programming problem requires analysis of the interaction between discretization, sparsity, and algorithm efficiency. The relative merits of various discretization techniques with regard to accuracy of the solution and efficiency of the sparse linear algebra and nonlinear program will be presented. Construction of gradient and Hessian information for the nonlinear program will be described. Issues affecting the nonlinear program algorithm strategy will be discussed.

A sparse nonlinear programming algorithm suitable for optimal control applications will be described. The method is based on a successive quadratic programming (SQP) technique, with quadratic programming subproblems solved using a Schur-complement approach, that exploits sparse symmetric indefinite linear algebra software. First and second derivative information is constructed using sparse finite differences. The cost of evaluating Hessian information is dictated by the number of differential equations (not the number of gridpoints), as well as the discretization formula. Finite difference Hessian information can be obtained for the trapezoidal discretization more efficiently than for higher order discretization formulas, and consequently is the preferred choice. Refined accuracy of the solution can be achieved by utilizing the method of deferred corrections in conjunction with efficient extrapolation techniques.

The implementation of these ideas will be illustrated in the solution of a continuous low-thrust orbit transfer problem. The dynamics of the vehicle have been implemented in equinoctial elements, and the transfer requires a significant orbital plane change achieved during a relatively long duration trajectory. The resulting problem involves 4352 variables, and 868 degrees of freedom, with over 3484 very nonlinear constraints.

---

[*]Invited Paper

[†]Applied Mathematics and Statistics Department, Research and Technology Division, Boeing Computer Services, P.O. Box 24346, MS 7L-21, Seattle, WA 98124-0346

The Direct Transcription Method

Preliminaries

The optimal control problem can be solved using a *direct transcription method* [1], [2], [5], [6], [7], [8], [11], [12], [14], [16]. Essentially the method can be described as follows:

1. Transcribe the optimal control problem into a nonlinear programming (NLP) problem by discretization

2. Solve the nonlinear program as follows:

   - Solve a sparse quadratic programming (QP) subproblem to estimate the NLP solution
   - If the solution estimate is acceptable–stop; otherwise update the NLP estimate and repeat the QP subproblem

3. Check the accuracy of the discretized solution

   - If the accuracy is acceptable–stop;
   - otherwise refine the discretization and return to step 1.

Transcription Into a Nonlinear Program

In general the optimal control problem requires finding the n_u-dimensional control vector $\boldsymbol{u}(t)$ to minimize the performance index $\phi[\boldsymbol{y}(t_f), t_f]$ evaluated at the final time t_f. The dynamics of the system are defined by the state equations

$$\dot{\boldsymbol{y}} = \boldsymbol{h}[\boldsymbol{y}(t), \boldsymbol{u}(t), t] \tag{1}$$

where y is the n_e dimension state vector. Initial conditions at time t_0 are defined by

$$\boldsymbol{\psi}[\boldsymbol{y}(t_0), \boldsymbol{u}(t_0), t_0] \equiv \boldsymbol{\psi}_0 = \boldsymbol{0}, \tag{2}$$

and terminal conditions at the final time t_f are defined by

$$\boldsymbol{\psi}[\boldsymbol{y}(t_f), \boldsymbol{u}(t_f), t_f] \equiv \boldsymbol{\psi}_f = \boldsymbol{0}. \tag{3}$$

In addition the solution must satisfy path constraints of the form

$$\boldsymbol{\Psi}_L \leq \boldsymbol{\Psi}[\boldsymbol{y}(t), \boldsymbol{u}(t), t] \leq \boldsymbol{\Psi}_U, \tag{4}$$

where $\boldsymbol{\Psi}$ is a vector of size n_p, as well as simple bounds on the state variables

$$\boldsymbol{y}_L \leq \boldsymbol{y}(t) \leq \boldsymbol{y}_U, \tag{5}$$

and control variables

$$\boldsymbol{u}_L \leq \boldsymbol{u}(t) \leq \boldsymbol{u}_U. \tag{6}$$

All transcription approaches divide the time interval into n_s segments

$$t_0 < t_1 < t_2 < \ldots < t_f = t_{n_s},$$

where the time points are referred to as mesh points, grid points, or nodes. Let us introduce the notation $y_j \equiv y(t_j)$ to indicate the value of the state variable at a grid point. In like fashion denote the control at a grid point by $u_j \equiv u(t_j)$. For the trapezoidal discretization, the NLP variables are

$$x = \left[y_0, u_0, y_1, u_1, \ldots, y_f, u_f, t_f\right]^\top . \tag{7}$$

The state equations are approximately satisfied by setting the *defects*

$$\zeta_j = y_j - y_{j-1} - \frac{1}{2}\kappa_j \left[h_j + h_{j-1}\right] \tag{8}$$

to zero for $j = 1, \ldots, n_s$. The step size is denoted by $\kappa_j \equiv t_j - t_{j-1}$, and the right hand side of the differential equations evaluated at grid point j is denoted by the vector h_j.

As a result of the transcription process the differential-algebraic system defining the optimal control problem is replaced by the NLP constraints

$$c_L \leq c(x) \leq c_U, \tag{9}$$

where

$$c(x) = \left[\zeta_1, \zeta_2, \ldots, \zeta_f, \psi_0, \psi_f, \Psi_0, \Psi_1, \ldots, \Psi_f\right]^\top \tag{10}$$

with

$$c_L = \left[0, \ldots, 0, \Psi_L, \ldots, \Psi_L\right]^\top \tag{11}$$

and a corresponding definition of c_U. The first $n_e n_s$ equality constraints require that the defect vectors from each of the n_s segments be zero thereby approximately satisfying the differential equations. The boundary conditions denoted by ψ are enforced directly as equality constraints, and nonlinear path constraints denoted by Ψ are imposed at the grid points. Note that nonlinear equality path constraints are accommodated by setting $\Psi_L = \Psi_U$

Solution of the Nonlinear Program

A successive quadratic programming (QP) approach is used to solve the nonlinear programming problem which results from transcription of the optimal control problem. Solution of the QP subproblem is used to define new estimates for the variables according to the formula

$$\bar{x} = x + \alpha p, \tag{12}$$

where the vector $\boldsymbol{p}$ is referred to as the *search direction*. The scalar α determines the step length and is typically set to one. The search direction $\boldsymbol{p}$ is found by minimizing the quadratic

$$\boldsymbol{g}^{\top}\boldsymbol{p} + \frac{1}{2}\boldsymbol{p}^{\top}\boldsymbol{H}\boldsymbol{p} \tag{13}$$

subject to the linear constraints

$$\boldsymbol{b}_{\ell} \leq \left[\begin{array}{c} \boldsymbol{G}\boldsymbol{p} \\ \boldsymbol{p} \end{array} \right] \leq \boldsymbol{b}_{u} \tag{14}$$

where $\nabla_{x}f(\boldsymbol{x}) = \boldsymbol{g}(\boldsymbol{x}) = \boldsymbol{g}$ is the N-dimensional gradient vector, and $\boldsymbol{G}$ is the $m \times N$ Jacobian matrix of constraint gradients and $\boldsymbol{H}$ is a symmetric $N \times N$ positive definite approximation to the Hessian matrix. The upper bound vector is defined by

$$\boldsymbol{b}_{u} = \left[\begin{array}{c} \boldsymbol{c}_{U} - \boldsymbol{c} \\ \boldsymbol{x}_{U} - \boldsymbol{x} \end{array} \right] \tag{15}$$

with a similar definition for the lower bound vector $\boldsymbol{b}_{\ell}$.

The positive definite Hessian matrix used in the QP subproblem is

$$\boldsymbol{H} = \boldsymbol{H}_{L} + \tau(1 + |\sigma|)\boldsymbol{I} \tag{16}$$

where the Hessian of the Lagrangian

$$\boldsymbol{H}_{L} = \nabla_{x}^{2}f - \sum_{i=1}^{m}\lambda_{i}\nabla_{x}^{2}c_{i}. \tag{17}$$

Since $\boldsymbol{H}_{L}$ is not necessarily positive definite, the Levenberg parameter τ is chosen such that $0 \leq \tau \leq 1$ and is normalized using the Gerschgorin bound σ for the most negative eigenvalue of $\boldsymbol{H}_{L}$. Quadratic convergence of the algorithm can only occur when $\tau = 0$, therefore, the parameter is adjusted at every iteration, and the rate of decrease is accelerated by monitoring the change in the norm of the projected gradient.

There are two issues which critically determine the overall efficiency of this solution process

1. Calculation of the Hessian matrix $\boldsymbol{H}$ and,

2. Solution of the sparse QP subproblem

These issues are addressed in the following sections.

Sparse Differences

The gradient and Hessian information required by the optimization algorithm can be computed efficiently using sparse finite differencing. Define the matrix of first derivatives of the vector $q = (c^\top, f)^\top$ by

$$D = \begin{bmatrix} G \\ g^\top \end{bmatrix}, \tag{18}$$

Sparse differencing requires partitioning the column indices of D into subsets Γ^k. Each subset of the columns of D has the property that there is at most one nonzero element in each *row*. Then define the perturbation direction vector by

$$\Delta^k = \sum_{j \in \Gamma^k} \delta_j e_j \tag{19}$$

where δ_j is the perturbation size for variable j, and e_j is a unit vector in direction j. Then for each nonzero row i of a column $j \in \Gamma^k$, the central difference approximation is

$$D_{ij} \approx \frac{1}{2\delta_j}[q_i(x + \Delta^k) - q_i(x - \Delta^k)]. \tag{20}$$

Denote the total number of index sets Γ^k needed to span the columns of D by γ. Then, a central difference estimate requires evaluations at 2γ perturbed points. Second derivatives can be computed using the *same* index sets. Thus, *all* first and second derivative information can be computed using $\gamma(\gamma + 3)/2$ perturbations. Specifically, it is necessary to evaluate the quantities $q(x + \Delta^\imath)$ for $\imath = 1, \ldots, \gamma$; $q(x - \Delta^\imath)$ for $\imath = 1, \ldots, \gamma$; and $q(x + \Delta^\imath + \Delta^\jmath)$ for $\imath = 1, \ldots, \gamma$, $\jmath = 1, \ldots, \gamma$, with $\jmath > \imath$. Both the Jacobian and Hessian matrix which result from the transcription of an optimal control problem are very sparse, and for a sparse matrix the number of index sets can be significantly less than the number of variables, i.e. $\gamma \ll N$. Consequently the cost of evaluating a Jacobian using sparse differences can be substantially less than standard finite difference methods.

Schur-Complement QP

The efficient solution of a sparse quadratic program can be achieved using a method proposed by Gill, Murray, Saunders, and Wright [13]. The method derives its efficiency from two facts. First, the large sparse symmetric indefinite KT system

$$\begin{bmatrix} H & A^\top \\ A & 0 \end{bmatrix} \begin{bmatrix} -p \\ \pi \end{bmatrix} \equiv K_0 \begin{bmatrix} -p \\ \pi \end{bmatrix} = \begin{bmatrix} g \\ 0 \end{bmatrix}. \tag{21}$$

is factored only *once* using a very efficient *multifrontal* algorithm for symmetric indefinite systems [3]. Second, subsequent changes to the QP active set can be computed using a *solve* with the previously factored KT matrix K_0 and a *solve* with a small dense "Schur-complement" matrix. Since the factorization of the KT matrix is significantly more expensive than the "solve" operation, the overall method is quite effective.

The Impact of the Hessian Matrix

The manner in which the Hessian matrix is computed and used has a significant impact on the overall transcription method, as well as the sparse nonlinear programming algorithm.

When sparse finite differences are used to compute the Hessian the cost is proportional to the square of the number of index sets γ^2. In [7] it is shown that the number of index sets γ is determined by two things:

1. the number of state and control variables and,

2. the discretization method.

The number of index sets for the trapezoidal method is approximately half the number for the Hermite-Simpson discretization scheme. When compared to a Runge-Kutta discretization, the trapezoidal method requires approximately half the number of evaluations of the differential equations. Therefore, to reduce the cost of evaluating the Hessian, the preferred discretization method is trapezoidal.

Because the number of perturbations needed to compute the Hessian are determined by the number of state and control variables, they are not related to the number of mesh points. Thus the overall approach is limited by the number of dynamic variables, i.e. state and control, but not the number of NLP variables. In contrast, many methods for solving large sparse nonlinear programming problems rely on having a small number of degrees of freedom in the problem.

Because the trapezoidal method is the discretization technique preferred to reduce the Hessian evaluation cost, it is necessary to address the approximation error for this approach. In particular, it is important to consider mesh refinement techniques to avoid introducing a large number of grid points, and therefore NLP variables. This issue is addressed in [10].

Since a finite difference estimate of the Hessian computed at an arbitrary point is not necessarily positive definite, it may be necessary to modify the Hessian in order to provide a well posed QP subproblem. This is accomplished using a Levenberg parameter, which is updated from iteration to iteration using an accelerated trust region strategy. Ultimately, when the Hessian can be used without modification, the iterations converge quadratically. Quadratic convergence is very important for large NLP problems, since the convergence rate is not proportional to either the number of variables and/or the number of degrees of freedom. In contrast, Hessian approximations generated using a recursive quasi-Newton update yield algorithms whose convergence rate is dictated by the problem size.

Finally, the accuracy of the Hessian at an arbitrary point is affected by the Lagrange multiplier estimates. For nonlinearly constrained problems the multiplier estimates are more accurate near the active constraint surface. Therefore, the preferred NLP strategy is to locate a feasible point, and then stay "close" to feasibility during the course of optimization. In so doing the multiplier estimates are reasonable, which in turn yields an accurate Hessian approximation, and ultimately accelerated convergence.

The Low Thrust Orbit Transfer Problem

Preliminaries

The issues discussed in the previous section can be illustrated by solving a low thrust orbit transfer problem [9]. The equations of motion for a vehicle expressed in an ECI (Earth Centered Inertial) coordinate system are:

$$\dot{r} = v \tag{22}$$

$$\dot{v} = g(r) + \frac{T}{m}e \tag{23}$$

$$\dot{w} = \frac{T}{I_{sp}} \tag{24}$$

where r is the position vector, v is the velocity vector, g is the gravitational acceleration vector, e is a unit vector defining the thrust direction in the ECI frame where T/m is the acceleration due to the thrust T. The weight of the vehicle is denoted by w, the mass by m, and the specific impulse of the motor by I_{sp}. When the thrust on the vehicle is zero, and a spherical earth model is used, the equations of motion can be solved analytically. In this case the orbit is completely described by a constant set of orbital elements. When the perturbing forces on the body are small relative to the dominant central body gravitational force it is appealing to represent the equations of motion using a variation of parameters approach. Specifically, [15] proposes replacing the translational equations of motion with

$$\dot{z} = M[z(t), t]f + \overline{m}[z] \tag{25}$$

where the equinoctial elements $z \equiv (a, h, k, p, q, F)^{\top}$ are uniquely defined functions of the position and velocity (r, v). Computation of the 6×3 matrix $M[z(t), t]$ and the vector $\overline{m}^{\top} = (0, 0, 0, 0, 0, m_6)^{\top}$ is summarized in [9]. The perturbing acceleration

$$f = \hat{u}\frac{T}{m} + \Delta g \tag{26}$$

consists of a propulsive acceleration term defined by its magnitude T/m and unit direction vector $\hat{u}$ *in the equinoctial frame*, as well as a gravitational acceleration perturbation. Notice that f must include all contributions to the acceleration except the gravitational acceleration due to a spherical earth. As such, oblate earth effects, as well as thrust contributions can be accommodated. Observe that when there are no perturbing forces, $f = 0$, which implies that the first five equinoctial elements are constant.

The perturbing forces must be expressed in the equinoctial coordinate frame defined by the following three unit vectors:

$$\hat{f} = \frac{1}{(1 + p^2 + q^2)} \begin{pmatrix} 1 - p^2 + q^2 \\ 2pq \\ -2p \end{pmatrix} \tag{27}$$

$$\hat{g} = \frac{1}{(1+p^2+q^2)} \begin{pmatrix} 2pq \\ 1+p^2-q^2 \\ 2q \end{pmatrix} \tag{28}$$

$$\hat{w} = \frac{1}{(1+p^2+q^2)} \begin{pmatrix} 2p \\ -2q \\ 1-p^2-q^2 \end{pmatrix} \tag{29}$$

Thrust in the Equinoctial Frame

For engineering purposes, it is convenient to define the thrust direction relative the inertial velocity vector. That is, define

$$\boldsymbol{e} = u_x\hat{\boldsymbol{i}} + u_y\hat{\boldsymbol{j}} + u_z\hat{\boldsymbol{k}} \tag{30}$$

in a rotating frame where the unit vectors are defined by

$$\hat{\boldsymbol{i}} = \frac{\boldsymbol{v}}{\|\boldsymbol{v}\|} \tag{31}$$

$$\hat{\boldsymbol{j}} = \frac{\boldsymbol{r} \times \boldsymbol{v}}{\|\boldsymbol{r} \times \boldsymbol{v}\|} \tag{32}$$

$$\hat{\boldsymbol{k}} = \hat{\boldsymbol{i}} \times \hat{\boldsymbol{j}} \tag{33}$$

Notice that the primary axis is oriented along the velocity vector, so that for a coplanar transfer it is reasonable to expect the thrust to be primarily in the direction defined by $\hat{\boldsymbol{i}}$, that is $\boldsymbol{e} = (1,0,0)^{\top}$.

The position and velocity vector are given by

$$\boldsymbol{r} = X\hat{\boldsymbol{f}} + Y\hat{\boldsymbol{g}} \tag{34}$$

and

$$\boldsymbol{v} = \dot{X}\hat{\boldsymbol{f}} + \dot{Y}\hat{\boldsymbol{g}} \tag{35}$$

where the coefficients X, Y, $\dot{X}$, and $\dot{Y}$ are functions of the equinoctial elements, defined in [9]. Consequently it can be shown that

$$\hat{\boldsymbol{i}} = a_1\hat{\boldsymbol{f}} + a_2\hat{\boldsymbol{g}} \tag{36}$$

$$\hat{\boldsymbol{j}} = \hat{\boldsymbol{w}} \tag{37}$$

$$\hat{\boldsymbol{k}} = a_2\hat{\boldsymbol{f}} - a_1\hat{\boldsymbol{g}} \tag{38}$$

where $a_1 = \dot{X}(\dot{X}^2+\dot{Y}^2)^{-\frac{1}{2}}$, and $a_2 = \dot{Y}(\dot{X}^2+\dot{Y}^2)^{-\frac{1}{2}}$. Then combining (30), (36), (37), and (38) gives

$$\boldsymbol{e} = (u_x a_1 + u_z a_2)\hat{\boldsymbol{f}} + (u_x a_2 - u_z a_1)\hat{\boldsymbol{g}} + u_y\hat{\boldsymbol{w}} \tag{39}$$

Oblate Gravity Perturbations in Equinoctial Frame

Perturbations to the spherical earth model can be incorporated provided they are also expressed in the equinoctial coordinate frame. Oblate gravity models are typically defined in a local horizontal reference frame. Let us define the unit vectors

$$\tilde{\boldsymbol{k}} = \frac{-\boldsymbol{r}}{\|\boldsymbol{r}\|} \tag{40}$$

$$\tilde{\boldsymbol{i}} = \frac{\check{\boldsymbol{i}}}{\|\check{\boldsymbol{i}}\|} \tag{41}$$

where

$$\check{\boldsymbol{i}} = \begin{pmatrix} 0 \\ 0 \\ 1 \end{pmatrix} - \tilde{k}_3 \tilde{\boldsymbol{k}} \tag{42}$$

Then the gravitational perturbation force is given by

$$\boldsymbol{g} = g_x \tilde{\boldsymbol{i}} + g_z \tilde{\boldsymbol{k}}. \tag{43}$$

In the equinoctial frame the gravitational perturbation is given by

$$\boldsymbol{g} = \alpha_1 \hat{\boldsymbol{f}} + \alpha_2 \hat{\boldsymbol{g}} + \alpha_3 \hat{\boldsymbol{w}} \tag{44}$$

where by comparing (43) and (44) we find

$$\alpha_1 = g_x \tilde{\boldsymbol{i}}^\top \hat{\boldsymbol{f}} + g_z \tilde{\boldsymbol{k}}^\top \hat{\boldsymbol{f}} \tag{45}$$

$$\alpha_2 = g_x \tilde{\boldsymbol{i}}^\top \hat{\boldsymbol{g}} + g_z \tilde{\boldsymbol{k}}^\top \hat{\boldsymbol{g}} \tag{46}$$

$$\alpha_3 = g_x \tilde{\boldsymbol{i}}^\top \hat{\boldsymbol{w}} + g_z \tilde{\boldsymbol{k}}^\top \hat{\boldsymbol{w}} \tag{47}$$

If the tesseral harmonics are ignored and only the zonal harmonics are included in the geopotential function the gravitational accelerations (cf [4]) are given by

$$g_x = -\frac{\mu \cos \phi}{r^2} \sum_{k=2}^{\infty} \left(\frac{a_e}{r}\right)^k P_k' J_k \tag{48}$$

$$g_z = -\frac{\mu}{r^2} \sum_{k=2}^{\infty} (k+1) \left(\frac{a_e}{r}\right)^k P_k J_k \tag{49}$$

where ϕ is the geocentric latitude, μ the gravitational constant, a_e is the equatorial radius of the earth, $r = \|\boldsymbol{r}\|$ is the radius, $P_k(\sin \phi)$ is the k-th order Legendre polynomial with corresponding derivative P_k', and the zonal harmonic coefficients are given by J_k.

Boundary Conditions

The preceding sections have defined the equations of motion which determine the vehicle trajectory. It is necessary to specify boundary conditions to complete a meaningful definition of the problem. Typically the initial and final orbits are specified in terms of classical orbit elements. To illustrate the techniques, we consider a transfer from a low altitude circular park orbit to a highly inclined (Molniya) eccentric orbit. The transfer is initiated at the ascending node of a circular, 150 nm altitude park orbit with inclination 28.5 deg. ($a = 21837080$ ft, $e = 0$, $i = 28.5$ deg). The desired final orbit has an inclination $i = 63.4$ deg, argument of perigee $\omega = 270$ deg, apogee altitude of 21450 nm, and perigee altitude of 350 nm ($a = 87155322$ ft, $e = 0.73550321$). The longitude of ascending node for the park orbit can be chosen arbitrarily without altering the results as long as the longitude of ascending node for the mission orbit is unspecified. For convenience the park orbit is fixed at $\Omega = -180$ deg.

The eccentricity is related to the equinoctial elements according to

$$e = \sqrt{h^2 + k^2} \tag{50}$$

and the argument of perigee is given by

$$\omega = \arctan\left(\frac{h}{k}\right) - \arctan\left(\frac{p}{q}\right). \tag{51}$$

The eccentricity constraint can be imposed directly, however it is desirable to reformulate the argument of perigee constraint. In particular, the angles computed by the inverse tangent function are defined on a specific range (e.g. $-\pi \leq \theta \leq \pi$), and consequently the function does not behave smoothly in the neighborhood of $\pm\pi$. Taking the tangent of both sides, and then using the formula for the difference of two angles it can be shown that

$$\sin\omega = \frac{N}{\sqrt{N^2 + D^2}} \tag{52}$$

and

$$\cos\omega = \frac{D}{\sqrt{N^2 + D^2}} \tag{53}$$

where $N = hq - pk$ and $D = kq + hp$. Therefore, if $\omega = 270$ deg., $\cos\omega = 0$, and $\sin\omega = -1 \leq 0$, which can be enforced by setting

$$D = kq + hp = 0 \tag{54}$$

and

$$N = hq - pk \leq 0 \tag{55}$$

Notice also that it is preferable to set $\cos\omega = 0$ as an equality constraint because at the root the *slope* of this constraint is nonzero. Imposing the constraint be setting

$\sin \omega = -1$ is not recommended because at the root, the slope of this constraint is zero, and linear convergence can be expected from the optimization algorithm. The inclination is given by the expression

$$i = 2 \arctan \sqrt{p^2 + q^2}. \tag{56}$$

As with the argument of perigee constraint, discontinuities caused by the inverse tangent function are avoided by rewriting the constraint

$$\tan \frac{i_d}{2} = \sqrt{p^2 + q^2} \tag{57}$$

where i_d is the desired value of the inclination.

Low Thrust Optimal Control Problem

The results of the preceding subsections can be combined to produce an expression for the total perturbing force in the equinoctial frame. The components of $\hat{u}$ follow by inspection from (39). Similarly the components of Δg are defined by (44). The state variables for the problem are $y(t) = (z, w)^\top = (a, h, k, p, q, F, w)^\top$, and the control variables are the thrust direction vector $u(t) = (u_x, u_y, u_z)$ and the parameter T. The objective is to maximize the final weight, i.e. minimize the performance index

$$\phi = -y_7(t_f) = -w(t_f) \tag{58}$$

evaluated at the final time t_f. During the transfer $0 \leq t \leq t_f$, the dynamics are defined by the system of differential-algebraic equations

$$\dot{z} = M[z] \left[\frac{T}{m} \begin{pmatrix} u_x a_1 + u_z a_2 \\ u_x a_2 - u_z a_1 \\ u_y \end{pmatrix} + \begin{pmatrix} \alpha_1 \\ \alpha_2 \\ \alpha_3 \end{pmatrix} \right] + \overline{m}[z] \tag{59}$$

$$\dot{w} = \frac{T}{I_{sp}} \tag{60}$$

$$0 = \|u\| - 1. \tag{61}$$

Initial conditions at time t_0 are defined for all state variables, that is $y(t_0) \equiv y_0$. The terminal value for a is defined by the mission orbit, and the values for the other state variables must satisfy the mission orbit constraints (50), (54), (55), and (57).

The manner in which the thrust is formulated is significant in two ways. First, the magnitude of the thrust is not allowed to become zero. It is well known (cf. [17]) that if throttling is permitted the optimal solution will approximate an impulsive orbit transfer, with a large number of burns made on successive orbit revolutions. To preclude trajectories which Zondervan refers to as 'geometrically similar' to impulsive transfers, the formulation is designed to use a single burn. Second, the direction of the thrust is specified by treating the three components of the vector

$\boldsymbol{u}$ as control variables, and then imposing the constraint $\|\boldsymbol{u}(t)\| = 1$. A more common alternative is to simply specify the direction of the thrust vector by two angles, e.g. a yaw and pitch in some coordinate frame. While the latter approach is intuitively appealing it leads to an undesirable optimization formulation. The angular formulation leads to multiple non-unique solutions, since the angles $\theta = \theta_1 \pm 2k\pi$ all yield the *same* thrust direction vector for $k = 0, 1, 2, \dots$. In contrast, there is a *unique* set of control variables $\boldsymbol{u}$ corresponding to any thrust direction.

Computational Results

Solutions to the problem were obtained using two, four, six, and eight orbital revolutions with an initial guess for the variables constructed by linearly interpolating between the initial and final time. Details of the solution are given in [9]. A perspective view of the transfer is shown in Figure 1.

Ref. Iter.	1	2	Total
N	1502	4352	–
NDOF	298	868	–
m	1204	3484	–
NGRID	150	435	–
NRHS	1658550	314940	1973490
$w(t_f)$	.22029	.22019	–
NFE	11057	724	11781
NFC	641	10	651
NGC	227	11	238
NHC	98	7	105
Time (sec)	2568.55	839.53	3408.08

Table 1: Optimization Iteration Summary

Table 1 presents more detailed information on the sparse nonlinear programming algorithm performance. For this case two mesh refinement iterations were run in order to obtain a relative error of 10^{-4}. The corresponding NLP data is shown in columns 2 and 3 of the table, and column 4 contains totals for the run. The NLP problem size is defined by the number of variables N, the number of degrees of freedom $NDOF$, and the number of active constraints m. The trapezoidal transcription method is defined by the number of grid points $NGRID$, and the number of times the right hand side of the differential equations were evaluated $NRHS$. The total number of function evaluations needed to solve the problem is given by NFE, where a "function evaluation" consists of a single evaluation of the objective and constraint functions. The number of function calls needed

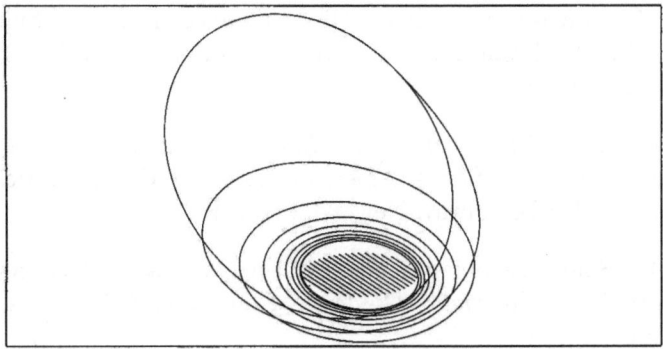

Figure 1: Orbit Transfer (8 revs)

for the NLP line search is given by NFC, and the number of Jacobian/gradient evaluations is given by NGC. The number of Hessian evaluations is tabulated as NHC. The total run time required to solve the NLP on a Sun IPX workstation is listed in the final row of the table.

Summary and Conclusions

This paper describes a number of issues in the direct transcription method for solving optimal control problems which determine algorithm efficiency. The interaction between discretization technique, its impact on sparsity, and ultimately on the design of a nonlinear programming algorithm are discussed. The methods are illustrated in the solution of a low thrust orbit transfer problem. Vehicle dynamics are defined using equinoctial coordinates, and the resulting optimal control problem requires determination of the thrust pointing direction for the duration of the trajectory. Solutions were obtained with oblate earth perturbations, for trajectories of up to eight revolutions of the earth. The resulting nonlinear programming problem required the determination of 4352 variables, with over 3484 very nonlinear constraints.

References

[1] Ascher, U., Christiansen, J., and Russell, R., 'Collocation Software for Boundary-Value ODEs,' *ACM Transactions on Mathematical Software*, Vol. 7, No. 2, June 1981, pp. 209–222.

[2] Ascher, U., Mattheij, R., and Russell, R.D., *Numerical Solution of Boundary Value Problems for Ordinary Differential Equations*, Prentice Hall, Englewood Cliffs, N.J., 1988.

[3] Ashcraft, C.C., and Grimes, R.G., 'The Influence of Relaxed Supernode Partitions on the Multifrontal Method,' Boeing Computer Services Technical Report ETA-TR-60, 1988.

[4] Battin, R.H. *An Introduction to the Mathematics and Methods of Astrodynamics*, AIAA Education Series, American Institute of Aeronautics and Astronautics, Inc., 1633 Broadway, New York, NY 10019.

[5] Betts, J.T., 'Sparse Jacobian Updates in the Collocation Method for Optimal Control Problems,' *Journal of Guidance, Control, and Dynamics*, Vol. 13, No. 3, May–June, 1990.

[6] Betts, J.T., and Huffman, W.P., 'Trajectory Optimization on a Parallel Processor,' *Journal of Guidance, Control, and Dynamics*, Vol. 14, No. 2, March–April, 1991.

[7] Betts, J.T., and Huffman, W.P., 'The Application of Sparse Nonlinear Programming to Trajectory Optimization,' *Journal of Guidance, Control, and Dynamics*, Vol. 15, No. 1, Jan–Feb, 1992.

[8] Betts, J.T., and Huffman, W.P., 'Path Constrained Trajectory Optimization Using Sparse Sequential Quadratic Programming,' *Journal of Guidance, Control, and Dynamics*, Vol. 16, No. 1, Jan–Feb, 1993, pp. 59–68.

[9] Betts, J.T., 'Using Sparse Nonlinear Programming to Compute Low Thrust Orbit Transfers,' *Journal of the Astronautical Sciences*, to appear.

[10] Betts, J.T. and Huffman, W.P., 'Accuracy Refinement in Transcription Methods for Optimal Control,' BCSTECH-93-006, Boeing Computer Services, P.O. Box 24346-0346, MS 7L-68, Seattle, WA 98124.

[11] Dickmanns, E.D., 'Efficient Convergence and Mesh Refinement Strategies for Solving General Ordinary Two-Point Boundary Value Problems by Collocated Hermite Approximation,' 2nd IFAC Workshop on Optimisation, Oberpfaffenhofen, Sept. 15–17, 1980.

[12] Enright, P.J., and Conway, B.A. 'Optimal Finite-Thrust Spacecraft Trajectories Using Collocation and Nonlinear Programming,' *Journal of Guidance, Control, and Dynamics*, Vol. 14, No. 5, 1991, pp. 981–985.

[13] Gill, P.E., Murray, W., Saunders, M.A., and Wright, M.H., 'A Schur-Complement Method for Sparse Quadratic Programming,' Report SOL 87-12, Department of Operations Research, Stanford University.

[14] Hargraves, C.R., and S.W. Paris, 'Direct Trajectory Optimization Using Nonlinear Programming and Collocation,' *J. of Guidance, Control, and Dynamics*, Vol. 10, No.4, July–Aug, 1987, p. 338.

[15] Kechichian, J.A., 'Trajectory Optimization with a Modified Set of Equinoctial Orbit Elements,' Paper AAS 91-524, AAS/AIAA Astrodynamics Specialist Conference, Durango, CO, Aug. 19–22, 1991

[16] Kraft, D., 'On Converting Optimal Control Problems Into Nonlinear Programming Problems,' NATO ASI Series Vol. F15, *Computational Mathematical Programming*, Springer-Verlag, 1985.

[17] Zondervan, K.P., Wood, L.J., and Caughy, T.K, 'Optimal Low-Thrust, Three-Burn Orbit Transfers with Large Plane Changes,' *The Journal of the Astronautical Sciences*, Vol. 32, No. 3, July–Sept., 1984, pp. 407–427.

International Series of Numerical Mathematics, Vol. 115, © 1994 Birkhäuser Verlag Basel

Optimization in Control of Robots*

F.L. Chernousko[†]

Abstract

Optimization of control algorithms and parameters of industrial robots is regarded as an effective means for improving their operational characteristics. Various possible versions of optimization problems for robots were considered in the scientific literature over the last 20 years. The paper presents a brief survey of research carried out in this field.

Introduction

A manipulation robot is a multi-purpose machine able to carry out various motions depending on the chosen control programs. When planning a technological process involving a robot one is to choose the motion of the manipulator for accomplishing a prescribed operation. A natural and reasonable approach to this choice is to optimize the motion of the manipulator with respect to some performance criterion that depends on parameters of the robot, a technological task, and environment. Thus, we come to optimal control problems for robots. The number of scientific publications on optimization in robotics is large and grows rapidly. These papers differ in mathematical models of robots, choice of constraints and performance indices, and methods of solution. In this paper we present a brief survey of research carried out in this field (see also [11, 17]).

Mathematical Models and Optimal Control Problems

The dynamics of robots can be described by Lagrangian equations

$$\frac{d}{dt}\frac{\partial T}{\partial \dot{q}_i} - \frac{\partial T}{\partial q_i} = P_i(q, \dot{q}) + Q_i, \quad i = 1, \ldots, n \tag{1}$$

Here, q_i are generalized coordinates of the robot, T is the kinetic energy, P_i are generalized forces including gravity, friction, etc., and Q_i are control forces. The kinetic energy is a quadratic form

$$T = \tfrac{1}{2}(A(q)\dot{q}, \dot{q}) \tag{2}$$

---

*Invited Paper

[†]Institute for Problems in Mechanics, Russian Academy of Sciences, Vernadskogo 101, Moscow, 117526, Russia

where $A(q)$ is a symmetric positive-definite $n \times n$-matrix. The forces P_i can be specified functions of coordinates and velocities, and may include also some uncertain bounded disturbances. The control Q_i are torques and/or forces for revolute and/or prismatic joints, respectively. They are subject to constraints

$$|Q_i| \leq Q_i^0, \quad i = 1, \ldots, n \tag{3}$$

where Q_i^0 are given constants. Substituting (2) into (1), we obtain

$$\begin{aligned} A(q)\ddot{q} + b(q, \dot{q}) &= Q \\ q &= (q_1, \ldots, q_n), \\ Q &= (Q_1, \ldots, Q_n) \end{aligned} \tag{4}$$

Here, the vector b includes all forces (except controls Q) as well as inertial terms of equations (1).

In general terms, the optimal control problem for robots implies finding the controls Q_i that satisfy the imposed constraints and drive the system (4) from an initial state

$$q_i(t_0) = q_i^0, \qquad \dot{q}_i(t_0) = (\dot{q}_i)^0 \tag{5}$$

to the prescribed terminal state

$$q_i(t_*) = q_i^*, \qquad \dot{q}_i(t_*) = (\dot{q}_i)^* \tag{6}$$

so that the given functional (performance index) J attains its minimal (maximal) possible value.

The imposed constraints may include, besides (3), some other constraints related to robot configuration, avoidance of obstacles, restrictions on energy consumption, etc. The performance index J is usually either time ($J = t_* - t_0$) or energy consumption.

By solving the optimal control problem for a given initial state (5), we obtain the open-loop optimal controls $Q_i(t)$. Much more difficult task is to find the closed-loop (feedback) controls $Q_i(q, \dot{q})$ which can be applied in case of any initial or current states.

The model described by equation (1) is purely mechanical and does not take into account equations of drives. More detailed models include equations of electric or hydraulic drives. Under some reasonable simplifications, equations describing a robot with electric DC drives can be reduced to the form of Lagrangian equations (4). In some cases, for sufficiently accurate models, it is also necessary to take into account the elastic compliance of robot joints and/or links. Thus, there exists a wide variety of optimal control problems for robots.

Optimal Motion with a Prescribed Gripper Trajectory

Here, the problem is stated as follows: find the control law that drives the gripper of a manipulator along a prescribed trajectory from a given initial position to a given terminal one and minimizes a given performance index. This problem is natural for technological operations that require the gripper to move along the prescribed trajectory with a definite orientation at each point. Such requirements are typical for welding, cutting, inspecting and other operations. The basic idea here is to reduce the problem to an optimal control problem for a single-degree-of-freedom system. The length of the trajectory arc is usually considered as a sole generalized coordinate.

In [10] the time-optimal control algorithm for a multi-link non-redundant manipulator is developed. The initial problem is reduced to optimal control of a point mass driven along the straight line by a control force under constraints imposed on the state and control. Under certain assumptions, at each time instant, the acceleration of the gripper assumes either maximal or minimal admissible values. The algorithm for determining the switching points of the optimal control is suggested. Another technique for treating the same problem is used in [43]. The described approach is widespread in scientific publications on robotics. In [9, 22, 27, 52], the algorithm proposed in [10] is used for more complicated optimization problems, in particular, for trajectory optimization under collision avoidance conditions.

In some papers, trajectories of special kind and performance indices other than time are considered. For example, the paper [37] deals with trajectories consisting of straight segments and circular arcs. In [53], instead of a continuous trajectory, certain points are specified which the robot must pass through during its motion. The combined integral functional taking into account the time, the averaged kinetic energy and the driving torques is considered in [44].

Optimization of Manipulator Trajectories

This direction is connected with the conventional optimal control problem stated as follows: find the control driving a manipulator from the state (5) to the state (6) and minimizing the prescribed cost functional. In the general case, both the state and control variables are constrained. Here, the trajectory is not specified beforehand but is determined as a result of the optimization. Below, we discuss three approaches to the optimization of robot trajectories, namely the direct use of optimal control methods, the parametric optimization, and the separation of motions.

Direct use of optimal control methods. This way is most efficient for systems with a small number of degrees of freedom. General mathematical theorems establishing the existence and structure of optimal control in different robotic motions are proved in [1, 16].

A number of publications are devoted to the time-optimal control of a two-link planar manipulator. In [3, 13, 14, 21, 23, 39, 41, 56, 58], time-optimal motions for a two-link manipulator are calculated numerically by methods based on Pontryagin's maximum principle. Note that in [39, 41, 56], the optimal control ensuring the point-to-point motion of the gripper is found together with the optimal placement of the robot itself.

The authors of [42] consider cylindrical- and spherical-coordinate robots controlled by restricted forces and torques. The possibility of a chattering time-optimal control law with infinite number of switches is established. In [15], such chattering controls are rigorously studied in detail.

A considerable number of publications deal with the numerical calculation of optimal motions of manipulation robots. These publications contain the description of computational algorithms used for solving optimal control problems (see, e.g. [45, 54]) as well as numerous examples of optimal trajectories for robots of various kinematic structures (see [23, 25, 33, 34, 50]). Some optimal control problems regarding the energy consumption for the motion of a manipulator are considered in, e.g., [5, 51].

Parametric optimization. In this approach, the initial admissible set of controls is replaced by a more narrow one, namely, by a parametric family with a comparatively small number of parameters. Hence, the optimal control problem is reduced to the minimization of a function of several variables. This approach is used quite often because it simplifies the computational procedure substantially. Some authors seek optimal control in the class of bang-bang functions, the switch instants being regarded as the parameters to be found (see, e.g. [7, 21, 31, 38]). Ritz's method is used in [36, 49]. The optimal trajectory is presented as a linear combination of some basic functions with coefficients determined from the optimality conditions. A number of authors use the parametric optimization combined with the method [10] related to a prescribed trajectory. In this approach, the parametric family of trajectories is specified, and for each trajectory the optimal control is obtained. The parameters are determined that correspond to the minimal value of the performance criterion, see [9, 22].

Separation of motions. In this approach, the original nonlinear system of coupled equations is replaced by a more simple system consisting of independent subsystems. The separation of motions can be achieved by an appropriate change of variables in the equations of motion, by ignoring small terms, by using asymptotic perturbation methods, by an appropriate choice of control and also by a combination of these techniques.

The authors of [28] consider the time-optimal control problem for a three-link anthropomorphic manipulator. They linearize the equations of motion in the vicinity of the terminal state and then, by using an appropriate change of variables, reduce the initial problem to three independent optimal control problems for single-degree-of-freedom systems. After that, the optimal control is obtained analytically in the feedback form.

In [18–20], a new method of control is proposed for a general Lagrangian system with controlled degrees of freedom. The system is reduced to the set of linear subsystems, the nonlinear and coupling terms being regarded as disturbances. If the level of the disturbances for each degree of freedom is less than the maximal admissible value of the corresponding control, the feedback control law is obtained that brings the original system to the terminal state in finite time.

The control is based on game theory, it is time-suboptimal and robust. If the disturbances due to the dynamical interaction of different degrees of freedom are small, they are often ignored when constructing the control. A limiting level for the disturbances is established that still guarantees the arrival of the system at the desired state for such control laws.

To conclude this section, we mention some publications on optimal trajectories for manipulation robots of most used designs. Different approaches (analytical and numerical) to time-optimal control for a two-link anthropomorphic manipulator are discussed in [21]. Both open-loop and feedback controls are obtained. In a series of papers [33, 34, 50], time-optimal trajectories and open-loop controls are constructed and investigated numerically for an anthropomorphic manipulator [33], a polar coordinate manipulator [34], and a SCARA-type robot [50]. Time-optimal and near-optimal controls for a cylindrical-coordinate manipulator with irreversible gear trains are constructed in [4]. The authors of [8] calculated time-optimal transport motions of a gantry robot. The authors of [52, 60] constructed optimal trajectories of manipulation robots in the presence of obstacles in their working zones.

The papers [2, 5, 6, 12, 46–48] are devoted to optimal control of manipulation robots with electric drives.

Optimization of Redundant Manipulators

If the number of degrees of freedom of a manipulator exceeds that of the manipulation object, the manipulator is called kinematically redundant. For such robots an infinite number of the manipulator's configurations match each position of the manipulation object, and many trajectories in the configuration space of the manipulator correspond to each trajectory of the gripper in the physical space. The kinematic redundancy makes a robot more versatile and allows to avoid collisions with obstacles by the choice of appropriate configurations. It also implies the possibility to optimize the manipulator motion, provided that the motion of the gripper is prescribed.

In the majority of publications, the optimization problem for redundant manipulators is stated as follows: given a prescribed motion of the robot's gripper, find the motion of the manipulator in order to minimize either a scalar function of the state variables at each time instant (the local optimization) or a functional that depends on the motion as a whole (the global optimization). The local optimization is considered, for example, in [24, 26, 35, 55, 57, 59]. As performance criteria, the sum of squared generalized velocities of the manipulator or its kinetic energy are used. The global optimization is treated, for example, in [29, 30, 32, 40].

Time-optimal motions of kinematically redundant manipulators are considered in [3] under the assumption that each degree of freedom moves independently, according to the prescribed law. This assumption, verified experimentally for certain types of robots, entails that the motion of the robot as a whole is determined, if the initial and terminal configurations of the manipulator are specified. Optimal configurations and motions of kinematically redundant robots are obtained.

Conclusion

Since robots are highly nonlinear systems with many degrees of freedom, optimal control problems for robots are very difficult. Most papers devoted to these problems deal only with open-loop optimal controls. These optimal motions are useful because they help to estimate possibilities of robotic systems. At the same time, closed-loop controls, optimal as well as suboptimal, are very important for practical applications. It seems that a lot of research is still to be done in this direction.

References

[1] Ailon, A. and Langholz, G., On the existence of time-optimal control of mechanical manipulators, *J. of Optimization Theory and Applications*, 46, 1, 1985.

[2] Akulenko, L. D. and Bolotnik, N. N., Synthesis of optimal control of transport motions of manipulation robots, *Mechanics of Solids*, 21 (4), 18, 1986.

[3] Akulenko, L. D., Bolotnik, N. N., Chernousko, F. L., and Kaplunov, A. A., Optimal control of manipulation robots, in *Proc. of the Ninth Triennial World Congress of IFAC, Volume 1*, Gertler, J., and Keviczky, L., Eds., Pergamon Press, Oxford, 1985, 331.

[4] Akulenko, L. D., Bolotnik, N. N., and Kaplunov, A. A., Some control modes for industrial manipulators, *Izvestiya AN SSSR. Tekhnicheskaya Kibernetika*, No. 6, 44, 1985.

[5] Avetisyan, V. V., Akulenko, L. D., and Bolotnik, N. N., Optimization of control modes of manipulation robots with regard of the energy consumption, *Izvestiya AN SSSR.* an electromechanical manipulator with a high degree of positioning accuracy, *Mechanics of Solids*, 25 (5), 32, 1990.

[6] Avetisyan, V. V., Bolotnik, N. N., Suboptimal control of an electromechanical manipulator with a high degree of positioning accuracy, *Mechanics of Solids*, 25 (5), 1990.

[7] Avetisyan, V. V., Bolotnik, N. N., and Chernousko, F. L., Optimal programmed motion of a two-link manipulator, *Soviet J. Computer and Systems Sciences*, 23 (5), 65, 1985.

[8] Berbyuk, V. E. and Yanchak, Ya. I., Minimum-time optimization of transport motions of a gantry robot, *Izvestiya AN SSSR. Tekhnicheskaya Kibernetika*, No. 1, 126, 1991 (in Russian).

[9] Bobrow, J. E., Optimal robot path planning using the minimum-time criterion, *IEEE J. Robotics and Automation*, 4, 443, 1988.

[10] Bobrow, J. E., Dubowsky, S., and Gibson, J. S., Time-optimal control of robotic manipulators along specified paths, *Int. J. Robotics Research*, 4, 3, 1985.

[11] Bolotnik, N. N. and Chernousko, F. L., Optimization of manipulation robot control, *Soviet J. Computer and Systems Sciences*, 28 (5), 127, 1990.

[12] Bolotnik, N. N., Gorbachev, N. V., and Shukhov, A. G., Optimization of control of an electromechanical system with respect to the minimax performance index, *Izvestiya AN SSSR. Mekhanika Tverdogo Tela*, No. 6, 30, 1992 (in Russian).

[13] Bolotnik, N. N. and Kaplunov, A. A., Optimal rectilinear transfer of a load by means of a two-link manipulator, *Izvestiya AN SSSR. Tekhnicheskaya Kibernetika*, No. 1, 160, 1982 (in Russian).

[14] Bolotnik, N. N. and Kaplunov, A. A., Optimization of control and configurations of a two-link manipulator, *Izvestiya AN SSSR. Tekhnicheskaya Kibernetika*, No. 4, 144, 1983 (in Russian).

[15] Borisov, V. F. and Zelikin, M. I., Modes with switchings of increasing frequency in the problem of controlling a robot, *J. Appl. Maths. Mechs.*, 52, 731, 1988.

[16] Chen, Y., Existence and structure of minimum-time control for multiple robot arm handling a common object, *Int. J. Control*, 53, 855, 1991.

[17] Chernousko, F. L., Models and optimal control of robotic systems, *System Modelling and Optimization, Proc. 14th IFIP-Conference, Lecture Notes in Control and Information Sciences*, Sebastian, H.-J., and Tammer, K., Eds., 143, 1, 1990.

[18] Chernousko, F. L., Decomposition and suboptimal control in dynamical systems, *J. Appl. Maths. Mechs.*, 54, 727, 1990.

[19] Chernousko, F. L., Decomposition and synthesis of control in dynamical systems, *Soviet J. Computer and Systems Sciences*, 29 (5), 126, 1991.

[20] Chernousko, F. L., Decomposition and suboptimal control in dynamic systems, *Optimal Control Applications and Methods*, 14, 1993 (to appear).

[21] Chernousko, F. L., Akulenko, L. D., and Bolotnik, N. N., Time-optimal control for robotic manipulators, *Optimal Control Applications and Methods*, 10, 293, 1989.

[22] Dubowsky, S., Norris, M. A., and Shiller, Z., Time optimal trajectory planning for robotic manipulators with obstacle avoidance: A CAD approach, in *Proc. of IEEE Int. Conf. on Robotics and Automation*, San Francisko CA, 1986.

[23] Geering, H. P., Guzzella, L., Hepner, S. A. R., and Onder, C. H., Time-optimal motions of robots in assembly tasks, *IEEE Trans. on Automatic Control*, AC-31, 512, 1986.

[24] Hanafusa, H., Yoshikawa, T., and Nakamura, Y., Analysis and control of articulated robot arm with redundancy, in *Prepr. of the 8th IFAC World Congress*, 1981, August, XIV-78.

[25] Heimann, B., Loose, H., Schmidt, K. D., Rothe, H., and Lyubushin, A. A. Dynamics and optimal control of manipulation robots, *Advances in Mechanics*, 7, 1984 (in Russian).

[26] Hollerbach, J. M. and Suh, K. C., Redundancy resolution of manipulators through torque optimization, *IEEE J. Robotics and Automation*, RA-3, 308, 1987.

[27] Huang, H. P. and McClamroch, N. H., Time-optimal control for a robotic contour following problem, *IEEE J. Robotics and Automation*, 4, 140, 1988.

[28] Kahn, M. E. and Roth, B., The near-minimum-time control of open-loop articulated kinematic chains, *Trans ASME. J. Dynamic Syst. Measurem. and Control*, 93, 165, 1971.

[29] Kazerounian, K. and Wang, Z., Global versus local optimization in redundancy resolution of robotic manipulator, *Int. J. Robotics Research*, 7, 3, 1988.

[30] Khadem, S. E. and Dubey, R. V., A global Cartesian space obstacle avoidance scheme for redundant manipulators, *Optimal Control Applications and Methods*, 12, 279, 1991.

[31] Kiriazov, P. and Marinov, P., Robot control synthesis in conjunction with moving workpieces, in *Prepr. 6th CISM-IFToMM Symp. on Theory and Practice of Robots and Manipulators, Ro. Man. Sy. '86*, Crakow, 1986, 284.

[32] Kobrinskii, A. A. and Kobrinskii, A. E., *Manipulation Systems of Robots*, Nauka, Moscow, 1984 (in Russian).

[33] Konzelmann, J., Bock, H. G., and Longman, R. W., Time-optimal trajectories of elbow robots by direct methods, in *Proc. of the AIAA Guidance, Navigation, and Control Conference*, Boston, August 1989, 895.

[34] Konzelmann, J., Bock, H. G., and Longman, R. W., Time-optimal trajectories of polar robot manipulators by direct methods, *Modeling and Simulation*, 20, 1933, 1989.

[35] Liegeois, A., Automatic supervisory control of the configuration and behavior of multi-body mechanisms, *IEEE Trans. Sysyt. Man. Cybern.*, SMC-7, 868, 1977.

[36] Loose, H., Rothe, H., and Schmidt, C.-D., Verfahren und Programme zur

Optimalen Steuerung von Industrierobotern, *Z. Angew. Math. und Mech,* 64, M 476, 1984.

[37] Luh, J. Y. S. and Lin, C. S., Optimum path planning for mechanical manipulators, *Trans. ASME. J. Dynamic Syst. Measurem. and Control,* 102, 142, 1981.

[38] Marinov, P. and Kiriazov, P., A direct method for optimal control synthesis of manipulator point-to-point motion, in *Prepr. 9th IFAC World Congress, Budapest, Hungary, 1984,* Volume IX, Mac Farlane, A. G. J., and Rauch, H. E., Volume Eds, Budapest, 1984, 219.

[39] Meier, E. B. and Bryson, A. E., Efficient algorithm for time-optimal control of a two-link manipulator, *J. Guidance,* 13, 859, 1990.

[40] Nakamura, Y. and Hanafusa, H., Optimal redundancy control of robot manipulators, *Int. J. Robotics Research,* 6, 32, 1986.

[41] Oberle, H. J., Numerical computation of singular control functions for a two-link robot arm, in *Proc. of the Conference on Optimal Control and Variational Calculus,* Bulirsch, R., Miele, A., Stoer, J., and Well, K. H., Springer Verlag, Berlin, 1987, 244.

[42] Osipov, S. N. and Formalskii, A. M., The problem of the time-optimal turning of a manipulator, *J. Appl. Maths. Mechs.,* 52, 725, 1988.

[43] Pfeiffer, F., Geometrical solution of a manipulator optimization problem, in *Control Applications of Nonlinear Programming and Optimization 1989, Proc. of the 8th IFAC Workshop,* Siguerdidjane, H. B., and Bernhard, P., Eds, Pergamon Press, Oxford, 1989, 83.

[44] Pfeiffer, F. and Johanni, R., A concept for manipulator trajectory planning, *IEEE J. Robotics and Automation,* RA-3, 115, 1987.

[45] Sahar, G. and Hollerbach, J. M., Planning of minimum-time trajectories for robot arms, *Int. J. Robotics Research,* 5, 90, 1986.

[46] Sato, O., Shimojima, H., and Kitamura, Y., Minimum-time control of a manipulator with two degrees of freedom, *Bull. JSME,* 26, 1404, 1983.

[47] Sato, O., Shimojima, H., Kitamura, Y., and Yoinara, H., Minimum-time control of a manipulator with two degrees of freedom (2nd Report, Dynamic characteristics of gear train and axes), *Bull. JSME,* 28, 959, 1985.

[48] Sato, O., Shimojima, H., and Kitamura, Y., Minimum-energy control of a manipulator with two degrees of freedom, *Bull. JSME,* 29, 573, 1986.

[49] Schmitt, D., Soni, A. H., Srinvasan, V.,and Nagamathan, G., Optimal motion programming of robot manipulators, *J. of Mechanisms, Transmissions and Automation in Design,* 107, 239, 1985.

[50] Steinbach, M., Bock, H. G., and Longman, R., Time optimal control of SCARA robots, in *Proc. of the AIAA Guidance, Navigation, and Control Conference*, Portland, Oregon, 1990.

[51] Stepanenko, Yu. A., Problem of optimal control of a manipulator, in *Theory of Machines of Automatic Action*, Nauka, Moscow, 1970 (in Russian).

[52] Takano, M., and Susaki, K., Time optimal control of PTP motion of a robot with collision avoidance, in *Proc. of the* 3-rd *Conf. on Robotics, ICAR 87*, Versailles, 1987.

[53] Troch, I., Time-suboptimal quasi-continuous path generation for industrial robots, *Robotica*, 7, 297, 1989.

[54] Vukobratovic, M. and Kircanski, M., A method for optimal synthesis of manipulation robotic trajectories, *Trans ASME J. Dynamic Syst. Measurem. and Control*, 102, 69, 1980.

[55] Vukobratovic, M. and Kircanski, M., A dynamic approach to nominal trajectory synthesis for redundant manipulators, *IEEE Trans. Syst. Man. Cybernet*, SMC-14, 1984.

[56] Weinreb, A. and Bryson, A. E., Optimal control systems with hard control bounds, *IEEE J. Automatic Control*, AC-30, 1135, 1985.

[57] Whitney, D. E., The mathematics of coordinated control of prosthetic arms and manipulators, *Trans. ASME. J. Dynamic Syst. Measurem. and Control*, 94, 303, 1972.

[58] Wie, B., Chuang, C. H., and Sunkel, J., Minimum-time pointing control of a two-link manipulator, *J. Guidance*, 13, 867, 1990.

[59] Yashi, O. S. and Ozgoren, K., Minimal joint motion optimization of manipulators with extra degrees of freedom, *Mechanism and Machine Theory*, 19, 325, 1984.

[60] Zhang, W. and Wang, R. K. C., Collision-free time optimal control of a two-link manipulator, *Int. J. Robotics and Automation*, 1, 96, 1986.

International Series of Numerical Mathematics, Vol. 115, © 1994 Birkhäuser Verlag Basel

Large-scale SQP Methods
and their Application in Trajectory Optimization*

Philip E. Gill†   Walter Murray‡   and   Michael A. Saunders‡

Abstract

Optimal trajectory problems are representative of a class of constrained optimization problems in which the constraints are discrete versions of continuous constraints (including ordinary or partial differential equations) and the variables define a discrete approximation to a continuous solution. This class is characterized by a large number of variables and sparse problem derivatives. Optimal trajectory problems have the additional property that the problem derivatives, or their approximations, are very expensive to evaluate.

SQP methods solve a nonlinear optimization problem by a sequence of quadratic programming (QP) subproblems. These methods converge rapidly and with great reliability when properly applied to optimal trajectory problems. However, as the problem size increases beyond a few hundred variables and constraints, the linear algebra required to solve each QP subproblem becomes significant.

Practical and theoretical aspects of extending SQP algorithms to large optimal trajectory problems are discussed. Attention is focussed on problems with relatively few degrees of freedom at the solution. In this case it is possible to formulate and analyze methods based upon a quasi-Newton approximation of a *reduced* Hessian, as well as sparse-matrix techniques for the Jacobian.

Keywords. Nonlinear programming, sequential quadratic programming methods, optimal control, optimal trajectory calculations.

Introduction

The flight dynamics of a spacecraft or aircraft are completely determined by initial and terminal conditions on the flight trajectory and by (differential) equations of motion. The equations of motion include control functions that can be chosen to optimize some performance criterion, such as the amount of fuel burned, or the elapsed time of the mission. Such *optimal trajectory problems* may be formulated

---

*Invited Paper

†Department of Mathematics, University of California at San Diego, La Jolla, CA 92093-0112, USA.

‡Systems Optimization Laboratory, Department of Operations Research, Stanford University, Stanford, CA 94305-4022, USA.

in the following general form. Associated with each problem is the definition of an
N-stage dynamical system, whose ith stage is

$$\dot{x}^i(t) = \psi^i(x(t), u(t), d, t), \qquad t \in [t_i, t_{i+1}], \tag{1}$$

where $x(t)$ and $u(t)$ are vector-valued functions of state and control variables, and
d is a vector of *design variables* (e.g., planform area, shape parameters, rocket
nozzle diameter, etc.). At the start of each stage, a boundary condition of the
form

$$x(t_i) = b^i(v_i, t_i, d) \tag{2}$$

is applied, where v_i is a vector of variables that determine the position of the
trajectory at t_i. The initial and final values of t_i are denoted by t_0 and t_f. In
general, t_0 is given, but other t_i's may be variables to be optimized.

The *optimal trajectory problem* is to find a local minimizer of a nonlinear
function $\psi^0(t_f, x(t_f))$, subject to the differential constraints (1) and (2), and the
so-called *path constraints*

$$b_l(t) \leq \phi(x(t), u(t), d, t) \leq b_u(t), \qquad t \in [t_0, t_f], \tag{3}$$

where $b_l(t)$ and $b_u(t)$ are prescribed vector functions of upper and lower bounds.

Recently, an approach based on Hermite collocation and the SQP algorithm
NPSOL has been employed within the optimal trajectory package OTIS (see Hargraves and Paris [HP87]). The resulting system has had a significant impact on
space-vehicle design, and is being used in the calculation of trajectories for the
National Aerospace Plane, the Mars Lander, and the single-stage-to-orbit test vehicle. From our viewpoint, OTIS will be regarded as a *problem generator* for finite-
dimensional optimization. OTIS has been successful, in part, because it generates
finite-dimensional problems that are well suited to optimization.

Here we give a highly simplified description of the discretization scheme used
in OTIS. The reader is referred to Hargraves and Paris [HP87] for further details.
(For related approaches, see [Kra93, Bet93].) The idea is to obtain a discrete set
x_1, x_2, x_3, ..., whose elements are the values of $x(t)$ at a set of so-called *nodes*
τ_1, τ_2, τ_3, ... The x-values are adjusted by the optimization routine, and they
must satisfy a set of functional equalities and inequalities that represent discretized
versions of the continuous constraints.

To illustrate the ideas involved, we describe the discretization of the state
equations (1). Consider the simplest case, with scalar-valued state equations $\dot{x} = \psi(x, t)$ on an interval $[a, b]$. The function $x(t)$ is approximated by a piece-wise
cubic Hermite polynomial $w(t)$. In our example, we consider a discretization that
involves two nodes, at a and at b. Suppose that x_a and x_b are given approximate
values of $x(a)$ and $x(b)$ (x_a and x_b will be the variables of the discretized problem).
The four conditions

$$w(a) = x_a, \qquad\qquad w(b) = x_b,$$
$$\dot{w}(a) = \psi(x_a, a), \qquad\qquad \dot{w}(b) = \psi(x_b, b), \tag{4}$$

completely define the piecewise cubic Hermite polynomial $w(t)$. The values x_a and x_b are now adjusted by the optimization method to force the differential equation to be satisfied at the midpoint $m = (a + b)/2$. This is done by defining a quantity $d(x_a, x_b)$, known as the *defect*, which is the error $\psi(w(m), m) - \dot{w}(m)$ at the midpoint of the interval. The required nonlinear constraint is then the equality $d(x_a, x_b) = 0$. This collocation can be defined for any subinterval of $[t_0, t_f]$. If a node coincides with a point at which $x(t)$ is specified (such as t_0 or t_f), the exact value is used in the definition of the appropriate $w(t)$.

The path constraints (3) may be discretized in a similar way, giving the finite-dimensional problem

$$\text{NP} \qquad \underset{x \in \Re^n}{\text{minimize}} \qquad f(x)$$

$$\text{subject to} \quad \ell \leq \begin{pmatrix} x \\ Ax \\ c(x) \end{pmatrix} \leq u,$$

where x is a set of variables, $f(x)$ is a nonlinear function, A is an $m_L \times n$ matrix, and $c(x)$ is an m_N-vector of nonlinear functions.

SQP methods require the specification of the derivatives of the nonlinear functions $f(x)$ and $c(x)$. When derivatives are not available a common approach is to approximate the derivatives by finite differences. In general, derivatives of ψ and ϕ (and hence of $c(x)$ and $f(x)$) are not provided. Fortunately, the problem derivatives are *sparse*, since each variable is associated with a particular node that is linked only to nodes at nearby points. In OTIS the sparsity pattern and the structure implied by the discretization is exploited to reduce the number of function evaluations required to approximate the derivatives. Nevertheless, the derivatives are very expensive to approximate.

This formulation treats both $x(t)$ and $u(t)$ as independent variables, i.e., there is no attempt to follow the traditional approach of using the differential constraints (1) to eliminate the state variables from the problem. An important consequence of this approach is that the differential constraints are satisfied only as the discretized variables converge to their optimal values. This increases the importance of the reliability of the optimizer, since termination at a point other than a solution will not provide a valid trajectory.

SQP Methods

Sequential quadratic programming (SQP) methods solve a nonlinearly constrained optimization problem using a sequence of quadratic programming subproblems. (For a survey of SQP methods, see [Pow83, GMSW89].) Many researchers have studied SQP (or SQP-related) methods; see, e.g., [Big72, Han76, Han77, Pow77, Pow83, Fle85, Fle87, Sch81, CC82a, CC82b, CC84]. SQP methods have a "two-level" structure of *major* and *minor* iterations. The major iterations generate a sequence of iterates $\{x_k\}$ such that

$$x_{k+1} = x_k + \alpha_k p_k, \tag{5}$$

where x_k is the current iterate, the non-negative scalar α_k is the *steplength*, and p_k is the *search direction*. The defining feature of an SQP method is that the search direction p_k is the solution of a quadratic programming (QP) subproblem. The minor iterations are then those of the method used to solve the quadratic program. The steplength α_k is usually chosen to yield a sufficient decrease in a *merit function* (a suitable combination of the objective and constraint functions). For a discussion of several important merit functions and their properties, see, e.g., [Fle87, GMSW89].

NPSOL is an SQP method that utilizes an *augmented Lagrangian merit function*. Under suitable conditions, the iterates will converge from any starting point to a point that satisfies the first-order necessary conditions for a solution of NP. Furthermore, the augmented Lagrangian merit function does not impede superlinear convergence, in the sense that when the iterates converge superlinearly with $\alpha_k = 1$, the unit step gives a sufficient decrease in the merit function.

Nodes	n	m	n_z	Majors	n_f	cpu secs
20	184	232	7	19	32	60.5
40	364	472	22	31	35	562.9
60	544	712	40	39	43	3002.9
80	724	934	55	45	49	9292.7
100	904	1192	74	46	50	18843.1

Table 1: NZOPT applied to the F4 minimum-time-to-climb.

In Table 1 we give the results of applying the SQP method NZOPT[1] to a standard optimal trajectory problem: the F4 minimum-time-to-climb. In this problem, a pilot cruising in an F-4 at Mach 0.34 at sea level wishes to ascend to 65,000 feet and be at Mach 1.0 in minimum time. The problem has two path constraints: a constraint on the maximum altitude and a constraint on the maximum dynamic pressure. The table gives the details of six runs with increasing numbers of nodes. The columns "n" and "m" give the number of variables and constraints (excluding the simple upper and lower bounds). The columns "Majors" and "n_f" give the total number of major iterations and function evaluations in each case. Note that at each major iteration, OTIS approximates the gradients of the problem functions. The gradient approximations dominate the computational effort for small n. The column "n_z" gives the number of *degrees of freedom* associated with the solution of

---

[1]NZOPT is a special version of NPSOL [GMSW86] that has been developed in conjunction with McDonnell Douglas Space Systems for optimal trajectory problems.

each problem. (n_z is the dimension of the underlying *unconstrained* problem, i.e., the number of variables n less the number of binding constraints at the solution; see Section 'SQP Methods for Sparse Problems'.) All runs were performed on a Sun SPARCstation 10/41.

The results of Table 1 indicate that the dense SQP algorithm used by NZOPT is extremely effective for problems with a small to moderate number of nodes. These results are typical of those obtained on practical problems, i.e., NZOPT converges *rapidly* and with great *reliability*. The problem functions, which are very expensive to evaluate, are only computed during the major iterations and so the remarkably few major iterations needed makes NZOPT very well suited to trajectory calculations provided n is not too large. Note that only 50 evaluations of the problem functions were necessary to solve a problem with almost 1000 variables. This property gives SQP methods like NZOPT an advantage over the method of MINOS [MS82, MS93], which is based upon solving a sequence of linearly constrained problems (see [Rob72]). MINOS requires function and gradient evaluations during both the major and minor iterations, and hence it is handicapped on problems such as the trajectory problem where the problem functions are expensive to compute.

When the Jacobian is treated as a dense matrix, the cost of solving the QP subproblems grows as a cubic function of the number of variables, whereas the total cost of evaluating the problem functions grows quadratically with the size of the discretized problem. Inevitably a point will be reached when the linear algebra required to solve each QP subproblem becomes significant. For example, the solution of the 100-node discretization of the F4 minimum-time-to-climb required approximately 5.2 cpu hours. Much of the time came from treating the Jacobian (and the Hessian of the Lagrangian) as dense matrices.

In this paper we discuss some recent approaches to extending SQP methods to large problems with a sparse Jacobian. The F4 minimum-time-to-climb is a typical trajectory problem in the sense that the number of degrees of freedom (the value of n_z in Table 1) is small compared to n. The main focus is on SQP methods that exploit this feature by using a quasi-Newton approximation of the *reduced* Hessian.

Some of the discussion also relates to methods that use second derivatives. The provision of second derivatives can confer several practical and theoretical benefits on a calculation. On the theoretical side, the potential exists of generating an iterative sequence that converges to a point satisfying the *second-order* necessary conditions for optimality. On the practical side, second-derivative methods have been observed to be significantly more robust and to require fewer iterations than methods that use only first derivatives. These benefits may be tempered by the sometimes time-consuming and error-prone task of providing subroutines to calculate higher derivatives. However, effective automatic differentiation packages are now becoming available, and second derivatives are easily defined for many important classes of nonlinear problems.

Finally, we note that we know of only one other significant non-SQP method in the area of general large-scale constrained optimization. Recently, a preliminary version of the LANCELOT package has been released [CGT92]. The method used in LANCELOT may be categorized as a *sequential augmented Lagrangian method*, since all constraints other than simple bounds are included in an augmented Lagrangian function, which is minimized subject to the bounds. In its current form, LANCELOT appears to be most suited to sparse problems with very few binding constraints at the solution.

SQP Methods for Sparse Problems

The constrained optimization problem is assumed to have the following *standard form*:

$$\text{NP} \qquad \underset{x\in\Re^n}{\text{minimize}} \quad f(x) \qquad \text{subject to} \quad c(x)=0, \quad x\geq 0, \qquad (6)$$

where $f(x)$ and $\{c_i(x)\}$ are prescribed nonlinear functions with continuous second derivatives. For simplicity, general linear constraints are not included in (6), since they appear without alteration in the associated subproblem (see (11) below). Further, all nonlinear constraints appear in (6) as *equalities*. A constraint $c_i(x) \geq 0$ may be converted to an equality by adding a bounded slack variable. (Provided the proper sparse-matrix techniques are used, there is no significant penalty with the increased problem size.) It is important that any practical algorithm be designed to treat variables with both upper and lower bounds, or no bounds at all. In the algorithms to be considered, the treatment of these cases is straightforward both in theory and in practice. We omit them for simplicity.

The optimality conditions for problem (6) involve the Lagrangian,

$$L(x,\pi) = f(x) - \pi^T c(x), \qquad (7)$$

where π is an m-vector. Let $g(x)$ and $A(x)$ denote the gradient of $f(x)$ and the Jacobian of $c(x)$ respectively. Let a_j denote the j-th column of A and assume for simplicity that A has full rank. The Hessian of the Lagrangian is $W(x,\pi) = \nabla^2 f(x) - \sum_{i=1}^{m} \pi_i \nabla^2 c_i(x)$.

It is convenient to define a matrix Q that facilitates the concise statement of the optimality conditions for problem (6). Consider the partition of A into nonbasic, basic and superbasic columns N, B and S (see Murtagh and Saunders [MS78]). If A has full rank, this partition may be written as $AP = (\ B\quad S\quad N\ )$, where P is a permutation matrix, B is square and nonsingular, and the n_N columns of N define variables that are temporarily frozen at their current values. Matrices and vectors associated with the partitions of A are denoted by appropriate subscripts. Let Q denote the $n \times n$ nonsingular matrix

$$Q = P \begin{pmatrix} B^{-1} & -B^{-1}S & \\ & I_S & \\ & & I_N \end{pmatrix}. \qquad (8)$$

It is easy to verify that

$$AQ = (\ I_B \quad 0 \quad N\ ),\tag{9}$$

and if the columns of Q are partitioned to match those of AQ above, then

$$Q = (\ Y_B \quad Z \quad Y_N\ ),\tag{10}$$

and the columns of Z form a basis for the null space of the matrix A augmented by the gradients of the temporarily frozen constraints.

Let x be a point at which the partition (B S N) gives an associated vector π satisfying $B^T\pi = g_B$. Then x is a *Karush-Kuhn-Tucker (KKT) point* for problem (6) if: (i) $g_j - a_j^T\pi \geq 0$ for all j associated with N; (ii) $Z^Tg(x) = 0$; and (iii) $Z^TW(x,\pi)Z$ is positive semidefinite. If an appropriate constraint qualification holds, these conditions must be satisfied at a local minimizer of NP (see, e.g., Powell [Pow74]).

Definition of the Subproblem

To a large extent, the choice of QP subproblem depends on whether an *equality QP* (EQP) or *inequality QP* (IQP) approach is taken in the SQP method. Typically, an IQP subproblem has the form

$$\begin{aligned}
&\underset{p\in\Re^n}{\text{minimize}} &&\varphi(p) \equiv g^Tp + \tfrac{1}{2}p^THp\\
&\text{subject to} &&c(x_k) + A(x_k)p = 0, \qquad x_k + p \geq 0,
\end{aligned}\tag{11}$$

where g and H are constant. The matrix H is defined from an approximation to the Hessian of the Lagrangian (see Section 'Definition of the QP Hessian').

With an EQP strategy, only major iterations are defined. At each iteration, a prediction is made of the nonbasic set at the solution and a QP subproblem analogous to (11) is solved in which these variables are temporarily fixed on their bounds. Persuasive arguments can be given for and against both approaches, and neither strategy is likely to be consistently better on all problems. It seems unlikely that an EQP strategy would require fewer major iterations than an IQP strategy. Therefore an EQP strategy is not likely to be preferable for problems whose function evaluations are expensive. (See Murray and Wright [MW82] for further discussion.)

The SQP algorithm of Prieto [Pri89] lies "in between" the IQP and EQP strategies, and is intended to reduce the calculation needed for a search direction. Prieto's algorithm follows an IQP strategy, except that only a *partial* solution of the QP subproblem is found. It is shown that active-set methods for the subproblem (11) generate information that may be used to construct a direction of sufficient descent for the augmented Lagrangian merit function. Other compromises for quasi-Newton SQP methods have been considered by [MW78, Sch81, NO85, Gur86, FdlM87].

Other SQP-like methods cannot be categorized as either IQP or EQP. Two important methods are based on the definition of an unconstrained subproblem whose objective is the sum of a quadratic model of the Lagrangian and a penalty term involving the violations of the *linearized* constraints. The $S\ell_1 QP$ of Fletcher [Fle85, Fle87] utilizes an ℓ_1 penalty function and has a major/minor iteration structure similar to an IQP method. Powell [Pow83] has considered use of the quadratic penalty function in this context and has shown that the resulting "$S\ell_2 QP$" algorithm is equivalent to the recursive quadratic programming method of Biggs [Big72]. Since the linearized constraints are preassigned to the penalty term in this case, an EQP-like iteration structure is obtained. (See Section 'Definition of the QP Hessian' for a further discussion of these methods.)

Computation of the Search Direction

Whatever the choice of QP subproblem, it can be shown that all algorithms either implicitly or explicitly solve symmetric indefinite *KKT systems* of the form

$$\begin{pmatrix} \mathcal{H} & \mathcal{A}^T \\ \mathcal{A} & 0 \end{pmatrix} \begin{pmatrix} u \\ v \end{pmatrix} = \begin{pmatrix} r \\ s \end{pmatrix}, \qquad (12)$$

where $\mathcal{H}$ is related to the Hessian of the Lagrangian and $\mathcal{A}$ is derived from the constraint gradients. (In some cases a diagonal matrix with small elements may occupy the $2,2$ block.) The symmetric matrix K associated with this system is called the *Karush-Kuhn-Tucker (or just KKT) matrix*. The definition of $\mathcal{H}$ and $\mathcal{A}$, and the frequency with which the KKT system must be solved, depends on the method. In an EQP approach, a new system is solved every iteration. For an IQP method, many related KKT systems may be solved each linearization. In both cases, $\mathcal{H}$ and $\mathcal{A}$ are the columns of the Hessian and Jacobian associated with basic and superbasic variables.

Definition of the QP Hessian

Research on SQP methods has concentrated mainly on the case in which H is a *positive-definite quasi-Newton* approximation to the Hessian of the Lagrangian. For large problems, attention has been focussed on methods that approximate only the *reduced* Hessian $Z^T W Z$ (see Section 'Quasi-Newton Methods for Large Problems').

Despite the utility of quasi-Newton methods, there are many applications where first *and* second derivatives of the objective and constraint functions are explicitly available. It might appear that a second-derivative SQP method would be straightforward to define and analyze. Given a "best known" multiplier estimate π_k, SQP methods can be defined in which H in (11) is taken as $W(x_k, \pi_k)$, the "exact" Hessian of the Lagrangian. In a neighborhood of a solution, such a method should display the desirable behavior of Newton's method. However, although the *reduced* Hessian of the Lagrangian must be positive semidefinite in a neighborhood of a solution, $W(x_k, \pi_k)$ may be indefinite at every iterate. The possibility of an indefinite Hessian may lead to difficulties that do not arise in the quasi-Newton case—namely, QP subproblems with unbounded or non-unique (local) solutions.

Because of its similarity to an EQP method, the $S\ell_2QP$ method is more readily implemented for large problems with known second derivatives. In this case, the search direction and multipliers satisfy a system similar to (12) except that the 2, 2 block contains a diagonal matrix that goes to zero as ρ goes to infinity. If the dimension of the reduced Hessian is a large fraction of the number of variables, it is efficient to factorize the KKT matrix explicitly using a symmetric indefinite solver (see Gould [Gou86b]). Coleman and Hemple [CH87] describe a similar method based on minimizing the $S\ell_2QP$ penalty function subject to a quadratic trust-region constraint.

Despite recent progress in algorithms for solving general quadratic programs [GMSW90, GMSW91, Gou86a], there are still unresolved issues associated with methods that utilize an IQP subproblem. If the size of the QP subproblem necessitates an explicit sparse factorization of KKT system, it is not immediately clear how to define an initial nonsingular KKT system with an associated nontrivial positive-definite reduced Hessian. In practice, this task is considerably complicated by the desirability of specifying a "target" initial basis. Forsgren and Murray [FM93] show that the identification of a suitable initial KKT system is equivalent to the computation of a feasible direction of negative curvature from an explicit KKT system—a problem of considerable difficulty.

Quasi-Newton Methods for Large Problems

Given a definition of the nonbasic set and its associated matrices B, S and Q (9)–(10), each iteration of an EQP method uses the *null-space equations*

$$Z^THZp_Z = -Z^Tg - Z^THYp_Y, \tag{13}$$

where Y is formed from Y_B and Y_N of (10) When all constraints are linear, the term involving Z^THY is not present, and methods for linear constraints need retain only the reduced Hessian, as in [MS78]. In the nonlinear case, many difficult issues arise if one wishes to use only reduced matrices, since no information has been retained concerning the one-sided reduced matrix Z^THY.

If an IQP strategy is used, a complication of updating the reduced Hessian is that the QP subproblem requires the definition of *all* of H. This implies that if more than one minor iteration is required to solve the QP and a nonbasic variable becomes superbasic, H_Z must be "unprojected" into an $n \times n$ matrix in order to compute the QP gradient $g + Hp$ and the new column of Z^THZ. Techniques for doing so are suggested in [NO85, Gur86]. Neither of these approaches has been completely satisfactory in a general algorithm because of the open issue of global convergence.

Recently, Eldersveld [Eld91] has described a large-scale SQP algorithm in which the QP Hessian is defined as

$$H = Q_0^{-T} \begin{pmatrix} I_B & & \\ & R_0^T R_0 & \\ & & I_N \end{pmatrix} Q_0^{-1}, \qquad (14)$$

where Q_0 denotes the matrix Q of (9) evaluated at the initial point of the QP subproblem and R_0 is the Cholesky factor of the initial reduced-Hessian approximation. With this definition of H, the reduced Hessian for the first minor iteration is $R_0^T R_0$. Eldersveld describes various choices for Q and shows how the resulting QP subproblem may be solved efficiently. A preliminary Fortran code incorporating many of these ideas has been developed as a modification of MINOS. Comparisons with NPSOL and MINOS on approximately 100 small and large problems indicate that reduced-Hessian SQP algorithms can be robust and efficient for many important applications.

Recently, the SQP code SNOPT has been developed for large problems with sparse constraints (see Gill *et al.* [GMS93]). SNOPT employs a QP Hessian similar to (14), but updates an approximation to a *transformed* Hessian that includes the reduced Hessian as a principal submatrix. When a QP variable is made nonbasic (i.e., fixed on its bound) the matrix Z shrinks by a column, and the curvature associated with the exiting column is removed from the reduced Hessian. Conversely, if a variable changes from nonbasic to superbasic, Z grows by a column and the relevant parts of $Z^T H Z$ must be re-estimated in subsequent iterations. If the optimal QP nonbasic set changes significantly close to a solution, the convergence of a reduced Hessian method can be significantly impaired. If a variable repeatedly enters and leaves the nonbasic set from one major iteration to another, its contribution to the reduced Hessian must be continually re-estimated. The use of a transformed Hessian eliminates this difficulty, since curvature can be maintained for nonbasic variables as well as superbasic variables. The use of a (orthogonally) transformed Hessian update is largely responsible for the excellent robustness and efficiency of NPSOL.

The Lagrangian is a linear and separable function of the slack variables. Consequently, it is known that the Hessian of the Lagrangian has a zero row and column corresponding to a slack variable. The approximation of (14) can be modified to reflect this property, with the hope that an improved approximation will result in fewer iterations as well as smaller intermediate reduced Hessian approximations.

Closing Remarks

There are other highly successful applications of SQP methods to practical problems (see, e.g., Burchett *et al.* [BHV84] and Ringertz [Rin88]). We have emphasized the OTIS project because it illustrates an inevitable consequence of the introduction of a successful new technology: *substantial success is rapidly followed by a substantial increase in expectation.* Design engineers now see numerical optimization as the means by which the *real-time* control of flight dynamics may be achieved. As SQP methods gain acceptance as a reliable numerical tool, the size and difficulty of the problems attempted will inevitably increase. It is crucial that new SQP methods be developed for problems that cannot yet be solved efficiently using existing technology.

There is much research remaining to be done on SQP methods for large problems. For example, an important topic is the derivation of improved quasi-Newton updates for the reduced Hessian. Numerous issues arise—for example, it is unclear which multiplier estimates should be included in the definition of the Lagrangian gradient. Our experience is that SNOPT usually requires more major iterations than NZOPT, especially on large problems. Since the QP nonbasic set generally settles down on the smaller problems, the difference in performance cannot be due solely to the loss of information inherent in a reduced Hessian update. Given the disparity between the performance of SNOPT and NZOPT, the potential gain from improving the existing quasi-Newton update is substantial.

Acknowledgement. The material contained in this report is based upon research partially supported by Department of Energy Grant DE-FG03-92ER25117, National Science Foundation Grants ECS-8715153 and DDM-9204547, and Office of Naval Research Grant N00014-90-J-1242.

References

[Bet93]   J. T. Betts. Issues in the direct transcription of optimal control problems to sparse nonlinear programs. In R. Bulirsch and D. Kraft, editors, *Control Applications of Optimization*, Basel, 1993. Birkhauser.

[BHV84]   R. C. Burchett, H. H. Happ, and D. R. Vierath. Quadratically convergent optimal power flow. *IEEE Transactions on Power Apparatus and Systems*, PAS-103, 3267–3275, 1984.

[Big72]   M. C. Biggs. Constrained minimization using recursive equality quadratic programming. In F. A. Lootsma, editor, *Numerical Methods for Nonlinear Optimization*, pages 411–428. Academic Press, London and New York, 1972.

[CC82a]   T. F. Coleman and A. R. Conn. Nonlinear programming via an exact penalty function: asymptotic analysis. *Mathematical Programming*, 24, 123–136, 1982.

[CC82b]   T. F. Coleman and A. R. Conn. Nonlinear programming via an exact penalty function: global analysis. *Mathematical Programming*, 24, 137–161, 1982.

[CC84]   T. F. Coleman and A. R. Conn. On the local convergence of a quasi-Newton method for the nonlinear programming problem. *SIAM J. on Numerical Analysis*, 21, 775–769, 1984.

[CGT92]   A. R. Conn, N. I. M. G. Gould, and Ph. L. Toint. Large-scale nonlinear constrained optimization. Report 92/2, Département de Mathématique, Facultés Universitaires de Namur, 1992.

[CH87]   T. F. Coleman and C. Hempel. Computing a trust region step for a penalty function. Report 87-847, Department of Computer Science, Cornell University, 1987.

[Eld91]   S. K. Eldersveld. *Large-scale sequential quadratic programming algorithms*. PhD thesis, Department of Operations Research, Stanford University, 1991.

[FdlM87]   R. Fletcher and E. Sainz de la Maza. Nonlinear programming and nonsmooth optimization by successive linear programming. Numerical Analysis Report NA 100, Department of Mathematical Sciences, University of Dundee, Scotland, 1987.

[Fle85]   R. Fletcher. An ℓ_1 penalty method for nonlinear constraints. In P. T. Boggs, R. H. Byrd, and R. B. Schnabel, editors, *Numerical Optimization 1984*, pages 26–40, Philadelphia, 1985. SIAM.

[Fle87]   R. Fletcher. *Practical Methods of Optimization*. John Wiley and Sons, Chichester, New York, Brisbane, Toronto and Singapore, second edition, 1987.

[FM93]   A. L. Forsgren and W. Murray. Newton methods for large-scale linear equality-constrained minimization. *SIAM J. on Matrix Analysis and Applications*, 14, 560–587, 1993.

[GMS93]   P. E. Gill, W. Murray, and M. A. Saunders. Reduced-Hessian SQP methods for large scale optimization. To appear, 1993.

[GMSW86]   P. E. Gill, W. Murray, M. A. Saunders, and M. H. Wright. User's guide for NPSOL (Version 4.0): a fortran package for nonlinear programming. Report SOL 86-2, Department of Operations Research, Stanford University, 1986.

[GMSW89]   P. E. Gill, W. Murray, M. A. Saunders, and M. H. Wright. Constrained nonlinear programming. In G. L. Nemhauser, A. H. G. Rinnooy Kan, and M. J. Todd, editors, *Handbooks in Operations Research and Management Science, Volume 1. Optimization*, chapter 5, pages 171–210. North Holland, Amsterdam, New York, Oxford and Tokyo, 1989.

[GMSW90] P. E. Gill, W. Murray, M. A. Saunders, and M. H. Wright. A Schur-complement method for sparse quadratic programming. In M. G. Cox and S. J. Hammarling, editors, *Reliable Numerical Computation*, pages 113–138. Oxford University Press, 1990.

[GMSW91] P. E. Gill, W. Murray, M. A. Saunders, and M. H. Wright. Inertia-controlling methods for general quadratic programming. *SIAM Review*, 33, 1–36, 1991.

[Gou86a] N. I. M. Gould. An algorithm for large scale quadratic programming. Report CSS 219, Computer Science and Systems Division, AERE Harwell, Oxford, England, 1986.

[Gou86b] N. I. M. Gould. On the accurate determination of search directions for simple differentiable penalty functions. *IMA J. on Numerical Analysis*, 6, 357–372, 1986.

[Gur86] C. B. Gurwitz. Sequential quadratic programming methods based on approximating a projected Hessian matrix. Report 219, Department of Computer Science, New York University, New York, 1986.

[Han76] S. P. Han. Superlinearly convergent variable metric algorithms for general nonlinear programming problems. *Mathematical Programming*, 11, 263–282, 1976.

[Han77] S. P. Han. A globally convergent method for nonlinear programming. *J. Optimization Theory and Applications*, 22, 297–309, 1977.

[HP87] C. R. Hargraves and S. W. Paris. Direct trajectory optimization using nonlinear programming and collocation. *J. of Guidance, Control, and Dynamics*, 10, 338–348, 1987.

[Kra93] D. Kraft. TOMP—Fortran modules for optimal control calculation. To appear in *ACM Transactions on Mathematical Software*, 1993.

[MS78] B. A. Murtagh and M. A. Saunders. Large-scale linearly constrained optimization. *Mathematical Programming*, 14, 41–72, 1978.

[MS82] B. A. Murtagh and M. A. Saunders. A projected Lagrangian algorithm and its implementation for sparse nonlinear constraints. *Mathematical Programming Study*, 16, 84–117, 1982.

[MS93] B. A. Murtagh and M. A. Saunders. Minos 5.4 user's guide. Report SOL 83-20R, Department of Operations Research, Stanford University, 1993.

[MW78] W. Murray and M. H. Wright. Projected lagrangian methods based on the trajectories of penalty and barrier functions. Report SOL 78-23, Department of Operations Research, Stanford University, 1978.

[MW82]   W. Murray and M. H. Wright. Computation of the search direction in constrained optimization algorithms. *Mathematical Programming Study*, 16, 62–83, 1982.

[NO85]   J. Nocedal and M. L. Overton. Projected Hessian updating algorithms for nonlinearly constrained optimization. *SIAM J. on Numerical Analysis*, 22, 821–850, 1985.

[Pow74] M. J. D. Powell. Introduction to constrained optimization. In P. E. Gill and W. Murray, editors, *Numerical Methods for Constrained Optimization*, pages 1–28, London and New York, 1974. Academic Press.

[Pow77] M. J. D. Powell. A fast algorithm for nonlinearly constrained optimization calculations. Report 77/NA 2, Department of Applied Mathematics and Theoretical Physics, University of Cambridge, England, 1977.

[Pow83] M. J. D. Powell. Variable metric methods for constrained optimization. In A. Bachem, M. Grötschel, and B. Korte, editors, *Mathematical Programming: The State of the Art*, pages 288–311. Springer Verlag, London, Heidelberg, New York and Tokyo, 1983.

[Pri89] F. J. Prieto. *Sequential quadratic programming algorithms for optimization*. PhD thesis, Report SOL 89-7, Department of Operations Research, Stanford University, 1989.

[Rin88] U. T. Ringertz. *A mathematical programming approach to structural optimization*. PhD thesis, Department of Aeronautical Structures and Materials, Royal Institute of Technology, 1988.

[Rob72] S. M. Robinson. A quadratically-convergent algorithm for general nonlinear programming problems. *Mathematical Programming*, 3, 145–156, 1972.

[Sch81] K. Schittkowski. The nonlinear programming method of Wilson, Han, and Powell with an augmented Lagrangian type line search function. *Numerische Mathematik*, 38, 83–127, 1981.

International Series of Numerical Mathematics, Vol. 115, © 1994 Birkhäuser Verlag Basel

Solving Optimal Control and Pursuit-Evasion Game Problems of High Complexity*

Hans Josef Pesch[†]

Abstract

Optimal control problems which describe realistic technical applications exhibit various features of complexity. First, the consideration of inequality constraints leads to optimal solutions with highly complex switching structures including bang-bang, singular, and control- and state-constrained subarcs. In addition, also isolated boundary points may occur. Techniques are surveyed for the computation of optimal trajectories with multiple subarcs. If the precise computation of the switching structure holds the spotlight, the indirect multiple shooting method is top quality. Second, the differential equations describing the dynamics may be so complicated that they have to be generated by a computer program. In this case, direct methods such as direct collocation are generally superior. Third, the task is often given in applications to solve many optimal control problems, either for parameter homotopies in the course of the solution process itself or for sensitivity investigations of the solutions with respect to various design parameters. Closely related to optimal control problems, pursuit-evasion game problems require, in a natural way, the solution of often thousands of boundary-value problems, in order to synthesize the open-loop controls by feedback strategies. In these cases, efficient homotopy methods must be used in connection with vectorized or parallelized versions of the aforementioned methods.

These three degrees of complexity in the solution of optimal control or pursuit-evasion game problems, respectively, are discussed in this survey paper by means of three examples: the abort landing of a passenger aircraft in the presence of a varying down burst, the time- and energy-optimal control of an industrial robot, and a pursuit-evasion game problem between a missile and a fighter aircraft.

Introduction

For the numerical solution of complicated optimal control problems, several numerical methods have proved to be reliable and efficient: the indirect multiple shooting method (see, e.g., [12], [23], [32], [33], [45], [46], [52]) turns out to be superior when highly accurate optimal solutions are to be computed and a scrutiny

---

*Invited Paper

[†]Department of Mathematics, Munich University of Technology, D–80290 München

of the necessary conditions of optimal control theory is desired. If these strong requirements are withdrawn for an easier handling of the method, the direct collocation method (see, e.g., [3], [4], [30], [53], [54]) turns out to be more favorable. In between these two methods, the direct simple shooting method [36], [37] and the direct multiple shooting method [5] are established.

The present investigations and the results obtained for various real-life optimal control problems (cf. [13]–[19], [21], [38], [45], [51], [55]) clearly show the trend that control problems are treated of which the mathematical models become closer and closer to reality. This causes an increase of complexity of optimal control problems by an increasing number of constraints, by more complicated differential equations, and finally also by an increasing number of unknowns that are involved in the model. In particular, these difficulties arise from multiple subarcs due to control and/or state variable inequality constraints, bang-bang, and singular controls or from the high complexity of the dynamic equations themselves, which even have to be generated by a computer program. In addition, special tasks require the solution of often thousands of related problems, for example, when optimal strategies of pursuit-evasion game problems are to be computed; see [6]–[10].

Especially, the performance of the indirect multiple shooting method and the direct collocation method is discussed in this paper with respect to the various difficulties arising in the solution of highly complicated real-life optimal control and pursuit-evasion game problems.

Optimal Control Problems with Multiple Subarcs

If optimal control problems include side conditions in form of control or state variable inequality constraints, the optimal solutions usually consist of several subarcs, which are related to the time intervals in which the constraints are non-active or active. In general, the sequence of the so-called junction points in which a constraint becomes active or inactive, is not known a priori, but must be computed as part of the solution of the problem.

When using an indirect method, it is recommended to solve the unconstrained problem firstly and then to introduce the constraint step by step by a homotopy method; see, e.g., [18] and [19]. Here we tacitly assume that both the unconstrained and the constrained problem is solvable. Note that this is not true, for example, for problems for which the objective function as well as the equations of motion are linear in the control variables. In this important case, the subarcs are related either to bang-bang or to singular optimal controls. We will not review this type of control problems here, but refer to [18] and [19], too.

When using a direct method, the (infinite dimensional) inequality constraints give rise to a finite number of inequality constraints in the resulting nonlinear programming problem which is then solved by an appropriate optimization method, e.g., the SQP-method of [28]. Here, one has not to care about the switching structure, i.e., the sequence of the junction points and the associated control laws between these junction points according to necessary conditions. However, knowledge of the switching structure helps to get more accurate results; see [53].

In the so-called hybrid approach of [54], which amalgamates direct collocation and multiple shooting by estimating the adjoint variables for the indirect approach, the transition from the solution obtained by a direct collocation method to the solution of the multipoint boundary-value problem, resulting from necessary conditions, by means of the indirect multiple shooting method may fail for particular problems. A failure particularly occurs when the domain of convergence of the multiple shooting method is too small; see, e.g., [20] and [51].

Hence, there is still a strong demand for the solution of constrained optimal control problems via the indirect approach. Moreover, only the indirect approach allows a scrutiny of the various necessary conditions known in optimal control theory; see, e.g., [11], [31], [34], and [43]. In the survey paper [50], recipes for many practically important classes of optimal control problems are given by which the multipoint boundary-value problems can be established via these necessary conditions. In the following example from civil aviation, the use of special sign conditions is demonstrated for the construction of a correct switching structure.

Abort Landing of a Passenger Aircraft in a Windshear

If strong downbursts, e.g., caused by thunderstorms, are present when a passenger aircraft is coming in for landing, the large difference between initial headwind and terminal tailwind leads to an additional acceleration of the aircraft and results very often in disastrous crashes. The safest flight through such a windshear is an abort landing maneuver during which the ground clearance is maximized, or in other words the minimum altitude is maximized. Optimal control problems of this type are so-called Chebyshev problems. The mathematical formulation of this problem is briefly described as follows; compare [18] and [44].

The performance index is given by

$$\max_{u \in U} \min_{0 \le t \le t_f} h(t) . \tag{1}$$

Here U denotes the set of all admissible control variables u, and $[0, t_f]$ is the flight time interval. Instead, we can also minimize the peak value of the altitude drop, that is, the difference between a constant reference altitude h_R and the instantaneous altitude

$$I[u] := \max_{0 \le t \le t_f} \{h_R - h(t)\} \stackrel{!}{=} \min . \tag{2}$$

Here, the reference altitude has to be chosen so as to satisfy

$$h_R \ge h(t) , \quad \text{for all } t \in [0, t_f].$$

Because of the relation between the Hölder norm and the Chebyshev norm, see [18], the optimal trajectory of the Chebyshev functional (2) can be approximated by the optimal trajectory of the Bolza functional

$$J[u] := \int_0^{t_f} (h_R - h(t))^q \, dt \stackrel{!}{=} \min , \qquad (3)$$

where q must be a sufficiently large even integer; here $q = 6$ as in [44].

The equations of motion are

$$\dot{x} = V \cos\gamma + W_x , \qquad (4a)$$

$$\dot{h} = V \sin\gamma + W_h , \qquad (4b)$$

$$\dot{V} = \frac{T}{m} \cos(\alpha + \delta) - \frac{D}{m} - g \sin\gamma - (\dot{W}_x \cos\gamma + \dot{W}_h \sin\gamma) , \qquad (4c)$$

$$\dot{\gamma} = \frac{T}{mV} \sin(\alpha + \delta) + \frac{L}{mV} - \frac{1}{V} g \cos\gamma + \frac{1}{V} (\dot{W}_x \sin\gamma - \dot{W}_h \cos\gamma) , \qquad (4d)$$

$$\dot{\alpha} = u . \qquad (4e)$$

The state variables are the horizontal distance x, the altitude h, the relative velocity V, the relative path inclination γ, and the relative angle of attack α. The time derivative of α is chosen as control variable u. For simplicity, the power setting is assumed to be a prescribed function of time; see [18] or [44].

These equations are supplemented by the approximations of the aerodynamic forces, thrust $T = T(t, V)$, drag $D = D(V, \alpha)$, and lift $L = L(V, \alpha)$, which also can be found in [18] and [44]. In addition, the windshear model is given by its wind velocity components[1],

$$W_x = k A(x) , \qquad (5a)$$

$$W_h = k \frac{h}{h_*} B(x) \qquad (5b)$$

with

$$A(x) = \begin{cases} -50 + ax^3 + bx^4 , & \text{for} \quad 0 \le x \le 500 , \\ \frac{1}{40}(x - 2300) , & \text{for} \quad 500 \le x \le 4100 , \\ 50 - a(4600 - x)^3 - b(4600 - x)^4 , & \text{for} \quad 4100 \le x \le 4600 , \\ 50 , & \text{for} \quad 4600 \le x , \end{cases}$$

$$B(x) = \begin{cases} dx^3 + ex^4 , & \text{for} \quad 0 \le x \le 500 , \\ -51 \exp[-c(x - 2300)^4] , & \text{for} \quad 500 \le x \le 4100 , \\ d(4600 - x)^3 + e(4600 - x)^4 , & \text{for} \quad 4100 \le x \le 4600 , \\ 0 , & \text{for} \quad 4600 \le x . \end{cases}$$

---

[1]Note there is a misprint in the term $B(x)$ for $4100 \le x \le 4600$ in [18].

Here, the parameter k characterizes the intensity of the windshear/downdraft combination. Its value $k = 1$ corresponds to a horizontal wind velocity difference (maximum tailwind minus maximum headwind) of $100\,ft\,s^{-1}$. In Fig. 1 the wind flow field is shown.

All other quantities not yet mentioned until now are constants the values of which can be found in [18], too.

For the treatment of the minimax optimal control problem, the following transformation technique to an optimal control problem of standard form is used; see [18] and the references cited therein.

By introducing the new (constant) state variable

$$\zeta(t) := \max_{0 \le \tau \le t_f} \{h_R - h(\tau)\} , \tag{6}$$

we have the new performance index

$$I[u] = \zeta(t_f) \overset{!}{=} \min . \tag{7}$$

Moreover, we have to add the constraints

$$\dot{\zeta} = 0 , \tag{8}$$

$$h_R - h(t) \le \zeta(t) . \tag{9}$$

Hence, we end up with an optimal control problem of Mayer's type.

For the numerical solution of the Chebyshev problem (2), the Mayer problem (7) and the Bolza problem (3) must be coupled,

$$\Im[u] := (1 - \Theta) \cdot \int_0^{t_f} (h_R - h(t))^6 \, dt + \Theta \cdot \zeta(t_f) . \tag{10}$$

This ansatz allows a homotopy from the Bolza problem to the Chebyshev problem by which the state variable inequality constraint (9) is introduced step by step when changing the homotopy parameter Θ from 0 to 1. Note that the state variable inequality constraint (9) is of order 3; see [18].

For the complete problem including an inequality constraint on the control variable u, a first-order state constraint on the angle of attack α, and the boundary conditions, it is referred to [18]. In this reference, the complete derivation of the necessary conditions, too, can be found which form a rather complicated multipoint boundary-value problem. This boundary-value problem is solved with the multiple shooting code given in [46]; see [19].

Starting with the solution of the Bolza Problem, i.e., with $\Theta = 0$, the optimal state-constrained solution of the coupled problem (10) exhibits first one touch point, then two touch points, and finally the first touch point splits into a boundary arc; compare Fig. 1. This process is described in detail in [19]. Note that the Hamiltonian is not regular here, so that a third-order state constraint may have

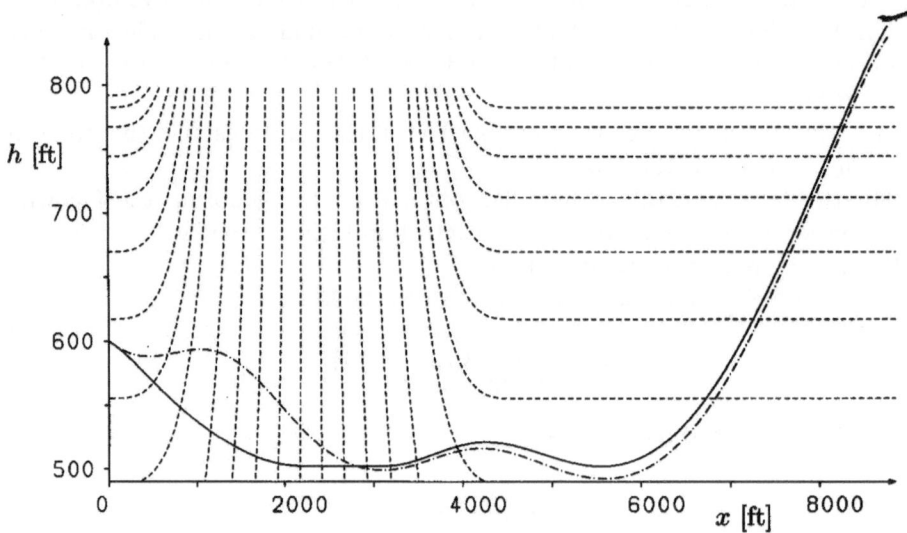

Figure 1: Wind flow field (dashed line) and altitude versus horizontal distance for Bolza (dashed-dotted line) and Chebyshev (solid line) functional.

both touch points and boundary arcs, contrary to the regular case; compare [34] and [43].

These changes of the switching structure can be detected only if necessary sign conditions as given in [31], [34] and [43], see also [19], are carefully obeyed. To explain this in more detail, we consider the Lagrange parameter μ by which the third total time derivative of the state constraint (9) is added to the Hamiltonian,

$$H(x, h, V, \gamma, \alpha, \zeta, \lambda_x, \lambda_h, \lambda_V, \lambda_\gamma, \lambda_\alpha, \lambda_\zeta, u) := (1 - \Theta)(h_R - h)^6$$

$$+ \lambda_x \dot{x} + \lambda_h \dot{h} + \lambda_V \dot{V} + \lambda_\gamma \dot{\gamma} + \lambda_\alpha \dot{\alpha} + \lambda_\zeta \dot{\zeta} + \mu \frac{d^3}{dt^3}(h_R - h - \zeta) .$$

Then the following necessary conditions due to [43], see also [31], must hold along the constrained subarc $[t_{entry}, t_{exit}]$,

$$\mu \geq 0 , \quad \dot{\mu} \leq 0 , \quad \mu^{(2)} \geq 0 , \quad \mu^{(3)} \leq 0 ; \tag{11a}$$

$$\mu(t_{exit}) = 0 , \quad \dot{\mu}(t_{exit}) = 0 . \tag{11b}$$

The numerical results published in [19] fulfill these conditions very precisely with the exception of the necessary condition $\dot{\mu}(t_{exit}) = 0$. The candidate optimal trajectory given in [19] fulfills $\dot{\mu}(t_{exit}) \approx -10^{-2}$ only. Taking into account the high precision attainable by the multiple shooting method, the trajectory is approximately optimal only. Hence, the switching structure must be changed in order to improve the objective function. Since state-constrained subarcs of problems having a non-regular Hamiltonian behave as singular subarcs, i.e., the switching function, which is λ_α here, vanishes along these subarcs, another singular subarc must follow the state-constrained subarc, and then a new touch point is released at the end of that boundary arc. This candidate optimal trajectory now fulfills all necessary conditions (11) very precisely, e.g., $\dot\mu(t_{exit}) \approx -10^{-13}$ now. However, the improvement in safety is about $3.\,10^{-6}\,ft$ only; see Fig. 2. The inserted diagram in Fig. 2 shows the state-constrained subarc followed by another touch point at the level of the minimax altitude.

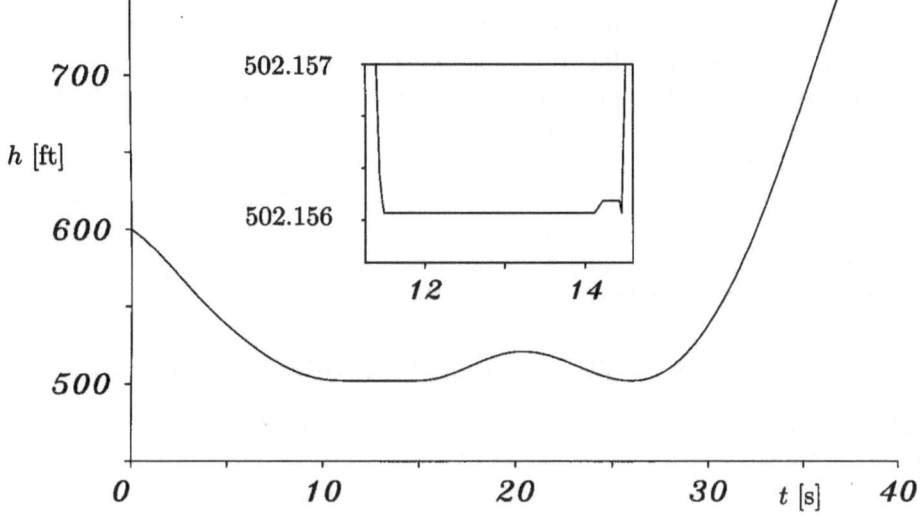

Figure 2: Altitude versus time.

Because of the petty improvement of the maximum minimal altitude, the result is of theoretical interest only, but nevertheless a challenge for a numerical method. Because of lack of appropriate theoretical results concerning the existence of boundary arcs and touch points depending on the order of the state constraint if the Hamiltonian is non-regular, it is interesting to simply report on the phenomenon of the changing of the switching structure depending on a homotopy parameter, here Θ. In [11], p. 123, the behavior of a second-order state variable inequality constraint is said to be typical for a problem with a regular Hamiltonian if the unconstrained trajectory is followed first by a trajectory with a touch point and then with a constrained subarc. However, this behavior is neither typical for such a second-order state constraint in the regular case, as can be seen from the reentry problem investigated in [49], nor for higher-order state constraints in the non-regular case, as can be seen here.

Results for different windshear models are investigated in [2], e.g., for the wind model (5), the considered aircraft, a Boeing 727, can survive a maximum windshear intensity of $k \approx 2$ which corresponds to a maximum wind velocity difference of about $200\,ft/s$ and a minimax altitude of $1\,ft$.

Summary

If precise numerical results must be obtained for optimal control problems exhibiting a complicated switching structure, the multiple shooting method is superior to direct methods. The limit of problems which today can be solved by the multiple shooting method can be seen, for example, from the investigations of ion-driven gravity-assisted mission to asteroids; see [13] and [14]. The multiple rendezvous mission to certain asteroids, for example, is modelled by an optimal control problem subject to several control and state variable inequality constraints. The fully optimized trajectory including a Moon swing-by and spiraling down to and up from low asteroid orbits exhibits more than 50 switching points. Moreover, the extremely different linear expansions between interplanetary flight and orbit around small asteroids make these problems ill-conditioned. For complex missions like this, the outstanding accuracy provided by the multiple shooting method is no longer an unnecessary by-product of an over-precise method, but of vital and decisive importance for the mission planning, and the high accuracy renders possible the computation of the optimal trajectory at all.

Optimal Control Problems with Complicated Dynamics

Difficulties of a different kind appear for problems with complicated equations of motion. Such equations occur mainly in robotics, vehicle dynamics, biomechanics, and chemical engineering. The following problem may be representative for this class.

Minimum-Time and Minimum-Energy Control of an Industrial Robot

The motion of a three-degrees-of-freedom robot, the mathematical model of which was developed in [47], can be described by the differential equations for a multibody system,

$$M(q)\,\ddot{q} = \chi^d(q, \dot{q}) + \chi^g(q) + u \ . \tag{12}$$

Here M is the 3×3 symmetric, positive definite matrix of the moments of inertia, and the vector q denotes the relative angles between adjacent arms. The vector functions χ^d and χ^g describe the moments due to Coriolis and centrifugal forces, and due to gravitational forces, respectively. The control vector u is proportional to the control input signals of the robot drives.

A more detailed impression of the complexity of the equations of motion can be obtained from the relation for $\chi^d(q, \dot{q})$,

$$\chi_i^d(q, \dot{q}) = \sum_{j=1}^{3} \left[\sum_{k=1}^{3} \Gamma_{jk}^i \, \dot{q}_k \right] \dot{q}_j$$

with the Christoffel symbols

$$\Gamma^i_{jk} := -\frac{1}{2}\left[\frac{\partial M_{ij}}{\partial q_k} + \frac{\partial M_{ik}}{\partial q_j} - \frac{\partial M_{jk}}{\partial q_i}\right]$$

where each element M_{ij} of the matrix M is a complicated function of q, e.g.,

$$M_{11} = c_1 \sin^2(q_2 + q_3) + c_2 \sin^2(q_2 + q_3) \sin q_2 + c_3 \sin^2 q_2$$

$$+c_4 \cos^2(q_2 + q_3) + c_5 \cos^2 q_2 + c_6 .$$

The full model including all data can be found in [47] and [51]. Eighteen control and state variable inequalities constraints are imposed on the model; see [51] and [55] for further information and detailed numerical results which have been obtained by a combination of direct collocation and multiple shooting [54]. The qualitatively very different motions of the minimum-time and minimum-energy trajectories make homotopies by means of the indirect multiple shooting method very tedious. Note that the right hand side of the adjoint differential equations, automatically generated with Maple [22], consists of more than 3,000 lines of code. See Fig. 3 for a certain point-to-point minimum-time (solid line, minimum final time $t_f = 1.32\,s$) and minimum-energy trajectory (dashed-dotted line, prescribed final time $t_f = 1.60\,s$).

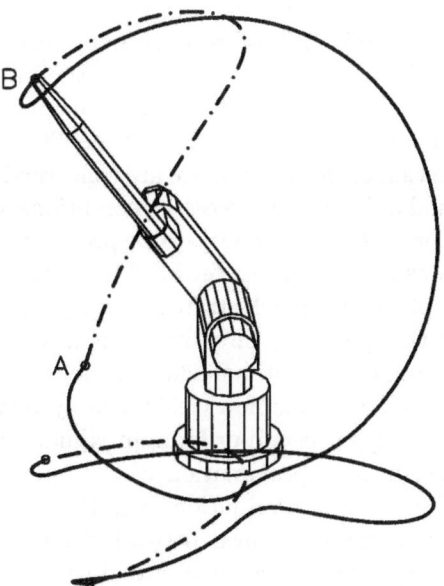

Figure 3: Minimum-time and minimum-energy trajectories of the Manutec r3.

Summary

Problems like this are typical for new control applications in robotics, vehicle dynamics, or chemical engineering. The equations of motion exhibit such a degree of complexity that they have to be established by appropriate software and, of course, the derivation of necessary conditions is then to be done either by symbolic differentiation [22] or automatic differentiation [29]. Hence, direct methods are preferable to indirect ones for problems like these, since one has not to be concerned with necessary conditions. However, appropriate general-purpose optimization software is not yet available for very large-scale optimal control problems [4], [27].

Pursuit-Evasion Game Problems

In practical applications there often arises the task to solve many related optimal control problems. Such tasks are typical for the computation of the optimal solution itself by homotopy techniques. These techniques enable the computation of optimal trajectories even in the presence of a small domain of convergence, see [24] and [25]. Especially for the indirect multiple shooting method, this domain of convergence is known to be very small, generally considerably smaller than the one of the direct collocation method, see [20], [51], and [53]–[55]. Furthermore, engineering design and sensitivity investigations require parameter soundings for which also many similar problems have to be solved. Finally, pursuit-evasion game problems, which are closely related to optimal control problems, demand the solution of often thousands of related multipoint boundary-value problems by nature. Only dynamic programming methods and indirect methods seem to be appropriate for the numerical solution of these problems because of the information structure [1] coined to pursuit-evasion games, i.e., the freedom of the players' decision to choose their controls instantaneously and independently; see [9].

Following the intention of this section, we are concerned here with the indirect multiple shooting method only. Again, necessary conditions for an equilibrium, i.e., for saddle-point trajectories, form boundary-value problems by which the so-called open-loop representations of the optimal feedback strategies [1] can be computed. For a synthesis of the optimal feedback strategies many boundary-value problems have to be solved. For this purpose, three methods have been developed so far: In [6], a continual updating of the optimal solution along saddle-point trajectories by means of the multiple shooting method yields reliably very precise optimal feedback strategies. Alternatively, a second-order approximation of the value function enables a sufficiently accurate approximation of the optimal feedback strategies even in real-time [7]. Finally, the information from the solution of many differential games can be used to train a neural network for the approximation of the optimal feedback strategies which is capable of real-time applications, too [26].

To describe the way to fight against this third kind of complexity problem, we consider a special air combat scenario, which is described in detail in [9], [10] and [42].

An Air Combat between a Fighter Aircraft and a Missile

The two players involved in the game, the pursuing missile and the evading aircraft, are both governed by the following equations of motion,

$$\dot{x} = V \cos\gamma , \tag{13a}$$

$$\dot{h} = V \sin\gamma , \tag{13b}$$

$$\dot{V} = g \left[\eta \, T_{\max}(t,h,V) - D(t,h,V,n)\right] / W(t) - g \sin\gamma , \tag{13c}$$

$$\dot{\gamma} = (g/V) \, (n - \cos\gamma) . \tag{13d}$$

Here x, h, V, and γ denote the horizontal position, the altitude, the velocity, and the flight path angle, respectively. The thrust $T := \eta \, T_{\max}$ is controlled by the power setting η, and γ by the load factor n. For more details, especially the complicated functions for the weight $W = W(t)$, the drag $D = D(t,h,V,n)$, and the lift $L = L(t,n)$ it is referred to [9]. Moreover, a state constraint, which limits the dynamic pressure, must be satisfied by the trajectory of the aircraft. This is described in [9], too.

The pursuit-evasion maneuver begins with some initial conditions

$$y(0) = y_0 \tag{14}$$

at time $t = 0$ and terminates as soon as the capture condition

$$F(y(t_f)) := (x_E(t_f) - x_P(t_f))^2 + (h_E(t_f) - h_P(t_f))^2 - d^2 = 0 \tag{15}$$

is fulfilled. This condition determines the capture time t_f. The constant d denotes the capture radius. If capture is not possible in finite time, we define $t_f := \infty$. The vector y combines the state variables of both the pursuer and the evader. The indices P and E denote variables associated with the pursuer and the evader, respectively.

The objective function of the game is defined by

$$J(u_P, u_E; t = 0, y = y_0) := t_f . \tag{16}$$

The pursuer P tries to drive, with his control vector u_P, the state y from the initial state y_0 to the terminal manifold (15) in minimum time. The evader E tries, with his control vector u_E, to avoid capture or, if escape is impossible, E tries to maximize capture time. Hence

$$\max_{u_E} \min_{u_P} J \equiv \max_{u_E} \min_{u_P} t_f \tag{17}$$

for all piecewise continuous functions u_P and u_E defined on the interval $[0, \infty[$. In Eq. (17) it is tacitly assumed that the operators min and max commute, i.e.,

$$\max_{u_E} \min_{u_P} J = \min_{u_P} \max_{u_E} J . \tag{18}$$

The above differential game is a so-called two-person zero-sum differential game with complete information [1]. The necessary conditions for saddle-point strategies in the capture zone and in the barrier, which separates capture zone from escape zone, are similar to those for optimal control functions in optimal control problems; see, e.g., [1] and [9]. However, open-loop strategies are of no significance in differential games [1], contrary to optimal (open-loop) controls for optimal control problems since they can be realized practically if, for example, appropriate (neighboring) feedback schemes are applied, see [39]–[41], [48], [49]. For differential games, however, one must compute optimal feedback strategies $U_P^*(t, y(t))$ and $U_E^*(t, y(t))$ which are related to the optimal open-loop realizations $u_P^*(t)$ and $u_E^*(t)$ by

$$u_P^*(t) := U_P^*(t, y(t)) ., \quad u_E^*(t) := U_E^*(t, y(t)) . \tag{19}$$

Because of this relation, one can synthesize the feedback strategies, if the open-loop realizations of the feedback strategies are known for each tuple (t_0, y_0) of the time-state space. In other words, the multipoint boundary-value problems based on the necessary conditions for a saddle-point trajectory must be solved for each initial point (t_0, y_0). In practice, this means that often thousands of multipoint boundary-value problems must be solved, the solutions of which then enable the approximation of the feedback strategies. An appropriate elaboration of this rough idea is still under research; see [6], [7], and [26].

It need not be mentioned that this procedure is exacting for the efficiency of numerical methods, especially if one has real-time computations in mind. As for parameter investigations in optimal control problems, efficient homotopy methods, e.g., [24] or [25], are valuable as well as implementations of the multiple shooting method for vector [35] or parallel computers. An implementation for the latter one is still under development. For some very first results, see [41].

This section is closed with a numerical result from [10]: Figure 4 shows the total flight times of pursuit-evasion maneuvers in the barrier of the game for a certain launch position of the pursuing missile, $x_P(0) = 0\,km$, $h_P(0) = 5\,km$, $V_P(0) = 250\,m/s$, whereas the evading aircraft takes initial values within its flight envelope. The flight envelope is the set of all meaningful initial values for altitude and velocity and given by the darkly shaded region in the $V_E(0)$-$h_E(0)$-plane of Fig. 4. The starting values for the related boundary-value problems have been obtained in [10] by extrapolation. Here, it is obviously possible to solve different boundary-value problems in a parallel way by proceeding according to the lines of the grid which discretizes the flight envelope. Changes of the switching structure due to the dynamic pressure constraint, however, prevent an automatic homotopy procedure.

By the way, differential game theory is a powerful tool also for the worst-case analysis of optimal control problems under uncertain disturbances [8].

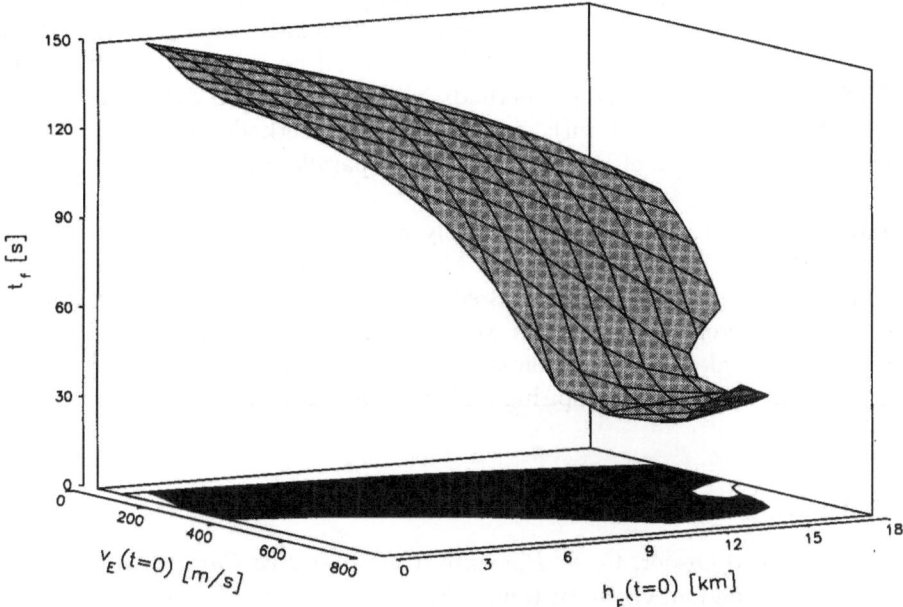

Figure 4: Total flight time over the flight envelope.

Summary

If the complexity is of the type that many related optimal control problems are to be solved, best results with respect to efficiency are obtained when making use of both homotopy techniques and special implementations for supercomputers. However, both approaches are not as well developed as the techniques for the other two degrees of complexity. The highest hurdle, for indirect methods even higher than for direct methods, is given by changing switching structures.

Conclusions

For the numerical solution of real-life optimal control and pursuit-evasion game problems, three degrees of complexity have been discussed. Firstly, optimal control problems with multiple subarcs can be computed very accurately by means of the indirect multiple shooting method. If one additionally cares much about a scrutiny of necessary conditions, the indirect multiple shooting is the only method appropriate for this task. However, if one prefers easier handling and is satisfied with moderate precision, the direct collocation method is top quality. This method is also preferable if the dynamic equations are very complicated which was denoted as the second complexity problem. Third, if many related problems are to be solved, the outcome of the comparison depends on the special task. Whereas for optimal control problems the direct collocation has some advantages because of its larger domain of convergence, only the multiple shooting method comes into question for pursuit-evasion games because it is compatible with the underlying information structure of the game. Finally, it should be noted that there is still a lack of more efficient homotopy methods in connection with either the indirect multiple shooting method or with direct collocation methods as well as a lack of efficient implementations of both approaches for parallel computers.

Acknowledgement. The author gratefully appreciates the contributions of his colleagues Michael Breitner, Francesco Montrone, and Oskar von Stryk to the solutions of the various problems discussed in this survey paper. He also is indebted to his students Peter Berkmann and Maximilian Schlemmer. Last but not least, a very special thank is due to Professor Roland Bulirsch who always took care, during so many years, of an inspiring and stimulating work atmosphere and who gave all facilities.

References

[1] Başar, T. and Olsder, G. J.: *Dynamic Noncooperative Game Theory*, Academic Press, London, Great Britain, 1982.

[2] Berkmann, P. and Pesch, H. J.: *Abort Landing under Different Windshear Conditions*, in preparation.

[3] Betts, J. T. and Huffman, W. P.: *Trajectory Optimization Using Sparse Sequential Quadratic Programming*, in: Optimal Control, Calculus of Variations, Optimal Control Theory and Numerical Methods, ed. by R. Bulirsch et. al., Birkhäuser (Inter. Series of Numer. Math. 111), Basel, Switzerland, 1993, 115–128.

[4] Betts, J. T. and Huffman, W. P.: *Path Constrained Trajectory Optimization Using Sparse Sequential Quadratic Programming*, AIAA J. of Guidance, Control, and Dynamics **16** (1993) 59–68.

[5] Bock, H. G. and Plitt, K. J.: *A Multiple Shooting Algorithm for Direct Solution of Optimal Control Problems*, in: Proc. of the 9th IFAC World Congress, Budapest, Hungary, 1984, Vol. IX, Colloquia 14.2, 09.2.

[6] Breitner, M. H.: *Construction of the Optimal Feedback Controller for Constrained Optimal Control Problems with Unknown Disturbances*, in: Control Applications of Optimization, München, Germany, 1992, ed. by R. Bulirsch and D. Kraft, Birkhäuser (Inter. Series of Numer. Math., this volume), Basel, Switzerland.

[7] Breitner, M. H.: *Real-Time Applicable Feedback Controller for Differential Games*, to appear in Proceedings of the Sixth International Symposium on Dynamic Games and Applications, St.-Jovite, Québec, Canada, 1994.

[8] Breitner, M. H. and Pesch, H. J.: *Re-entry Trajectory Optimization Under Atmospheric Uncertainty as a Differential Game*, in: Advances in Dynamic Games and Applications, ed. by T. Başar et al., Birkhäuser (Annals of the Inter. Society of Dynamic Games **1**), Basel, Switzerland, 1993.

[9] Breitner, M. H., Pesch, H. J., and Grimm, W.: *Complex Differential Games of Pursuit-Evasion Type with State Constraints. Part 1: Necessary Conditions for Optimal Open-Loop Strategies*, J. of Optim. Theory & Appl. **78** (1993), 419–441.

[10] Breitner, M. H., Pesch, H. J., and Grimm, W.: *Complex Differential Games of Pursuit-Evasion Type with State Constraints. Part 2: Numerical Computation of Optimal Open-Loop Strategies*, J. of Optim. Theory and Appl. **78** (1993), 443–463.

[11] Bryson, A. E. and Ho, Y.-C.: *Applied Optimal Control*, Rev. Printing, Hemisphere Publishing Corporation, New York, New York, 1975.

[12] Bulirsch, R.: *Die Mehrzielmethode zur numerischen Lösung von nichtlinearen Randwertproblemen und Aufgaben der optimalen Steuerung*, Oberpfaffenhofen, Germany, Report of the Carl-Cranz Gesellschaft, 1971; Reprint: Department of Mathematics, Munich University of Technology, München, Germany, 1993.

[13] Bulirsch, R. and Callies, R.: *Optimal Trajectories for an Ion Driven Spacecraft from Earth to the Planetoid Vesta*, in: Proc. of the AIAA Guidance, Navigation and Control Conference, New Orleans, Louisiana, 1991, AIAA Paper 91-2683 (1991).

[14] Bulirsch, R. and Callies, R.: *Optimal Trajectories for a Multiple Rendezvous Mission to Asteroids*, in: 42nd Inter. Astronautical Congress, Montreal, Canada, 1991, IAF-Paper IAF-91-342 (1991).

[15] Bulirsch, R. and Chudej, K.: *Ascent Optimization of an Airbreathing Space Vehicle*, in: Proc. of the AIAA Guidance, Navigation and Control Conference, New Orleans, Louisiana, 1991, AIAA Paper 91-2656 (1991).

[16] Bulirsch, R. and Chudej, K.: *Staging and Ascent Optimization of a Dual-Stage Space Transporter*, Zeitschrift für Flugwissenschaften und Weltraumforschung **16** (1992) 143–151.

[17] Bulirsch, R. and Chudej, K.: *Guidance and Trajectory Optimization under State Constraints*, in: Preprint of the 12th IFAC Symposium on Automatic Control in Aerospace — Aerospace Control 1992, München, Germany, 1992, ed. by D. B. DeBra and E. Gottzein, VDI/VDE-GMA, Düsseldorf, Germany, 1992, 533–538.

[18] Bulirsch, R., Montrone, F., and Pesch, H. J.: *Abort Landing in the Presence of Windshear as a Minimax Optimal Control Problem. Part 1: Necessary conditions*, J. Optim. Theory & Appl. **70** (1991) 1–23.

[19] Bulirsch, R., Montrone, F., and Pesch, H. J.: *Abort Landing in the Presence of Windshear as a Minimax Optimal Control Problem. Part 2: Multiple shooting and Homotopy*, J. Optim. Theory & Appl. **70** (1991) 223–254.

[20] Bulirsch, R., Nerz, E., Pesch, H. J., and von Stryk, O.: *Combining Direct and Indirect Methods in Optimal Control: Range Maximization of a Hang Glider*, in: Optimal Control, Calculus of Variations, Optimal Control Theory and Numerical Methods, ed. by R. Bulirsch et. al., Birkhäuser (Inter. Series of Numer. Math. 111), Basel, Switzerland, 1993, 273–288.

[21] Callies, R.: *Optimal Design of a Mission to Neptune*, in: Optimal Control, Calculus of Variations, Optimal Control Theory and Numerical Methods, ed. by R. Bulirsch et. al., Birkhäuser (Inter. Series of Numer. Math. 111), Basel, Switzerland, 1993, 341–349.

[22] Char, B. W., Geddes, K. O, Gonnet, G. H., Leong, B. L., Monagan, M. B., and Watt, S. M.: *Maple V, Language Reference Manual*, Springer, New York, New York, 1991.

[23] Deuflhard, P.: *A Modified Newton Method for the Solution of Ill-conditioned Systems of Nonlinear Equations with Application to Multiple Shooting*, Numerische Mathematik **22** (1974) 289–315.

[24] Deuflhard, P.: *A Stepsize Control for Continuation Methods and its Special Application to Multiple Shooting Techniques*, Numerische Mathematik **33** (1979) 115–146.

[25] Deuflhard, P., Pesch, H. J., and Rentrop, P.: *A Modified Continuation Method for the Numerical Solution of Nonlinear Two-Point Boundary Value Problems by Shooting Techniques*, Numerische Mathematik **26** (1976) 327–343.

[26] Gabler I., Miesbach S., Breitner M. H., and Pesch, H. J.: *Synthesis of Optimal Strategies for Differential Games by Neural Networks*, Schwerpunktprogramm der Deutschen Forschungsgemeinschaft, Anwendungsbezogene Optimierung u. Steuerung, Munich University of Technology, München, Germany, Report No. 468, 1993.

[27] Gill, P. E.: *Large-Scale SQP Methods and Their Application in Trajectory Optimization*, in: Control Applications of Optimization, München, Germany, 1992, ed. by R. Bulirsch and D. Kraft, Birkhäuser (Inter. Series of Numer. Math., this volume), Basel, Switzerland.

[28] Gill, P. E., Murray, W., Saunders, M. A., and Wright, M. H.: *User's Guide for NPSOL (Version 4.0)*, Department of Operations Research, Stanford University, California, Report SOL 86-2, 1986.

[29] Griewank, A.: *Automatic Evaluation of Discrete Adjoints with Logarithmic Increase in Storage*, in: Control Applications of Optimization, München, Germany, 1992, ed. by R. Bulirsch and D. Kraft, Birkhäuser (Inter. Series of Numer. Math., this volume), Basel, Switzerland.

[30] Hargraves, C. R. and Paris, S. W.: *Direct Trajectory Optimization Using Nonlinear Programming and Collocation*, AIAA J. of Guidance, Control, and Dynamics **10** (1987) 338–342.

[31] Hartl, R. F., Sethi, S. P., and Vickson, R. G.: *A Survey of the Minimum Principles for Optimal Control Problems with State Constraints*, Institut für Ökonometrie, Operations Research und Systemtheorie, Vienna University of Technology, Vienna, Austria, Report Nr. 153, 1992.

[32] Hiltmann P.: *Numerische Lösung von Mehrpunkt-Randwertproblemen und Aufgaben der optimalen Steuerung mit Steuerfunktionen über endlichdimensionalen Räumen*, Thesis, Munich University of Technology, München, Germany, 1990; see also Schwerpunktprogramm der Deutschen Forschungsgemeinschaft, Anwendungsbezogene Optimierung u. Steuerung, Munich University of Technology, München, Germany, Report No. 448, 1993.

[33] Hiltmann, P., Chudej, K., and Breitner, M. H.: *Eine modifizierte Mehrzielmethode zur Lösung von Mehrpunkt-Randwertproblemen — Benutzeranleitung*, Sonderforschungsbereich 255 der Deutschen Forschungsgemeinschaft, Transatmosphärische Flugsysteme, Munich University of Technology, München, Germany, Report No. 14, 1993.

[34] Jacobson, D. H., Lele, M. M., and Speyer, J. L.: *New Necessary Conditions of Optimality for Control Problems with State-Variable Inequality Constraints*, J. of Math. Anal. and Appl. **35** (1971) 255–284.

[35] Kiehl, M.: *Vectorizing the Multiple-Shooting Method for the Solution of Boundary-Value Problems and Optimal Control Problems*, in: Proc. of the 2nd Inter. Conference on Vector and Parallel Computing Issues in Applied Research and Development, Tromsø, Norway, 1988, ed. by J. Dongarra et. al., Ellis Horwood, London, Great Britain, 1989, 179–188.

[36] Kraft, D.: *FORTRAN Computer Programs for Solving Optimal Control Problems*, Report 80-03, Institute for Flight Systems Dynamics, German Aerospace Research Establishment DLR, Oberpfaffenhofen, Germany, 1980.

[37] Kraft, D.: *On Converting Optimal Control Problems into Nonlinear Programming Codes*, in: Computational Mathematical Programming, ed. by K. Schittkowski, Springer (NATO ASI Series 15), Berlin, Germany, 1985, 261–280.

[38] Kugelmann, B., Mihatsch, O., Mikulski, L., and Schmidt, W.: *Optimal Design of Elastic Arches in Combination with Bifurcation Theory*, submitted for publication; see also Schwerpunktprogramm der Deutschen Forschungsgemeinschaft, Anwendungsbezogene Optimierung u. Steuerung, Munich University of Technology, München, Germany, Report No. 477, 1993.

[39] Kugelmann, B. and Pesch, H. J.: *New General Guidance Method in Constrained Optimal Control. Part 1: Numerical Method*, J. of Optim. Theory & Appl. **67** (1990) 421–435.

[40] Kugelmann, B. and Pesch, H. J.: *New General Guidance Method in Constrained Optimal Control. Part 2: Application to Space Shuttle Guidance*, J. of Optim. Theory & Appl. **67** (1990) 437–446.

[41] Kugelmann, B. and Pesch, H. J.: *Serielle und parallele Algorithmen zur Korrektur optimaler Flugbahnen in Echtzeit-Rechnung*, in: Jahrestagung der Deutschen Gesellschaft für Luft- und Raumfahrt, Friedrichshafen, Germany, 1990, DGLR-Jahrbuch 1990 **1** (1990) 233–241.

[42] Lachner, R., Breitner, M. H., Pesch, H. J.: *Efficient Numerical Solution of Differential Games with Application to Air-Combat*, Schwerpunktprogramm der Deutschen Forschungsgemeinschaft, Anwendungsbezogene Optimierung u. Steuerung, Munich University of Technology, München, Germany, Report No. 466, 1993.

[43] Maurer, H.: *Optimale Steuerprozesse mit Zustandsbeschränkungen*, Habilitationsschrift, University of Würzburg, Würzburg, Germany, 1976.

[44] Miele, A., Wang, T., Melvin, and W. W.: *Optimal Abort Landing Trajectories in the Presence of Windshear*, J. of Optim. Theory & Appl. **55** (1987) 165–202.

[45] Oberle, H. J.: *Numerische Berechnung optimaler Steuerungen von Heizung und Kühlung für ein realistisches Sonnenhausmodell*, Habilitationsschrift, Munich University of Technology, München, Germany, 1982.

[46] Oberle, H. J. and Grimm, W.: *BNDSCO—A Program for the Numerical Solution of Optimal Control Problems*, Internal Report No. 515-89/22, Institute for Flight Systems Dynamics, German Aerospace Research Establishment DLR, Oberpfaffenhofen, Germany, 1989.

[47] Otter, M. and Türk, S.: *The DFVLR Models 1 and 2 of the Manutec r3 Robot*, DLR-Mitteilungen 88-13, Institute for Flight Systems Dynamics, German Aerospace Research Establishment DLR, Oberpfaffenhofen, Germany, 1988.

[48] Pesch, H. J.:*Real-time Computation of Feedback Controls for Constrained Optimal Control Problems. Part 1: Neighbouring Extremals*, Optimal Control Applications and Methods **10** (1989) 129–145.

[49] Pesch, H. J.:*Real-time Computation of Feedback Controls for Constrained Optimal Control Problems. Part 2: A Correction Method Based on Multiple Shooting*, Optimal Control Applications and Methods **10** (1989) 147–171.

[50] Pesch, H. J.: *Offline and Online Computation of Optimal Trajectories in the Aerospace Field*, in: Applied Mathematics in Aerospace Science and Engineering, 12th Course of the International School of Mathematics "G. Stampacchia", Erice, Italy, 1991, ed. by A. Miele and A. Salvetti, Plenum Publishing Corporation, New York, New York, 1994; see also Sonderforschungsbereich 255 der Deutschen Forschungsgemeinschaft, Transatmosphärische Flugsysteme, Munich University of Technology, München, Germany, Report No. 9, 1992.

[51] Pesch, H. J., Schlemmer, M., and von Stryk, O.: *Minimum-Energy and Minimum-Time Control of Three-Degrees-Of-Freedom Robots. Part 1: Mathematical Model and Necessary Conditions. Part 2: Numerical Methods and Results for the Manutec r3 Robot*, in preparation.

[52] Stoer, J. and Bulirsch, R.: *Introduction to Numerical Analysis*, 2nd Ed., Springer, New York, New York, 1993.

[53] von Stryk, O.: *Numerical Solution of Optimal Control Problems by Direct Collocation*, in: Optimal Control, Calculus of Variations, Optimal Control Theory and Numerical Methods, ed. by R. Bulirsch et. al., Birkhäuser (Inter. Series of Numer. Math. 111), Basel, Switzerland, 1993, 129–143.

[54] von Stryk, O. and Bulirsch, R.: *Direct and Indirect Methods for Trajectory Optimization*, Annals of Operations Research **37** (1992) 357–373.

[55] von Stryk, O. and Schlemmer, M.: *Optimal Control of the Industrial Robot Manutec r3*, in: Control Applications of Optimization, München, 1992, ed. by R. Bulirsch and D. Kraft, Birkhäuser (Inter. Series of Numer. Math., this volume), Basel, Switzerland.

2 Theoretical Aspects of Optimal Control and Nonlinear Programming

International Series of Numerical Mathematics, Vol. 115, © 1994 Birkhäuser Verlag Basel

Continuation Methods in Boundary Value Problems

Vassili V. Dikoussar*

Introduction

It is known that the solution of an optimal control problem can be reduced to the solution of a two-point boundary value problem. This problem includes solving two groups of nonlinear differential equations (direct and adjoint).

One of the most effective numerical techniques for the solution of optimal control problems is discretizing the differential equations. This approach combines a nonlinear programming problem with a discretization. The resulting problem is characterized by matrices which are large and sparse. The iteration routine requires a good initial guess for the adjoint variables. Such a treatment of the problem has been successfully utilized for applications [1], [2]. Two new techniques yielded extremely important contributions toward the numerical solution of nonlinear programming. One of the methods has been called the predictor-corrector or pseudo arc-lengths continuation method. The second method is the simplicial or piecewise linear method [3].

This paper extends the continuation method to ill-posed problems. Our approach utilizes the system dynamics and adjoint differential equations defined by the maximum principle. The resulting boundary value problem is characterized by Jacobi (or Hesse) matrices which are small and ill-conditioned. The method of introduction of a parameter helps to overcome the difficulties associated with adjoint initial values. Convergence of the iterations is often sensitive to the accuracy of the adjoint guess. For ill-posed linear systems we propose factor analysis methods. Typical factor methods include eigenvalue problem. We extend the effectiveness of the iteration methods for large scale systems. In this paper we present some basic iterative methods, discuss their practical implementation. We show how to extend Jacobi iterations, Gauss–Seidel iterations, and relaxation parameter techniques.

More details for iterative and direct methods can be found in papers [2], [5].

Factor Analysis for Solving Ill-Posed Linear Problems

Let us consider an arbitrary linear system

$$AX = B, \tag{2.1}$$

*Computer Center, Russian Academy of Sciences, Vavilov Str. 40, Moscow, 117967, Russia

where A is an real $m \times n$ matrix. A system (2.1) is said to be consistent if it has at least one solution and inconsistent if it has no solution.

Generally speaking, the system (2.1) is inconsistent. We transform the system as follows: multiply both sides of (2.1) by the transposed matrix A^T

$$A^T A Y = A^T B = B_0, \quad A^T A = A_0. \tag{2.2}$$

The system (2.2) is always consistent and can have nonunique solution. For (2.2) the regularization method could be reduced to solving an equation

$$(A_0 + \delta I) Z = B_0, \quad \delta > 0, \quad A_1 = A_0 + \delta I, \tag{2.3}$$

where δ is parameter of regularization. It is proved that for $\delta \to 0$ the solution $Z \to Y_0 \subseteq Y$; Y is the set of solution (2.2). The main difficulty in applying the regularization method lies in deducing algorithmic principles for choosing the regularization parameter δ.

Here is suggested a technique for calculating Z (2.3) for any δ, including $\delta = 0$.

Let us introduce the set of n-dimensional arithmetic vectors

$$Z_1 = \{1\,0\ldots0\}^T, \; Z_2 = \{0\,1\,0\ldots0\}^T, \ldots, Z_n = \{0\,0\,0\ldots1\}^T \tag{2.4}$$

and apply to (2.4) matrix operator A_1

$$A_1 Z_1 = B_1, \; A_1 Z_2 = B_2, \ldots A_1 Z_n = B_n.$$

Now we make up the observation matrix [4]

$$A_2 = (B_1 B_2 \ldots B_n) = A_1$$

which is real and symmetric; the rank A_1 is equal to n for $\delta > 0$.

The factor analysis (the method of principal components, or component analysis) [4] gives us

$$B_j = \sum_{i=1}^{p} a_{ji} F_i, \quad p \le n, \quad F_k = \sum_{i=1}^{n} \beta_{ki} B_i, \quad k = 1, \ldots, p, \; j = 1, \ldots, n. \tag{2.5}$$

Here F_k are the factors; a_{ij} and β_{ki} are the coefficients of the factor solution. Theoretically, the vectors F_k (2.5) form an orthonormal basis in R^n (R^n is n-dimensional real Euclidean space) for $\delta > 0$. Any vector B_0 (2.3) can be uniquely represented as a linear combination of the basis vectors F_k, $k = 1, \ldots, n$ ($p = n$)

$$B_0 = \gamma_1 F_1 + \gamma_2 F_2 + \cdots + \gamma_n F_n, \quad \gamma_k = (B_0, F_k) \tag{2.6}$$

Substituting the F_k from (2.5) into (2.6) we obtain

$$B_0 = b_1 B_1 + b_2 B_2 + \cdots + b_n B_n.$$

Proposition. The solution of system (2.3) is the linear combination of (2.4)

$$Z = b_1 Z_1 + b_2 Z_2 + \cdots + b_n Z_n. \tag{2.7}$$

Clearly, the smaller δ, the closer Z to the solution Y_0, so that taking very small δ seems to be appropriate. However the resulting quantities of the coefficients a_{ji} (2.5) are

$$a_{ji} = b_{ji}\sqrt{\lambda_i}/\sqrt{b_{1i}^2 + b_{2i}^2 + \cdots + b_{ni}^2},$$

where λ_i is an eigenvalue of the matrix A_1 and b_{1i}, $i = 1, \ldots, n$ are the coordinates of the eigenvector. Hence, the accuracy of the solution (2.3) is determined by the precision of calculating the eigenvalues λ_i. In the case of a singular matrix A_0 we have $\lambda_i = \delta$ for $i > r$, where r is the rank of A_0 (we consider that the eigenvalues λ_i are ordered by diminution).

Theorem 1. The solution (2.7) is a continuous function of the parameter δ.

If the system (2.1) has a unique solution then in (2.6) $p = n$. Otherwise $p < n$ and we have a generalized solution (2.7), where

$$B_0 = \gamma_1 F_1 + \gamma_2 F_2 + \cdots + \gamma_p F_p, \quad p < n.$$

Differential Equation Methods for Linear Systems

We consider a two-parameter dependent formula

$$[I + \delta A]X = \delta B + \alpha X, \quad 0 \le \alpha \le 1. \tag{3.1}$$

Here I is the identity matrix, δ is a small parameter, α is the parameter of the formula.

The code (3.1) defines the iteration

$$X_{n+1} = \delta[I + \delta A]^{-1}B + \alpha_n[I + \delta A]^{-1}X_n \tag{3.2}$$

We start with $\alpha_0 = 0$ and end with $\alpha_n = 1$. The main rule for choosing the sequence $\alpha_1, \ldots, \alpha_n$ is the principle of Runge-Kutta for the numerical solution of differential equations.

Now we can list the main advantages of the iterative method (3.2):

- it is "easy" to calculate the inverse matrix to control the local error;

- the matrix $I + \delta A$ is close to an orthogonal matrix;

- the method has "good" convergence properties even if the spectral radius of the matrix A is close to unity.

We can forecast α_{n+1} by processing the accumulated information about α_n, $n = 1, \ldots$.

The method (3.1) can be improved in different ways. Consider, for example a splitting

$$A = L + D + L^T, \tag{3.3}$$

where D is diagonal and L is strictly lower triangular. The result is

$$X_{n+1} = \delta[I + \delta D]^{-1} B + \alpha_n [I + \delta D]^{-1} [I - L] X_n - [I + \delta D]^{-1} L^T X_{n-1}$$

For a fixed value α_n we can construct a Gauss-Seidel iteration and introduce the relaxation parameter even for every component of the vector X. Practical schemes for estimating the optimum parameters are based on the principle of Runge-Kutta.

A Newton-like Method

Let us now consider the nonlinear vector equation

$$f(y) = 0, \quad f = (f_1, \ldots, f_n)^T, \quad y = (y_1, \ldots, y_n). \tag{4.1}$$

Together with (4.1) we introduce the vector-valued function

$$g(y) = BAf(y) + \lambda B(y - y_*)^T. \tag{4.2}$$

Here B and A are piecewise constant matrices, y_* is a root of equation (4.1), λ is a scalar parameter, $\lambda > 0$. The solution of the equation (4.2) can be obtained by the iteration

$$y_{k+1} = y_k + B_k A_k f(y_k) + \lambda_k B_k (y_{k+1} - y_k), \tag{4.3}$$

where the index k denotes the number of iterations. We assume that the derivative of $f(y)$ (4.1) exists. From the condition of maximum convergence we obtain (I is identity matrix)

$$B = -[Af_y(y) + \lambda I]^{-1}. \tag{4.4}$$

For $\lambda_k = 0$ we obtain Newton's method, provided that $f'_y(y_k)$ and A_k are non-singular.

Now we put $A_k = \mu_k [f'_y(y_k)]^T$, μ_k is a scalar parameter, $\mu_k > 0$. Then from (4.3), (4.4) we have

$$y_{k+1} = y_k + \mu_k [I - \lambda_k B_k]^{-1} [f'_y(y_k)]^T f(y_k). \tag{4.5}$$

The iteration (4.5) has two parameters λ_k and μ_k. By choosing μ_k and λ_k we can regulate the convergence of the iteration process (4.5).

Boundary Value Problem

The boundary value problem is as follows. It is required to find the solution of the differential equation

$$\dot{x} = f(x, \psi, t), \quad x(0) = x_0, \tag{5.1}$$

$$\dot{\psi} = g(x, \psi, t), \quad t \in [0, t_1] \tag{5.2}$$

satisfying the boundary conditions:

$$\psi_1(t_1) = 1, \quad x_i(t_1) = x_{i1}, \quad i = 2, \ldots, n. \tag{5.3}$$

Here x and ψ are n-dimensional vectors, x is he state variable, ψ the conjugate variable, t_1 is fixed. Without loss of generality, we consider x_1 as functional.

The solution (5.1), (5.2) depends on arbitrary parameters $\psi_i(0)$, $i = 1, \ldots, n$. We must choose $\psi_i(0)$ which satisfies (5.3).

Problem A

We consider the system of embedded intervals

$$[0, t_{11}], \ [0, t_{12}], \ \ldots, \ [0, t_{1k}], \quad t_{1k} = t_1, \quad t_{11} < t_{12} < \cdots < t_{1k}$$

and introduce the notations

$$\psi_1 = y_1, \quad x_i = y_i, \quad i = 2, \ldots, n, \quad \psi(0) = a$$

On the $[0, t_{11}]$ we consider the following problem.

Problem A_1: Find the minimum $R_1(a)$:

$$R_1 = \|y(t_{11}) - y(t_1)\| = \sum_{i=1}^{n} (y_i(t_{11}) - y_i(t_1))^2. \tag{6.1}$$

It is known that problem (6.1) can be reduced to the solution of the nonlinear equation (4.1), where $f(y)$ is the gradient of R_1.

In the iteration procedure (4.5) we consider λ_k as a large parameter and choose $\mu_k/\lambda_k < 1$. Therefore, we can "easily" calculate the inverse matrices (4.4), (4.5). The necessary derivatives are calculated by a factor analysis technique. The interval $[0, t_{11}]$ is sufficiently small and practically there are no problems in solving (6.1).

We denote the vector solution (6.1) as a_1. Now we continue this process to the problem A_k. As a first approximation for the problem A_2 we take a_1 from problem A_1. By using linear or quadratic extrapolation one can predict $a_3, \ldots, a_k$.

For solving the initial boundary value problem we must turn λ_k to zero. One usually succeeds to make λ_k equal to zero if (5.1)–(5.3) is a well-posed problem.

Problem B

To find the solution of the boundary value problem (5.1)–(5.3) we proceed in the following way. On $[0, t_{11}]$ one chooses an intermediate point τ_{11}. Thus we have two intervals: $[0, \tau_{11}], [\tau_{11}, t_{11}]$. One does a forward integration of (5.1), (5.2) on $[0, \tau_{11}]$ and a backward one on $[\tau_{11}, t_{11}]$. Let us consider the new problem.

Problem B_1: Find

$$\min_{a,b} R_{11}, \quad R_{11} = \sum_{i=1}^{2n} (u_i(\tau_{11}) - v_i(\tau_{11}))^2$$

$$a = \psi(0), \quad b = \psi(t_{11}), \quad \psi_1(t_{11}) = 1, \quad x_i(t_{11}) = x_{i1}, \quad i = 2, \ldots, n,$$

u_i and v_i are the results of integration (5.1), (5.2) on $[0, \tau_{11}], [\tau_{11}, t_{11}]$. This process can be continued by taking a new intermediate point τ_{12} on $[0, t_{12}]$ and so on.

The iterative solution of problem B gives us the sequence of well-posed problems B_k. By choosing intermediate points we can easily make λ_k equal to zero.

References

[1] Betts, J.T., and Huffman, W.P., "Trajectory Optimization on a Parallel Processor", Journal of Guidance, Control and Dynamics, Vol. 14, No. 2, March–April, 1991.

[2] Bulirsch, R., and Stoer, J., "Introduction to Numerical Analysis", Springer–Verlag, 1992.

[3] Allgover, E.L., and Georg, K., "Numerical Continuation Methods. An Introduction", Springer–Verlag, 1990.

[4] Harman, H., "Modern Factor analysis", The University of Chicago Press, 1960.

[5] Loan, Ch., "A Survey of Matrix Computations", Cornell Theory Center Technical Report, 1990.

International Series of Numerical Mathematics, Vol. 115, © 1994 Birkhäuser Verlag Basel

Second Order Optimality Conditions for Singular Extremals

A.V. Dmitruk*

Abstract

We consider the class of optimal control problems, linear in the control, with control bounded by linear inequalities, and with terminal equality and inequality constraints. Both control and state variables are multidimensional, and the examined control is totally singular.

For such problems we suggest quadratic-order necessary and sufficient conditions for a weak and a so-called Pontryagin minimum, which is a minimum of intermediate type between classic weak and strong minima. Necessary conditions transform into sufficient ones only by strengthening an inequality, what is similar to conditions in the classical analysis and calculus of variations (close pairs of conditions).

Keywords. Singular extremal, weak and Pontryagin minimum, quadratic order of estimation, necessary and sufficient conditions, third variation of Lagrange function.

Statement of the Problem

The problem of consideration is:

$$J = \kappa_0(p) \to \min, \quad K(p) = 0, \tag{1}$$

$$\kappa_i(p) \leq 0, \quad i = 1, \ldots, \nu, \tag{2}$$

$$\dot{x} = f(x, t) + F(x, t)u, \tag{3}$$

$$u(t) \in U(t). \tag{4}$$

Here $p = (x_0, x_1)$, $x_0 = x(t_0)$, $x_1 = x(t_1)$, the time interval $[t_0, t_1]$ is fixed; x is a Lipschitz function and u is a bounded measurable function, the dimensions of x, u, K are $d(x)$, $d(u)$, $d(K)$ respectively. The system (3) is linear in the control u, but nonlinear in the state variable x.

Assumptions.

A1) All functions κ, K, f, F are twice continuously differentiable in x and Lipschitzian in t.

A2) The set $U(t)$ is convex, continuous (in the Hausdorff metric) and uniformly solid in t.

*Central Economic and Mathematical Institute, Russian Academy of Sciences, Moscow 117418, ul. Krasikova, 32, Russia

Basic Notions

We denote by W the space of all pairs of functions $w = (x, u)$ of the abovementioned type. Let $w^o = (x^o, u^o)$ be an examined trajectory. We assume that $w^o(t)$ is totally singular over the whole $U(t)$, and $u^o(t)$ is continuous.

The Pontryagin Maximum Principle (MP) says that there exist Lagrange multipliers $\alpha = (\alpha_o, \dots, \alpha_\nu) \geq 0$, $c \in R^{d(K)}$, and a Lipschitz function $\psi(t)$ such that $|\alpha| + |c| = 1$, $\dot\psi = -H_x$, $\psi(t_0) = l_{x_o}$, $\psi(t_1) = -l_{x_1}$, and (because of the total singularity of w^o) $H_u = 0$, where $l(p) = \alpha \cdot \kappa(p) + c \cdot K(p)$,

$$H(x, u, t) = \psi[f(x, t) + F(x, t)u], \qquad \kappa = (\kappa_o, \dots, \kappa_\nu).$$

To simplify further considerations we assume here that such a collection $\lambda = (\alpha, c, \psi)$ is unique. (For a general case see [12–16].)

Without loss of generality we take $w^o(t) \equiv 0$, and $\kappa_i(0, 0) = 0$, $\forall i = 0, 1, \dots, \nu$, i.e. all indices are active.

It is generally known that the fulfillment of the MP does not guarantee the optimality of the given trajectory. In particular it is true for our class of problems (1)–(4). (One of the first examples: the famous Lawden's spiral.) Therefore, other optimality conditions are desirable, which may be obtained via investigations of the second order. Such investigations for this class of problems (in some special statements) have been done since the early 1960-ies by many authors: Kelley, Kopp, Moyer, Bryson, Robbins, Goh, Vapnyarsky, Bolonkin, Speyer, Jacobson, Bell, McDanell, Powers, Gabasov, Kirillova, Krener, Agrachiov, Gamkrelidze, Milyutin, Knobloch, Zelikin, Gurman, Dykhta, Lamnabhi-Lagarrigue, Stefani and others (see [2–10] and references therein). An overwhelming majority of works are devoted to higher-order *necessary* conditions which have *a pointwise* character (we call them Legendre conditions). A large number of such conditions have been obtained. But the question have been remained open: what is a full set of necessary conditions? Few works are devoted to *sufficient* conditions, but these conditions obtained till now are rather far from necessary ones (in particular because they include conditions of Frobenius type, which are not conditions of any finite order). We investigate this class of problems to obtain both necessary and close to them sufficient second order conditions of optimality.

First of all, we should define more accurately what means "optimality" and what are "second order" conditions.

Types of minimum. Speaking of optimality, we should point out a type of minimum under consideration. We consider the following two types of minimum, a weak and a so-called Pontryagin minimum (Π-minimum).

A weak minimum, known from the classics, is a minimum in the norm

$$\| w \| = \| x \|_\infty + \| u \|_\infty,$$

or, in other words, a minimum with respect to uniformly small variations. A Pontryagin minimum includes in addition so-called "needle-type" variations.

Definition. We say that $w^o \equiv 0$ is *a Pontryagin minimum point* in problem (1)–(4), if for all N there exists an $\varepsilon > 0$ such that w^o is a minimum point in problem (1)–(4) on the set

$$\| x \|_\infty < \varepsilon, \quad \| u \|_1 < \varepsilon, \quad \| u \|_\infty \leq N.$$

In other words, there can exist no sequence $w_n = (x_n, u_n)$ such that

$$\| x_n \|_\infty \to 0, \quad \| u_n \|_1 \to 0, \quad \| u_n \|_\infty \leq \mathcal{O}(1), \tag{5}$$

all constraints (1)–(4) are satisfied, and for all n, $J(p_n) < J(p^0)$.

We call *Pontryagin sequences* those satisfying (5); the set of all such sequences we denote by Π. A Pontryagin minimum (Π-minimum) obviously occupies an intermediate position between the classic weak and strong minima.

Quadratic order of estimation. Speaking of "second order" conditions, to be more accurate, we should point out a quadratic functional of estimation, regarding to which all the functions in the problem to be considered. For example, in the classical calculus of variations (where $\dot{x} = u$) the appropriate functional is

$$\gamma_0(w) = | x(t_o) |^2 + \int | u(t) |^2 \, dt. \tag{6}$$

(here and throughout the paper all integrals are taken over the whole interval $[t_0, t_1]$). This functional is adequate also for optimal control problems nonlinear in the control (see [11]), but obviously it is too rough for the problem (1)–(4), since the last is linear in u. It turns out that the adequate quadratic functional of estimation for this class of problems is

$$\gamma(w) = | x(t_0) |^2 + | y(t_1) |^2 + \int | y(t) |^2 \, dt, \tag{7}$$

where

$$\dot{y} = u, \qquad y(t_0) = 0 \tag{8}$$

(the latter is preserved throughout the paper). Note that the control u does not come as such in the quadratic order (9); it comes only through the new state variable y. Here we give conditions of this order γ.

Consider the *Lagrange function*

$$\Phi(w) = l(p) + \int ((\psi, \dot{x}) - H(x, u, t)) \, dt, \tag{9}$$

and it's second variation, the quadratic functional

$$\Omega(w) = (l''p, p) - \int ((H_{xx}x, x) + 2(x, H_{xu}u)) \, dt. \tag{10}$$

Define the matrices $A(t) = f_x(0,t)$, $B(t) = F(0,t)$, and the tensor $R(t) = F_x(0,t)$ in such a way that the equation (3) is reduced to

$$\dot{x} = A(t)x + B(t)u + (R(t)x, u) + \text{h.o.t.} \tag{11}$$

Let $\mathcal{K}$ be the so-called critical cone, consisting of all $w = (x,u)$ in W such that $\kappa'(0,0)p \leq 0$, $K'(0,0)p = 0$, and

$$\dot{x} = A(t)x + B(t)u. \tag{12}$$

Remind the well-known *Goh transformation*: $(x,u) \rightarrow (\xi,y,u)$, where $\xi = x - By$, and hence

$$\dot{\xi} = A\xi + B_1 y, \quad B_1 = AB - \dot{B}. \tag{13}$$

Under this transformation the functional (10) takes the form

$$\Omega(\xi,y,u) = g(\xi_0,\xi_1,y_1) + \int ((D\xi,\xi) + (P\xi,y) + (Qy,y) + (Vy,u)) \, dt, \tag{14}$$

where g is a terminal quadratic form, $Q(t)$ is a symmetric and $V(t)$ is a skew-symmetric Lipschitz matrices.

We also recall the Goh conditions (firstly proposed in [3]):

$$\forall t \quad a) \ \ V(t) = 0, \quad \text{and} \quad b) \ \ Q(t) \geq 0. \tag{15}$$

which obviously have a pointwise character. Now we pass to the optimality conditions obtained.

Conditions of a Weak Minimum

Consider firstly the case when $u^o(t)$ goes strictly inside $U(t)$, i.e. for some $\varepsilon > 0$, for every t the ε-neighborhood of $u^0(t)$ is contained in $U(t)$. It is clear that in this case constraint (4) is not essential, so we can neglect it.

Theorem 1 [2, 12]. *a) Let w^0 be a weak minimum point in problem (1)–(3). Then the Goh conditions (15) hold, and moreover*

$$\Omega(w) \geq 0 \quad \text{for all } w \in \mathcal{K}. \tag{16}$$

b) Suppose that the Goh conditions (15) hold, and for some $a > 0$

$$\Omega(w) \geq a\gamma(w) \quad \text{for all } w \in \mathcal{K}. \tag{17}$$

Then w^0 is a weak minimum point in problem (1)–(3).

As one can see, these necessary and sufficient conditions are close each to other; we call them a close pair of conditions. In this sense these conditions are precisely analogous to those in the analysis and the calculus of variations.

It is worth to note here (and it is known from the classics) that the full set of necessary conditions must definitely contain an inequality of the form (16), which is non-Legendre, and just by the strengthening of which necessary conditions transform into sufficient ones.

Now let us consider the general case, i.e. when $u^0(t)$ may contact the boundary of $U(t)$. Here we take some assumptions about the character of contacts:

*B*1) In a neighbourhood of contacts the control set $U(t)$ is a polyhedron.

*B*2) As before, the examined extremal $w^o(t)$ is totally singular with respect to the whole $U(t)$.

*B*3) Contacts with the bound of $U(t)$ have in some sense "good" character.

The critical cone in this case is $\mathcal{R} = \mathcal{K} \cap \mathcal{N}$, where the cone

$$\mathcal{N} = \{\bar{w} \in W : \ \bar{u}(t) \in N(t) = con(U(t) - u^0(t))\}$$

is generated by the pointwise cone $N(t)$.

Here the Goh conditions must be regarded with respect to the maximal linear subspace $l(t)$ in $N(t)$, i.e. if $P(t)$ is the projector onto $l(t)$, then

$$\forall t \quad a) \ \ PVP(t) = 0, \quad b) \ \ PQP(t) \geq 0. \tag{18}$$

Theorem 1 is still valid for this case, if we replace the Goh conditions (15) by (18). This was proved by A.A. Milyutin but is yet unpublished.

Pontryagin Minimum for a Free Control

Assume as before that $u^0(t)$ goes strictly inside $U(t)$, but now let us consider a Π-minimum. For a Π-minimum constraint (4) is essential and we cannot neglect it. Consider firstly the case, when (4) is absent, and so u is unbounded.

It is well known, that the first order conditions both for a weak and a Π-minimum are one and the same (for any problem convex in the control). Let us now pose the question: will conditions of the above- stated quadratic order γ be strengthened if we pass from a weak to a Π-minimum? Or in other words, *do Pontryagin (e.g. needle-type) variations bring some new optimality condition in addition to those provided by uniformly small variations?* The answer is that they do.

To give an accurate formulation, define the cubic functional

$$\rho(w) = \int [-(H_{uxx}x, x, u) + 2((Rx, u), H_{xu}y)] \, dt. \tag{19}$$

It is the third variation of the Lagrange function (9) at zero (in W) on equation (3) to within $o(\gamma)$ on Pontryagin sequences, see [13, 14].

Using the Goh transformation we reduce ρ to the form

$$\rho(w) = \int ((T_1\xi, \xi, u) + (T_2\xi, y, u) + (\mathcal{E}y, y, u)) \, dt. \tag{20}$$

Here the essential part is presented by the last term [14].

For all t^* we introduce the differential 1-form

$$\omega(t^*) = (\mathcal{E}(t^*)y, y, \, \mathrm{d}y) = \sum_{ijk} \mathcal{E}_{ijk}(t^*) y^i y^j \, \mathrm{d}y^k. \tag{21}$$

The new pointwise condition, provided by Pontryagin variations, is: for every t^* 1-form (21) is closed, i.e.,

$$\mathrm{d}\omega(t^*) = \sum_{ijk} \mathcal{E}_{ijk}(t^*)(y^i \, \mathrm{d}y^j + y^j \, \mathrm{d}y^i) \wedge \mathrm{d}y^k = 0 \tag{22}$$

(here t^* is a parameter, the differential is taken with respect to y).

Theorem 2. *a) Let w^0 be a Pontryagin minimum point in problem (1)–(3). Then both Goh conditions (15) and the new condition (22) hold, and as before, inequality (16) is valid.*

b) Suppose that both Goh conditions (15) and the new condition (22) hold, and as before, for some $a > 0$ inequality (17) is valid. Then w^0 is a Pontryagin minimum point in problem (1)–(3).

The proof is given in [14] and is based on a general theory of higher order conditions, developed recently by A.A. Milyutin and his co-workers [1].

Pontryagin Minimum for a Bounded Control

Now let the constraint (4) be present, but as before $u^0(t)$ goes strictly inside $U(t)$. In this case we must replace condition (22) by another one.

Consider the functional

$$L(y) = \int ((Qy, y) + (Vy, u) + (\mathcal{E}y, y, u)) \, \mathrm{d}t, \tag{26}$$

where the two first terms are from (14) and the last one is from (20). The new condition is $V(t) \equiv 0$, and for every t^* and every Lipschitz function $y(t)$, having $y(t_0) = y(t_1) = 0$ (we call such a function a cycle) and $\dot{y} = u \in U(t^*)$, the following inequality holds:

$$L[t^*](y) = \int ((Q(t^*)y, y) + (\mathcal{E}(t^*)y, y, u)) \, \mathrm{d}t \geq a \int (y, y) \, \mathrm{d}t, \tag{27}$$

where a is a real number. The functional in (27) is got by freezing the coefficients of (26) at the point t^* (with accounting that $V(t)$ vanishes); the set $U(t)$ is also frozen at t^*.

Theorem 3 [13, 16]. *a) Let w^0 be a Pontryagin minimum point in problem (1)–(4). Then both Goh conditions (15) and condition (27) with $a = 0$ hold, and as before, inequality (16) is valid.*

b) Suppose that the Goh conditions (15) hold, and for some $a > 0$ inequalities (27) and (17) are valid. Then w^0 is a Pontryagin minimum point in problem (1)–(4).

The proof is based on the general theory [1] and yet unpublished.

Condition (27) has a pointwise character, so it can be regarded as a new Legendre type condition. As one can see, it concerns not only the second variation of the Lagrange function, as usual, but the sum of the second and the third variations. In the case when $U(t) \equiv R^{d(u)}$, condition (27) with $a = 0$ decomposes onto (15,b) and (22).

But for an arbitrary convex $U(t^*)$ containing the origin in it's interior, (27) does not decompose, and we get an auxiliary problem: to determine all the $a \in R$ such that (27) holds for any cycle with $\dot{y} = u \in U(t^*)$. This problem has intrinsic interest, and it seems rather difficult. However, at the present time it's solution is known for three cases, when U is: a) the whole space (see above), b) a stripe $c \leq (m, u) \leq d$, where m is an arbitrary vector in $R^{d(u)}$ (A.A. Milyutin), c) an arbitrary ellipsis on the plane (A.A. Milyutin and the author).

Pontryagin Minimum for $u^0(t)$ Contacting the Bound of U(t)

As before, conditions for a Π-minimum are similar to those for a weak minimum, but instead of (27) a new condition must be introduced.

Denote by Leg $\Pi(U)$ the set of all Pontryagin sequences $w_n = (x_n, y_n, u_n)$, satisfying (4, 8, 12) and such that

$$| x_n(0) | + | y_n(1) | + \int | y_n(t) | \, \mathrm{d}t = o(\sqrt{\gamma_n}). \tag{29}$$

We call them Legendre sequences. A characteristic example is $y(t)$, having a triangle shape, based on an interval, tending to a point t^*.

Let $a \in R$, and, for any sequence from Leg $\Pi(U)$, the functional (26) satisfies the inequality:

$$\liminf_{n \to \infty} \frac{L[\lambda](w_n)}{\gamma(w_n)} \geq a. \tag{30}$$

If $u^0(t)$ goes strictly inside $U(t)$, this condition is reduced to (27).

Theorem 4 [16]. *a) Let w^0 be a Pontryagin minimum point in problem (1)–(4). Then both Goh conditions (28) and condition (30) with $a = 0$ hold, and as before, inequality (16) is valid.*

b) Suppose that the Goh conditions (28) hold, and for some $a > 0$ inequalities (30) and (17) are valid. Then w^0 is a Pontryagin minimum point in problem (1)–(4).

A more convenient form of condition (30) will be presented in one the forthcoming papers by the author.

Examples

Example 1. $\dot{x}_1 = u_1, \quad \dot{x}_2 = u_2 - bx_2u_1, \qquad x(0) = 0, \quad t \in [0, 1],$
the control u is unbounded, the functional is

$$J = \int (2x_1u_1 + 2x_2u_2 + x_1^2 + x_2^2)\, dt \;\rightarrow\; \min,$$

and $x^0 \equiv u^0 \equiv 0$. Here the critical cone $\mathcal{K}$ is given by $\dot{x} = u, \quad x(0) = 0$. The second variation of the Lagrange function is $\Omega[\lambda] = J$ and is equal to γ on $\mathcal{K}$, where

$$\gamma(y) = \; \gamma(y_1) + \gamma(y_2) \; = \mid y_1(1) \mid^2 + \mid y_2(1) \mid^2 + \int (\mid y_1(t) \mid^2 + \mid y_2(t) \mid^2)\, dt;$$

$$\rho[\lambda](w) = \; 2b \int y_2^2 u_1\, dt \; + \; o(\gamma),$$

$$\omega[\lambda] = 2by_2^2\, dy_1, \qquad d\omega[\lambda] = -4by_2\, dy_1 \wedge dy_2.$$

By Theorem 1, for each b there is a weak minimum at zero, but according to (22) and to Theorem 2 (a), only for $b = 0$ there is a Pontryagin minimum.

Now let us take $b \neq 0$ and add the constraint $\mid u \mid \leq r$ (a circle) to the problem. For which r there will be a Π-minimum? We know that since there is a weak minimum, obviously there is also a Π-minimum for all sufficiently small r. Theorem 3 allows us to determine the precise critical value of such r.

According to this Theorem, to answer the above question we must only check condition (27). Here the result is: $\max a = 1 - r \mid b \mid$, thus
    if $r \mid b \mid < 1,$ then a Π-minimum holds, and
    if $r \mid b \mid > 1,$ then a Π-minimum fails.

Consider the initial problem with another constraint $\mid u_1 \mid \leq r$ (a stripe). Here it can be shown that condition (27) holds for all $a \leq 1 - 2r \mid b \mid$, thus
    if $2r \mid b \mid < 1,$ then a Π-minimum holds, and
    if $2r \mid b \mid > 1,$ then a Π-minimum fails.
Note that the critical value of r for a stripe is two times less than that for a circle.

Let us now consider an ellipsis:

$$\left(\frac{u_1}{r_1}\right)^2 + \left(\frac{u_2}{r_2}\right)^2 \leq 1.$$

Here the critical value for a Π-minimum is $2\frac{r_1r_2}{r_1+r_2} \mid b \mid = 1.$
Note that if $r_1 = r_2$, it is reduced to the case of a circle, and if $r_2 \rightarrow \infty$, we get precisely the critical value for a stripe.

Consider yet another constraint to the initial problem, $u_1 \geq 0$. In this case the control u^0 lies entirely on the boundary of $U(t)$.

Let us check the new Legendre condition (30). Here it means that for any sequence of cycles $y^{(n)}$, having $\dot{y}_1^{(n)} = u_1^{(n)} \geq 0$ (in below we omit the index n), and such that

$$| y_1(1) | + | y_2(1) | + \int (| y_1(t) | + | y_2(t) |) \, dt = o(\sqrt{\gamma(y)}), \qquad (31)$$

the following inequality should be valid

$$L = \int (y_1^2 + y_2^2 + 2by_2^2 u_1) \, dt \geq (a - o(1)) \cdot \gamma(y). \qquad (32)$$

Note that since $u_1 \geq 0$, i.e. y_1 is monotone, for all t $y_1(t) \leq y_1(1)$, and hence we can take $\gamma(y_1) = | y_1(1) |^2$. Then (31) implies that

$$\gamma(y_1) = o(\gamma(y_2)), \qquad (33)$$

and condition (32) takes the form

$$\int (y_2^2 + 2by_2^2 u_1) \, dt \geq (a - o(1)) \int y_2^2 \, dt. \qquad (34)$$

Taking here the cubic term by parts, we get regarding (33) that

$$| \int y_2^2 u_1 dt | = | - \int y_1 y_2 u_2 dt | \leq$$

$$\leq \| y_1 \|_2 \cdot \| y_2 \|_2 \cdot \| u_2 \|_\infty \leq o(\sqrt{\gamma(y_2)}) \cdot \sqrt{\gamma(y_2)} \cdot \mathcal{O}(1) = o(\gamma(y_2)),$$

hence (34) is valid with $a = 1$, and by Theorem 4 there is a Π-minimum at w^0.

Example 2. Bilinear system:

$$\dot{x} = p + uAx + vBx, \qquad x(0) = q, \qquad (35)$$

$$J = (l, x(T)) \to \max, \qquad \dim x = n,$$

the controls u, v are scalars and the interval $[0, T]$ is fixed.

Let $u^0 = v^0 = 0$, $x^0 = q + pt$. Then MP yields

$$lAp = lAq = lBp = lBq = O. \qquad (36)$$

Assume it holds. Set $\dot{y} = u$, $y(0) = 0$, $\dot{z} = v$, $z(0) = 0$, $w = (y, z)$, $r = (u, v)$, so $\dot{w} = r$. Define matrices

$$P = \begin{pmatrix} lA^2 p & lBAp \\ lABp & lB^2 p \end{pmatrix}, \qquad Q = \begin{pmatrix} lA^2 q & lBAq \\ lABq & lB^2 q \end{pmatrix}.$$

The Goh equality condition $(15, a)$ yields:

$$l(AB - BA)p = l(AB - BA)q = 0. \qquad (37)$$

Assume it also holds (otherwise there is not even a weak minimum), which implies that matrices P and Q are symmetric. The second variation of the Lagrange function is:

$$\Omega = \frac{1}{2}((Q+PT)w(T), w(T)) + \int (Pw(T), w)\, \mathrm{d}t - \frac{1}{2}\int (Pw, w)\, \mathrm{d}t.$$

the Goh inequality condition $(15, b)$ means:

$P \leq 0$ (necessary), $\quad P < 0$ (sufficient).

Assume that $P < 0$. Let ρ_1, ρ_2 and θ_1, θ_2 are the eigenvalues of $-P$ and $-Q$ respectively.

Result: if $\theta_1 < 0$ or $\theta_2 < 0$, then $\Omega < 0$, and by Theorem 1 there is no weak minimum at the given extremal; if $\theta_1 > 0$ and $\theta_2 > 0$, i.e. $Q < 0$, then (17) is fulfilled and there is a weak minimum.

To analyze a Π-minimum assume that $P < 0$ and $Q < 0$.

Result: if

$$l[A, [B, A]]p = l[B, [B, A]]p,$$

$$l[A, [B, A]]q = l[B, [B, A]]q, \tag{40}$$

(where $[\ ,\ ]$ denotes a commutator) then condition (22) holds, and by Theorem 2 there is a Π-minimum for unbounded u, v.

Suppose that conditions (40) are not valid. Consider the additional constraint $-b \leq u \leq a$. Here the Legendre function (26) is

$$L = \int [G(y, z) + M(y, z)u]\, \mathrm{d}t,$$

where $\quad G = -\frac{1}{2}l(yA + zB)(yA + zB)p, \quad$ and

$$M(y, z) = l[A, [A, B]]x^0(t)yz + \frac{1}{2}l[B, [A, B]]x^0(t)z^2.$$

Using the above mentioned result by A.A. Milyutin we have to consider a quadratic form of three variables,

$$\Phi = a^{-1}G(y, z_1) + M(y, z_1) + b^{-1}G(y, z_2) - M(y, z_2). \tag{42}$$

Result: by Theorem 3 a Π-minimum implies that $\Phi \geq 0$, and $\Phi > 0$ implies that there is a Π-minimum at the examined extremal.

Thus, we have reduced the problem to a standard question of linear algebra: to check the nonnegativity and positivity of quadratic form (42). Another additional constraint, where (u, v) belong to an ellipsis, can also be considered in this way.

References

1. E.S. Levitin, A.A. Milyutin, N.P. Osmolovskii. Russian Math. Surveys, 1978, 33:6, pp. 85–148.

2. A.A. Milyutin. Quadratic conditions of an extremum in smooth problems with a finite-dimensional image, in "Metody teorii ekstremal'nyh zadach v ekonomike", "Nauka", Moscow, 1981, pp. 138–177, (in Russian).

3. B.S. Goh. SIAM J. on Control, 1966, 4:4, pp. 716–731.

4. J.P. McDanell, W.F. Powers. AIAA Journal, 1970, 8:8.

5. J.L. Speyer, D.H. Jacobson. J. Math. Analysis and Appl., 1971, 33:1.

6. R. Gabasov, F.M. Kirillova, "Singular optimal controls", "Nauka", Moscow, 1973.

7. D.J. Bell, D.H. Jacobson. "Singular Optimal Control Problems", Academic Press, NY, 1975.

8. H.W. Knobloch. Higher order necessary conditions in optimal control theory, Lecture Notes in Control and Information Sciences, Vol. 34, 1981.

9. M.I. Zelikin. Soviet Math. Doklady, 1982, 267:3 (in Russian).

10. F. Lamnabhi-Lagarrigue, G. Stefani. SIAM J. on Control and Optimization, 1990, 28:4, pp. 823–840.

11. N.P. Osmolovskii. Soviet Math. Doklady, 1988, 303:5 (in Russian).

12. A.V. Dmitruk. Soviet Math. Doklady, 1977, 18:2.

13. A.V. Dmitruk. Soviet Math. Doklady, 1983, 28:2.

14. A.V. Dmitruk. Mathematics of the USSR, Izvestija, 1987, 28:2   and 1988, 31:1.

15. A.V. Dmitruk. Siberian Math. Journal, 1990, 31:2.

16. A.V. Dmitruk. Lecture Notes in Control and Inf. Sci., 1992, v. 180, pp. 334–343.

International Series of Numerical Mathematics, Vol. 115, © 1994 Birkhäuser Verlag Basel

Synthesis of Adaptive Optimal Controls for Linear Dynamic Systems

R. Gabasov*   N.V. Balashevich*   and   F.M. Kirillova†

Abstract

Optimal feedback control of dynamic systems with adaptation to changing situations is the most complicated task of control theory (Bellman, 1961). In this contribution a synthesis method of an adaptive optimal regulator for a linear problem of terminal control is described.

Problem Formulation

In the class of piecewise continuous controls $u(t)$, $t \in T = [0, t^*]$, we consider the family of problems

$$c'x(t^*) \longrightarrow \max, \tag{1}$$

$$\dot{x} = Ax + bu + w, \quad x(\tau) = z, \tag{2}$$

$$Hx(t^*) = g, \tag{3}$$

$$\underline{d} \le u(t) \le \overline{d}, \quad t \in T_\tau = [\tau, t^*], \tag{4}$$

$$(x \in R^n, \quad u \in R, \quad g \in R^m, \quad \mathrm{rank} H = m < n),$$

depending on a totality $s = \{A, b, c, \underline{d}, \overline{d}, g, H, \tau, w, z\}$.

We call a totality s an admissible situation, a piecewise continuous function $u(t \mid s)$, $t \in T_\tau$, continuous function $x(t \mid s)$, $t \in T_\tau$, corresponding to s an admissible control and a trajectory if the relations (2)–(4) hold. The set of admissible situations is denoted by S. An admissible control $u^0(t \mid s)$, $t \in T_\tau$, and a trajectory $x^0(t \mid s)$, $t \in T_\tau$, are said to be the optimal program control and trajectory for $s \in S$ if the criterion (1) reaches its maximum value.

A piecewise continuous function $u^0(s)$, $s \in S$, is called an adaptive optimal feedback control if for every $s \in S$ the inequalities $\underline{d} \le u^0(s) \le \overline{d}$ hold and along the trajectory $x(t \mid s)$, $t \in T_\tau$, of the equation

$$\dot{x} = Ax + bu^0(A, b, c, \underline{d}, \overline{d}, g, H, \tau, w, x), \quad x(\tau) = z,$$

the identity $x(t \mid s) \equiv x^0(t \mid s)$, $t \in T_\tau$, is true. The classical optimal feedback control $u^0(\tau, z)$ represents a special case of the above one for fixed values of the

---

*Department of Mathematics, University of Minsk, Minsk
†Bjelorussian Academy of Sciences, Minsk

parameters $A = A_0$, $b = b_0$, $c = c_0$, $\underline{d} = \underline{d}_0$, $\overline{d} = \overline{d}_0$, $g = g_0$, $H = H_0$, $w = 0$. Below we use $u^0(s)$, $s \in S$, while s changes.

Consider the control process starting at $t = 0$ from $x(0) = x_0$, when the parameters of (1)–(4) are equal to $s(0) = \{A_0,\ b_0,\ c_0,\ \underline{d}_0,\ \overline{d}_0,\ g_0,\ H_0,\ \tau = 0,\ w_0,\ z = x_0\}$.

Suppose due to acting unknown perturbations the situation

$$s(t) = \{A(t),\ b(t),\ c(t),\ \underline{d}(t),\ \overline{d}(t),\ g(t),\ H(t),\ t,\ w(t),\ x(t)\}$$

is changing continuously and for $t \geq 0$ a real trajectory of the optimized system closed by feedback is described by the equation

$$\dot{x} = A(t)x + b(t)u^0(A(t), b(t), c(t), \underline{d}(t), \overline{d}(t), g(t), H(t), w(t), x),\quad x(0) = x_0,\quad (5)$$

$$\left(A(0) = A_0, b(0) = b_0, c(0) = c_0, \underline{d}(0) = \underline{d}_0,\right.$$

$$\left.\overline{d}(0) = \overline{d}_0, g(0) = g_0, H(0) = H_0, w(0) = w_0\right).$$

Denote by $x^*(t)$ a fixed trajectory of the equation (5) corresponding to a concrete realization $s^*(t)$, $t \in T$. The function

$$u^*(t) = u^0(A^*(t),\ b^*(t),\ c^*(t),\ \underline{d}^*(t),\ \overline{d}^*(t),\ g^*(t),\ H^*(t),\ t,\ w^*(t),\ x^*(t)),$$

$t \in T$, represents a realization of the adaptive optimal feedback control circulating in the closed system (5) of the concrete process.

A device which for every concrete process is able to construct $u^*(t)$, $t \in T$, in the real-time mode is called an *adaptive optimal controller*. A particular case of an adaptive optimal controller corresponding to the classical optimal feedback control is called the classical optimal controller.

Below we shall propose an algorithm of calculating the adaptive optimal controller when

1) the function $A^*(t)$, $t \in T$, takes the form $A^*(t) = A_0 + \alpha^*(t)A_1$, where $\alpha^*(t)$, $t \in T$, is a continuous scalar function,

2) there exists such a moment t^0, $0 < t^0 < t^*$, that the condition

$$\mathrm{rank}\{H^*(\tau)F_\tau(t^* - t)b^*(\tau),\ t \in T_\tau^*\} = m$$

holds for any $\tau \in [0, t^0]$, where

$$T_\tau^* = \{t \in T_\tau : u^0(t - 0|s) \neq u^0(t + 0|s)\},$$

$$\dot{F}_\tau = A^*(\tau)F_\tau,\quad F_\tau(0) = E,\tag{6}$$

and for every $t \in [t^0, t^*]$, the identities

$$\alpha^*(t) \equiv \alpha^*(t^0),\quad b^*(t) \equiv b^*(t^0),\quad c^*(t) \equiv c^*(t^0),\quad \underline{d}^*(t) \equiv \underline{d}^*(t^0),\quad \overline{d}^*(t) \equiv \overline{d}^*(t^0),$$

$$g^*(t) \equiv g^*(t^0),\quad H^*(t) \equiv H^*(t^0),\quad w^*(t) \equiv w^*(t^0)$$

are fulfilled,

3) the controller at every current moment $\tau \in T$ knows exact values of the state $x^*(\tau)$ and parameters $\alpha^*(\tau)$, $b^*(\tau)$, $c^*(\tau)$, $\underline{d}^*(\tau)$, $\overline{d}^*(\tau)$, $g^*(\tau)$, $H^*(\tau)$, $w^*(\tau)$.

Optimal Adaptive Controller

According to (Gabasov, Kirillova, 1984) the optimal program control corresponding to the admissible situation $s(\tau) \in S$ gets the form

$$u^0(t \mid s(\tau)) = \frac{1}{2}(\underline{d}^*(\tau) + \overline{d}^*(\tau) + (\overline{d}^*(\tau) - \underline{d}^*(\tau))\mathrm{sign}\Delta^0(t \mid s(\tau))), \quad t \in T_\tau,$$

where

$$\Delta^0(t \mid s(\tau)) = (c^*(\tau) - H^{*\prime}(\tau)y(s(\tau)))F_\tau(t^* - t)b^*(\tau), \; t \in T_\tau,$$

is an optimal co-control, $y(s(\tau))$ is an optimal potential vector of the problem (1)–(4).

The optimal program control is completely determined by the totality

$$t_i(s(\tau)), \; i \in P = \{1, 2, \ldots, p\}; \;\; y(s(\tau)), \tag{7}$$

consisting of the zeroes $t_1(s(\tau)) < \ldots < t_p(s(\tau))$ of the co-control $\Delta^0(t \mid s(\tau))$, $t \in T_\tau$, and the potential vector.

The elements in (7) satisfy the system of equations

$$f(s(\tau); \; t_i(s(\tau)), \; i \in P) = 0, \tag{8}$$

$$q_j(s(\tau); \; t_i(s(\tau)), \; i \in P; \; y(s(\tau))) = 0, \;\; j \in P;$$

where

$$f(s(\tau); \; t_i, \; i \in P) = \sum_{i=0}^{p} \int_{t_i}^{t_{i+1}} H(\tau, t)(b^*(\tau)k_i + w^*(\tau))d\tau + H(\tau, \tau)z - g^*(\tau),$$

$$t_0 = \tau, \; t_{p+1} = t^*, \; H(\tau, t) = H^*(\tau)F_\tau(t^* - t),$$

$$k_i = \frac{1}{2}(\underline{d}^*(\tau) + \overline{d}^*(\tau) + (\overline{d}^*(\tau) - \underline{d}^*(\tau))\mathrm{sign}\Delta^0(t_i + 0 \mid s(\tau))),$$

$$q_j(s(\tau); \; t_i, \; i \in P; \; y) = (c^{*\prime}(\tau) - y'H^*(\tau))F_\tau(t^* - t_j)b^*(\tau).$$

The system of equations (8) will be called *defining equations* of the adaptive optimal controller.

A numerical method of solving the defining equations (8) in the real-time mode is analogue to the method of solving the defining equations for the classical optimal controller (Gabasov, Kirillova, Kostyukova, 1991).

In addition we shall justify a method of calculating the function $F_\tau(t)$, $t \in T$. To calculate $F_\tau(t^* - t_j)$, $j \in P$, and integrals containing $F_\tau(t)$, $t \in T$, one can use asymptotic expansion of the function

$$F_\tau(t), \; t \in T : \;\; F_\tau(t) = F_0(t) + \alpha(\tau)F_1(t) + \ldots.$$

From (6) we get that $F_0(t)$, $F_1(t)$, ... satisfy the equations

$$\dot{F}_0 = A_0 F_0, \quad F_0(0) = E,$$

$$\dot{F}_i = A_0 F_i + A_1 F_{i-1}, \quad F_i(0) = 0, \quad i = 1, 2, \ldots$$

The values of $F_i(t^* - t_j(s(0)))$, $j \in P$, $i = \overline{0, l}$, can be calculated before the controller starts.

Realizition of the Controller

Introduce the parameter $\nu > 0$ characterizing the limit frequency of switching controls produced by the controller. The controller starts its work at the moment $t = 0$ and produces the value

$$u^*(0) = u^0(0 \mid A_0, b_0, c_0, \underline{d}_0, \overline{d}_0, g_0, H_0, 0, w_0, x_0),$$

where

$$u^0(t \mid A_0, b_0, c_0, \underline{d}_0, \overline{d}_0, g_0, H_0, 0, w_0, x_0)$$

is the optimal program control which can be calculated by finite methods (Gabasov, Kirillova, 1984) before turning on the controller. Let the controller had worked at the interval $[0, \tau[$. Denote by $u^*(t)$, $t \in [0, \tau[$ the control produced, by τ the last switching point of the control $u^*(t)$, $t \in [0, \tau[$. We propose the following rule for the controller: at the moment τ the controller produces the control

$$u^*(\tau) = \{ \begin{array}{ll} u^*(\tau + 0), & \tau - \underline{\tau} < \nu, \\ u^0(\tau \mid s^*(\tau)), & \tau - \underline{\tau} \geq \nu, \end{array}$$

where $u^0(\tau \mid s^*(\tau))$ is the control constructed as a result of numerically solving the defining equations (8).

Examples

Example 1

Consider the problem

$$\int_0^{t^*} u(t) dt \longrightarrow \min,$$

$$\ddot{x} + a(t)x = b(t)u, \quad x(0) = x_{10}, \quad \dot{x}(0) = x_{20},$$

$$x(t^*) = g_1(t),$$

$$\dot{x}(t^*) = g_2(t),$$

$$0 \leq u(t) \leq d(t), \ t \in T = [0, t^*];$$

$$a(0) = b(0) = d(0) = 1, \quad g_1(0) = g_2(0) = 0.$$

	$t_i, i = \overline{1,p}$	$u^*(+0)$	R	J^0
Unperturbed problem	0.333650, 0.935552, 6.61685, 7.218737	0	10^{-7}	1.203803
Controller 1	0.31, 0.8, 6.18043, 7.398571	0	10^{-3}	0.998618
Controller 2	0.25, 1.1, 5.78331, 6.381178	0	10^{-4}	0.932717

Table 1: Results of Example 1

Assign the values of the parameters

$$t^* = 4\pi, \quad x_{10} = 0.702958, \quad x_{20} = -0.954865,$$

$$a(t) = 1 + 0.1\sin 2t, \quad b(t) = 1 + 0.5\sin 3t, \quad d(t) = 0.5 + 0.5e^{-t},$$

$$g_1(t) = 0.1\ln(t+1) + 0.2\sin 0.5t, \quad g_2(t) = 0.1\sin 3t,$$

$$w(t) = 0.1\sin 5t, \quad t \leq t^0 = 5.5.$$

According to the scheme described above, two controllers were constructed in which various approximations of the matrix function $F_\tau(t)$, $t \in T$, were used:

$$1) F_\tau(t) \approx F_0(t), \quad t \in T,$$

$$2) F_\tau(t) \approx F_0(t) + \alpha(\tau)F_1(t), \quad t \in T.$$

The solution to the unperturbed problem and the results of operating the controllers are given in Table 1.

Here t_i, $i = \overline{1,p}$, are switching points of the controls $u^0(t|s(0))$, $u^*(t)$, $t \in T$; $u^*(+0)$ is the value of the controls at the first interval, R is the norm of the violation of the terminal constraints, J^0 is the value of the control criterion after finishing of operating the controllers.

Example 2

The problem investigated above can be generalized. Let us regard that the real motion of the optimized system is described by the equation

$$\dot{x} = [A^*(t) + A^0(t)]x + [b^*(t) + b^0(t)]u + w^*(t) + w^0(t), \ t \in T,$$

$$x(0) = x_0,$$

and the control criterion and the constraints take the form

$$[c^*(t) + c^0(t)]'x(t^*) \longrightarrow \max,$$

$$[H^*(t) + H^0(t)]x(t^*) = g^*(t) + g^0(t),$$

$$\underline{d}^*(t) + \underline{d}^0(t) \le u(t) \le \overline{d}^*(t) + \overline{d}^0(t), \quad t \in T.$$

We suppose that at every current moment $\tau \in T$ the controller knows the values $A^*(\tau)$, $b^*(\tau)$, $c^*(\tau)$, $\underline{d}^*(\tau)$, $\overline{d}^*(\tau)$, $g^*(\tau)$, $H^*(\tau)$, $w^*(\tau)$, $x^*(\tau)$. Unknown perturbations $A^0(t)$, $b^0(t)$, $c^0(t)$, $\underline{d}^0(t)$, $\overline{d}^0(t)$, $g^0(t)$, $H^0(t)$, $w^0(t)$, $0 \le t \le \tau$, influence the state $x^*(\tau)$. While producing the control $u^*(\tau)$ the controller uses the problem (1)–(4). Thereby we suppose that $A^0(t) \equiv 0$, $b^0(t) \equiv 0$, $c^0(t) \equiv 0$, $\underline{d}^0(t) \equiv 0$, $\overline{d}^0(t) \equiv 0$, $g^0(t) \equiv 0$, $H^0(t) \equiv 0$, $w^0(t) \equiv 0$, $t \in [\tau, t^*]$.

Instationary Processes

Another interesting generalization in practice consists in transferring from constant A, b to variable $A(t)$, $b(t)$, $t \in T$. In this case the change $A(\tau)$, $b(\tau)$ by $A(t, \tau)$, $b(t, \tau)$, $\tau \le t \le t^*$, means the use of forecast got to the moment τ. From calculations presented in section 2, 2–3, one can see that the investigation of nonstationary cases does not require principal modifications of the technique described above.

References

Bellman, R.E. (1961). Adaptive Control Processes, A Guided Tour. Princeton University Press, Princeton, New Jersey.

Gabasov, R. and F.M. Kirillova (1984). Constructive methods of optimization, P.2. University Press, Minsk.

Gabasov, R., F.M. Kirillova and O.I. Kostyukova (1991). Constructing optimal feedback controls in linear problems. *Doklady AN SSSR, 320,* pp. 1294–1297.

International Series of Numerical Mathematics, Vol. 115, © 1994 Birkhäuser Verlag Basel

Control Applications of Reduced SQP Methods*

Ekkehard W. Sachs[†]

Abstract

Reduced Successive Quadratic Programming (SQP) methods have been applied to various areas in optimization. In this paper we consider mainly applications from optimal control and neighboring areas. The convergence analysis in Hilbert spaces is reviewed and the main ingredients are filtered out for the design of reduced SQP methods. As applications we chose examples from control and parameter identification with partial differential equations, process optimization in chemical engineering, and aerodynamic shape optimization.

Scope

In this paper we review some of the literature on reduced SQP methods and their applications in the area of optimal control. Since this is a large area of current research interest we concentrate on several but not all issues involved. Among those omitted are for example problems with inequality constraints and globalization strategies. Since optimal control problems are by nature infinite dimensional problems, we include this aspect in the motivation and citation of reduced SQP methods. In finite dimensions there is a large number of papers devoted to the convergence theory in this area of which we reference only a few, [8], [9], [4], [22], [13], [12].

In the following section we motivate reduced SQP-methods by a non standard approach, which does not use the quadratic approximation explicitly. Then we focus on various applications. In several papers applications of SQP methods have been considered. It is essential for applications in optimal control that the structure of such problems is used in some form because otherwise the dimension of the underlying spaces becomes too large. We mention among others papers [14], [2], [21], [10], [1] which use special versions of SQP methods to solve large sparse problems in connection with optimal control. In this paper applications are presented which are closely connected to the reduced SQP method.

---

*Invited Paper
†Universität Trier, FB IV - Mathematik, D-54286 Trier

Reduced SQP Methods

In this section we want to present the main ideas in the derivation of reduced SQP methods and discuss some of its convergence properties.

We consider an optimization problem in infinite dimensional spaces, i.e. let X and Y be Hilbert spaces. Let f and h be mappings between the following spaces

$$f : X \to \mathbb{R}, \quad h : X \to Y.$$

The minimization problem consists in finding x^* such that $h(x^*) = 0$ and

$$f(x^*) \leq f(x) \text{ for all } x \text{ with } h(x) = 0. \tag{1}$$

Most algorithms do not solve this minimization problem directly but rather a substitute, the equations describing the necessary optimality conditions. In order to formulate these, we make the following smoothness assumption on the functions describing the problem:

Assumption 1 *Let f and h be twice Fréchet-differentiable on a ball D which contains the solution x^* of (1). Furthermore let $f''(\cdot)$ and $h''(\cdot)$ be Lipschitz-continuous on D. Let $h'(x^*) : X \to Y$ be a surjective map.*

The first order necessary optimality conditions are given as

Theorem 1 *Let assumption (1) hold and let x^* be a solution of (1). If $h'(x^*)$ is surjective then we have*

$$f'(x^*)s = 0 \text{ for all } s \in \mathcal{N} = \{s \in X : h'(x^*)s = 0\}. \tag{2}$$

We can simplify (2) further by the use of a proper representation of the nullspace $\mathcal{N}$ of the linearized equality constraint. The type of representation depends heavily on the underlying problem structure. In general, we assume the following

Assumption 2 *Let W be a Hilbert space and for each $x \in D$ let*

$$\mathcal{T}(x) : W \to \mathcal{N}$$

be an isomorphism. Furthermore we assume that $\mathcal{T}(\cdot)$ is Fréchet-differentiable in D and that $\mathcal{T}'(\cdot)$ is Lipschitz-continuous in D.

For examples of choices of $\mathcal{T}$ see the following sections.

We can use the nullspace representation to rewrite the necessary optimality condition in the following form by using the adjoint operator $\mathcal{T}^*$ of $\mathcal{T}$:

Corollary 1 *Let the assumptions of Theorem 1 hold. Then the solution x^* solves the following system of nonlinear equations*

$$\Phi(x) = \begin{pmatrix} \Phi_1(x) \\ \Phi_2(x) \end{pmatrix} = \begin{pmatrix} \mathcal{T}(x)^* f'(x) \\ h(x) \end{pmatrix} = 0. \tag{3}$$

If one uses Newton's method to solve (3) then one obtains a SQP method. Let us consider the differentiation of the first equation. We introduce the Lagrangian associated with the optimization problem: Let $l \in Y^*$ be given, where Y^* denotes the dual space of Y. Then

$$L(x, l) = f(x) + l(h(x))$$

is called the Lagrangian with $L : X \times Y^* \to I\!R$. Note that for all $l \in Y^*$ and $x \in X$

$$\Phi_1(x) = T(x)^* f'(x) = T(x)^* L_x(x, l), \tag{4}$$

because the range of T is in $\mathcal{N}$ and hence $h'(x)T(x)w = 0$ for all $w \in W$ and

$$T(x)^* h'(x)^* l = 0 \text{ for all } l \in Y^*.$$

Upon differentiation of (4) one obtains as the derivative

$$\Phi_1'(x) = T'(x)^* L_x(x, l) + T(x)^* L_{xx}(x, l). \tag{5}$$

This expression can be simplified at the solution x^*. Using the Lagrangian we can express the necessary optimality conditions (2) in a different form, cf. [20]: There exists $l^* \in Y^*$ with

$$L_x(x^*, l^*) = f'(x^*) + l^*(h'(x^*)) = 0. \tag{6}$$

Hence the first term in (5) vanishes at x^* and

$$\Phi_1'(x^*) = T(x^*)^* L_{xx}(x^*, l^*).$$

We use this representation also for $x \in D$ where D is a small neighborhood of x^*. Then the computation of a Newton-step s requires the solution of

$$\Phi'(x)s = \begin{pmatrix} T(x)^* L_{xx}(x, l)s \\ h'(x)s \end{pmatrix} = - \begin{pmatrix} T(x)^* f'(x) \\ h(x) \end{pmatrix} \tag{7}$$

In many applications it is a big advantage that we do no longer need to solve (3) for $s \in X$ but rather (8) for $w \in W$, where the nullspace W is sometimes a much smaller space than X. This system can be solved more easily with the use of a right inverse of $h'(x)$.

Assumption 3 *For each $x \in D$ there exists an operator $\mathcal{R}(x) : Y \to X$ with*

$$\mathcal{R}(x) \in \mathcal{L}(Y, X), \quad h'(x)\mathcal{R}(x) = I_Y.$$

Furthermore let $\mathcal{R}(\cdot)$ be Fréchet-differentiable and $\mathcal{R}'(\cdot)$ be Lipschitz-continuous on D.

The following identity can be used to reduce the solution of (7)

$$\{s \in X : h'(x) = -h(x)\} = \{s \in X : s = \mathcal{T}(x)w - \mathcal{R}(x)h(x), w \in W\}.$$

Then (7) is equivalent to the following equation for w

$$\mathcal{T}(x)^* L_{xx}(x,l)\mathcal{T}(x)w = -\mathcal{T}(x)^* f'(x) + \mathcal{T}(x)^* L_{xx}(x,l)\mathcal{R}(x)h(x). \qquad (8)$$

This equation is solvable in a neighborhood of (x^*, l^*), if the usual second order sufficiency condition holds:

Assumption 4 *Let x^* be a solution of (1) and let l^* be the Lagrange multiplier which satisfies (6). Assume there exists some $m > 0$ such that*

$$< \xi, L_{xx}(x^*, l^*)\xi > \geq m\|\xi\|^2 \text{ for all } \xi \in \mathcal{N}.$$

Another issue that arises is the choice of the Lagrange multiplier $l \in Y$. If the new iterate $x \in X$ is chosen, the new multiplier $l = l(x)$ should be determined depending on x in such a way that for $l : X \to Y^*$ there exists some $\lambda > 0$ such that

$$\|l(x_1) - l(x_2)\| \leq \lambda \|x_1 - x_2\| \qquad x_1, x_2 \in D \qquad (9)$$

holds. This means that the choice of $l(x)$ should depend on x in a smooth way, in particular Lipschitz continuously. A typical choice in finite dimensional spaces is the least squares solution

$$\text{Minimize } \|f'(x) + h'(x)^* l\| \text{ over } l \in Y^*. \qquad (10)$$

Next we want to point out an important simplification in the algorithm which is often made when using secant approximation to the reduced Hessian $\mathcal{T}(x)^* L_{xx}(x,l)\mathcal{T}(x)$. Suppose we approximate the reduced Hessian by some operator $B \in \mathcal{L}(W,W)$, which is very convenient since only an update in the small space W has to be carried out. However, the right hand side of (8) still contains the term $\mathcal{T}(x)^* L_{xx}(x,l)\mathcal{R}(x)h(x)$ which involves L_{xx}. This term is often simply omitted in the formulation of a reduced secant SQP method.

Algorithm 1

(1) Choose $x_0 \in X, l_0 \in Y^*, B_0 \in \mathcal{L}(W,W)$ self-adjoint and positive definite

(2) Solve $B_k w_k = -\mathcal{T}(x_k)^* f'(x_k)$

(3) Select $y_k \in Y$

(4) If $w_k = 0$, set $B_{k+1} = B_k$ else update B_{k+1} by a BFGS update

$$B_{k+1} = B_k + \frac{1}{< y_k, w_k >} < y_k, \cdot > y_k - \frac{1}{< w_k, B_k w_k >} < B_k w_k, \cdot > B_k w_k$$

(5) Set $x_{k+1} = x_k + \mathcal{T}(x_k)w_k - \mathcal{R}(x_k)h(x_k)$

In Step (3) the choice of y_k has to be specified. According to the theory of secant approximations of the reduced Hessian we expect convergence if there exists a neighborhood $D_1 \subset D$ of x^* and a constant c_1 such that for all $x_k, x_{k+1} \in D_1$ we have

$$\|y_k - T(x^*)^* L_{xx}(x^*, l^*) T(x^*) w_k\| \leq c_1 \max\{\|x_k - x^*\|, \|x_{k+1} - x^*\|\} \|w_k\|. \quad (11)$$

This estimate holds (see [16]) for the following choice ([6])

$$y_k = T(x_k)^* (L_x(x_{k+}, l_k) - f'(x_k)), \quad \text{where } x_{k+} = x_k + T(x_k) w_k. \quad (12)$$

It was shown in [16] that one obtains a superlinear rate of convergence also in Hilbert space.

Theorem 2 *Let $x^* \in X$ be a solution of (1) and let the Assumptions 1 - 4 be satisfied. Suppose that (11) is true. Then for any $r \in (0, 1)$ there exist positive constants δ, η such that if*

$$\|x_0 - x^*\| < \delta \qquad and \qquad \|B_0 - T(x^*)^* L_{xx}(x^*, l^*) T(x^*)\| < \eta$$

and

$$B_0 - T(x^*)^* L_{xx}(x^*, l^*) T(x^*) \qquad is\ compact, \quad (13)$$

the sequence $\{x_k\}$ generated by Algorithm A1 is well-defined and converges to x^* at a 2-step superlinear rate

$$\lim_{k \to \infty} \frac{\|x_{k+1} - x^*\|}{\|x_{k-1} - x^*\|} = 0.$$

It is interesting to note that for infinite dimensional problem the approximation of the reduced Hessian does not only need to be close in norm but the compactness requirement is an additional assumption for superlinear convergence. This can also be observed for finite dimensional problem which are discretizations of infinite dimensional problems, see e.g. [15]. The reason for the 2-step convergence rate lies in the omission of the last term in (8).

Corollary 2 *If Step (2) in Algorithm 1 is replaced by*

$$Solve \quad B_k w_k = -T(x_k)^* f'(x_k) + T(x_k)^* L_{xx}(x_k, l_k) \mathcal{R}(x_k) h(x_k),$$

then under the assumptions of Theorem 2 we obtain a 1-step superlinear rate of convergence.

In summary, if reduced SQP methods are used for problems in applications and one wants to make use of the special structure this can be done most efficiently through a proper choice of

- null space representation,

- right inverse,

- Lagrange multiplier.

In the following some of the applications in problems of optimal control and related areas will be discussed.

Finite Dimensional Problems

First consider the finite dimensional case: $X = \mathbb{R}^N, Y = \mathbb{R}^M, W = \mathbb{R}^{N-M}$. The null space is of dimension $N - M$ and assume that the columns of a matrix Z span the null space:

$$T(x) = Z(x) \in \mathbb{R}^{N \times (N-M)}, \qquad h'(x)Z(x) = 0.$$

Then M columns stored in a matrix $Y(x)$ are added in such way that the resulting $N \times N$ matrix is regular:

$$Y(x) \in \mathbb{R}^{N \times M}, \qquad (Y(x) \quad Z(x)) \in \mathbb{R}^{N \times N} \quad \text{invertible.} \tag{14}$$

A right inverse $\mathcal{R}(x)$ of $h'(x)$ is easily constructed by

$$\mathcal{R}(x) = Y(x)(h'(x)Y(x))^{-1} \in \mathbb{R}^{N \times M}. \tag{15}$$

In order to obtain a Lagrange multiplier update, consider the necessary optimality condition (6). If this equation is multiplied by $Y(x)^T$, it can be solved for l to give the least squares multiplier update (10)

$$l(x) = -(Y(x)^T h'(x)^T)^{-1} Y(x)^T f'(x) = -\mathcal{R}(x)^T f'(x). \tag{16}$$

The most common approach (see [11] e.g.) to construct Z and Y is using the QR-decomposition of the Jacobian

$$h'(x)^T = Q(x)R(x) = (Q_1(x) \quad Q_2(x)) \begin{pmatrix} R_1(x) \\ 0 \end{pmatrix}$$

where $Q(x)$ is orthogonal, $Q_1(x) \in \mathbb{R}^{N \times M}, Q_2(x) \in \mathbb{R}^{N \times (N-M)}$ and $R_1(x) \in \mathbb{R}^{M \times M}$ is an upper triangular matrix. In this case the null space representation T is given by

$$T(x) = Q_2(x) \in \mathbb{R}^{N \times (N-M)}.$$

It is well known, that T in this instance does not necessarily satisfy the smoothness assumptions posed in Assumption 2. This and modifications of the QR-decomposition are discussed in [5], [7]. The right inverse and the Lagrange multiplier are given by

$$R(x) = Q_1(x)(R_1^T(x))^{-1} \in \mathbb{R}^{N \times M}, \qquad l(x) = -R_1^{-1}(x)Q_1^T(x)f'(x).$$

Since it is possible to compute all the necessary ingredients for a reduced SQP method with a QR decomposition this approach is quite popular. A serious drawback occurs for example, if the problem size is too large and the sparsity has to be used. However, the theory outlined in the previous section gives the requirements for other choices of T, R, l which are needed to obtain similar convergence results.

Sometimes the null space of the Jacobian of the constraint can be represented in a way which is obvious from the underlying structure of the minimization problem. Suppose that the variable x can be decomposed into $z \in \mathbb{R}^M$ and $y \in \mathbb{R}^{N-M}$ such that the partial derivatives of $h'(x)$ are

$$h'(x) = (h'_y(x) \ h'_z(x)), \quad h'_y \in \mathbb{R}^{M \times (N-M)}, \quad h'_z \in \mathbb{R}^{M \times M} \text{ invertible,}$$

then the null space is represented by

$$T = \begin{pmatrix} I_{(N-M) \times (N-M)} \\ -h_z(x)^{-1} h_y(x) \end{pmatrix}. \tag{17}$$

In order to find $Y(x) \in \mathbb{R}^{N \times M}$ to satisfy (14) possible choices are

$$Y(x) = \begin{pmatrix} 0 \\ I_{M \times M} \end{pmatrix} \quad \text{or} \quad Y(x) = \begin{pmatrix} h_y(x) h_z(x)^{-1} \\ I_{M \times M} \end{pmatrix}.$$

The right inverse $\mathcal{R}$ and the multiplier l can be chosen according to (15) and (16), respectively.

Process Optimization

In a series of papers, see e.g. [25], [24], reduced SQP methods were applied to problems of process optimization in chemical engineering. The description of the application involves initial value problems for ordinary differential equations and algebraic equations. For a more detailed description of the problems we refer to the references in the citations mentioned above.

From a numerical analysis viewpoint it should be mentioned that in this application emphasis is also put on the treatment of two additional aspects

- the inclusion of inequality constraints,

- approximations to the omitted term $T^* L_{xx} \mathcal{R} h$ in Algorithm 1.

Here we want to concentrate on the second aspect. As we noticed in the section on 'Reduced SQP Methods', Theorem 2 and Corollary 2, the 1-step superlinear rate of convergence is reduced to 2-step superlinear when the term

$$T(x_k)^* L_{xx}(x_k, l_k) \mathcal{R}(x_k) h(x_k) \tag{18}$$

is omitted in the right hand side of the system of linear equations. On the other hand it is in general not reasonable to compute this term exactly when the reduced Hessian is approximated by quasi Newton updates in order to avoid the evaluation of second derivatives.

One approach to alleviate this problem is to use a finite difference approximation

$$T(x_k)^* (L_x(x_k + \mathcal{R}(x_k) h(x_k), l_k) - L_x(x_k, l_k))$$

which results in an additional evaluation of L_x. In order to avoid the additional evaluation one can use Broyden update for this term:

$$
\begin{aligned}
A_{k+1} &= A_k + \frac{(\eta_k - A_k s_k)s_k^T}{s_k^T s_k^T}, \quad \text{where} \\
\eta_k &= T(x^k)(L_x(x_k + \mathcal{R}(x_k)h(x_k), l_k) - L_x(x_k, l_k)), \\
s_k &= x_{k+1} - x_k.
\end{aligned}
$$

A detailed discussion on issues of convergence of these methods can be found in [19] and [12]. In particular it is possible to recover theoretically the 1-step superlinear rate of convergence. However, since each iteration step is more costly it is important to consider also the numerical efficiency. The effect of these updates on the numerical performance for reduced SQP methods is studied for problems from process optimization in [19]. The Broyden update is reported to perform best in terms of computation time.

Shape Optimization in Aerodynamics

Shape optimization problems occur in many areas of engineering. Here a problem from aerodynamics, in particular the design of a wing of an airplane, is considered. Typically, these problems include in their original formulation a partial differential equation so that the size of the discretized problem becomes quite large.

The wing design problem has been considered at various places in the literature, here we confine ourselves to an approach presented in [23] which uses a reduced SQP method for the solution of the problem. The problem is to find the shape of the surface of an airfoil such that a desired pressure distribution is achieved.

Let $\Omega \in I\!R^2$ denote the area around the airfoil, Γ_s the surface of the airfoil, Γ_∞ the farfield boundary. If ϕ denotes the velocity potential then the following boundary value problem for the flow around the airfoil is considered in [23]

$$
\begin{aligned}
\nabla \cdot ((1 - \frac{\gamma - 1}{2}\|\nabla\phi\|^2)^{\frac{1}{1-\gamma}}\nabla\phi) &= 0 \quad \text{in } \Omega \\
\frac{\partial \phi}{\partial n} &= 0 \quad \text{on } \Gamma_s \\
\frac{\partial \phi}{\partial n} &= u_\infty \quad \text{on } \Gamma_\infty
\end{aligned}
\tag{19}
$$

Here u_∞ denotes the normal component of the freestream density.

In order to solve this problem numerically a Galerkin discretization is introduced. This leads to a set of nonlinear equations as equality constraints. Finally, one obtains a problem of the form

$$
\begin{aligned}
\text{Minimize} \quad & \int_{\Gamma_s}(p(\Phi) - p^*)^2 \, d\Gamma \\
\int_{\Omega_h}(1 - \frac{\gamma - 1}{2}(\Phi^T Q\Phi))^{\frac{1}{\gamma-1}} d\Omega &= \int_{\Gamma_{h\infty}} N(\rho_\infty u_\infty)\cdot n \, d\Gamma \\
p(\Omega, \Phi)_{TE_+} - p(\Omega, \Phi)_{TE_-} &= 0
\end{aligned}
\tag{20}
$$

p^* denotes the desired pressure on the airfoil, where $p(\Omega)$ denotes the computed pressure according to the design variables chosen. The last equation corresponds to the Kutta condition in terms of pressures at the trailing edge which in [23] is added to the objective function through a penalty term.

It is interesting to note that the number of variables for this problem has to be chosen fairly high since they come from the solution of a discretized elliptic differential equation in 2-dimensional space whereas the number of design variables for the airfoil is only three. This was the reason why in [23] a reduced SQP method with BFGS updates was used. The nullspace representation $\mathcal{T}$ was chosen according to a partitioning of the variables into state and design variables as in (17). The solution of a system of linear equations with coefficient matrix h_z can be carried out efficiently when using the sparsity of the matrix and the proper choice of preconditioners. The right inverse $\mathcal{R}$ and the Lagrange multiplier l were chosen the same as in the lines below (17). For the update formula a BFGS update was used with negligible cost because the size of the matrix was very small due to the number of the design variables.

Optimal Control Problems with ODE's

As a first example we consider optimal control problems with constraints given by initial value problems. Let $F : [0, T] \times \mathbb{R}^n \times \mathbb{R}^m \to \mathbb{R}$ and $\phi : \mathbb{R}^n \times \mathbb{R}^m \to \mathbb{R}^n$ be given. Then consider the following optimal control problem:

$$\begin{aligned}
\text{Minimize} \quad & \int_0^T F(t, y(t), u(t))dt \\
\text{subject to} \quad & \dot{y}(t) = \phi(y(t), u(t)) \quad \text{a.e. on } [0, T] \\
& y(0) = y_0.
\end{aligned}$$

In order to use the Hilbert space framework outlined below we assume that

$$F \quad \text{is quadratic in } u, \quad \phi \text{ is affine in } u.$$

Let $X = L_m^2[0, T] \times W_n^{1,2}[0, T], \quad Y = L_n^2[0, T] \times \mathbb{R}^n$. With $x = (u, y)$ define

$$\begin{aligned}
f(x) &= \int_0^T F(t, y(t), u(t))dt, \\
h(x) &= (\dot{y} - \phi(y, u), y(0) - y_0)
\end{aligned}$$

such that we obtain an optimization problem of the form (1). The Fréchet-derivative of h is given by

$$h'(x)(\delta x) = (\dot{\delta y} - \phi_y \delta y - \phi_u \delta u, \delta y(0)).$$

The obvious choice for the parameterization of the null space of $h'(x)$ is to choose δu as the independent variable and select δy as the solution of the following linear initial value problem:

$$\dot{\delta y} = \phi_y \delta y + \phi_u \delta u, \quad \delta y(0) = 0. \tag{21}$$

Hence the null space representation is given by $W = L^2_m[0,T]$ and

$$T(x)(\delta x) = (\delta u, \delta y), \quad \text{where } \delta u \in W.$$

The right inverse $\mathcal{R}(x) : Y \to X$ can be defined by

$$\mathcal{R}(x)(r,\rho) = (0,s), \quad (r,\rho) \in Y,$$

where $s \in W^{1,2}_n[0,T]$ solves

$$\dot{s} = \phi_y(y,u)s + r, \quad s(0) = \rho.$$

The update for the Lagrange multiplier can be defined as in the finite dimensional case in (16)

$$l(x) = \mathcal{R}(x)^* f'(x)$$

which amounts to the solution of an adjoint equation.

Parameter Identification Problems

Parameter identification problems occur in various areas of engineering such as oil recovery, ground water modelling, etc. Typically, the processes are described by differential equations and the mathematical problem is to recover the coefficients of the dynamical system from certain observations.

As an example we consider a problem with partial differential equations. Let Ω be a bounded open set on $I\!\!R^n, n \leq 3$ with a smooth boundary. The goal is to determine the unknown coefficient function q in the following elliptic boundary value problem:

$$\begin{aligned} -\nabla \cdot (q\nabla u) &= g \text{ in } \Omega \subset I\!\!R^n \\ u &= 0 \text{ on } \partial\Omega \end{aligned} \tag{22}$$

Let $g \in H^{-1}(\Omega)$ be fixed and denote by $u(q)$ the solution of (22) depending on q. To obtain well defined elliptic boundary value problems assume that the solution q^* is bounded away from 0 on Ω and that the starting data are close enough to the solution to disregard this inequality constraint. The 'correct' parameter function q^* is to be determined from a given observation z which corresponds to some solution u^* of the boundary value problem. Since the determination of q from data u is an ill posed problem the objective function is regularized by a penalty term in q:

$$\min \frac{1}{2}\|u(q) - z\|^2 + \frac{\beta}{2}\|q\|^2 \quad \text{over } q. \tag{23}$$

In order to phrase the problem in the formulation which we used above for reduced SQP methods, we rewrite it in the following way: Let

$$X = \tilde{H}(\Omega) \times H^1_0(\Omega), \quad Y = H^1_0(\Omega)$$
$$\tilde{H}(\Omega) = H^2(\Omega) \text{ if } n = 2,3 \text{ and } \tilde{H}(\Omega) = H^1(\Omega) \text{ if } n = 1.$$

Let $\Delta : H_0^1(\Omega) \to H^{-1}(\Omega)$ denote the Laplacian operator. Then define with $x = (q, u)$

$$\begin{aligned} h(x) = h(q, u) &= -\Delta^{-1}(\nabla \cdot (q\nabla u) + g) \\ f(x) = f(q, u) &= \tfrac{1}{2}(\|u(q) - z\|_{H_0^1}^2 + \beta \|q\|_{\tilde{H}}^2) \end{aligned} \tag{24}$$

In order to simplify the notation we define an operator

$$A(h) : H_0^1(\Omega) \to H^{-1}(\Omega)$$

by

$$A(h)u = -\nabla \cdot (h\nabla u).$$

Then $h(x) = \Delta^{-1}(A(q)u - g)$ and the Fréchet-derivative is given by

$$h'(x)(\delta x) = \Delta^{-1}(A(\delta q)u + A(q)\delta u).$$

Therefore, if δx belongs to the null space $\mathcal{N}$, i. e. $h'(x)(\delta x) = 0$, the equation can easily be solved for δu depending on δq:

$$\mathcal{T}(x) : W \to \tilde{H} \times H_0^1 \text{ with } W = \tilde{H}$$

and

$$\mathcal{T}(x)\delta q = (\delta q, -A(q)^{-1}A(\delta q)u) \quad \text{for } (q, u) \in \tilde{H} \times H_0^1.$$

Moreover, a right inverse

$$\mathcal{R}(x) : H_0^1 \to \tilde{H} \times H_0^1$$

of $h'(x)$ can be defined by

$$\mathcal{R}(x)y = (0, A(q)^{-1}\Delta y). \tag{25}$$

The smoothness assumptions for $\mathcal{T}, \mathcal{R}$ are checked in [15]. In the same reference one finds numerical results for a reduced secant SQP method which was applied to a finite dimensional discretization. Among others one could observe the influence of

- the compactness condition (13),

- perturbation of the data on the convergence rate,

- the bilinear nature of the nonlinearity of h given by (24),

- a refinement of the discretization.

Control of Nonlinear Diffusion Processes

Let us consider another application of reduced SQP methods to optimal control problems with partial differential equations. In [3] the following optimal control problem is addressed: If a probe such as a piece of ceramic inside a kiln has to be heated the engineers choose a firing curve which the temperature inside the probe should follow. The problem is to heat the kiln in such a way that this firing curve in the probe is met as close as possible.

Mathematically, the problem is modelled as follows: Let $y(x,t)$ denotes the temperature of the probe at time t and at the location $x \in [0,T]$, then the change in time is given by a nonlinear heat equation with appropriate boundary conditions.

$$
\begin{aligned}
C(y(x,t))\frac{\partial y}{\partial t}(x,t) - \nabla(\lambda(y(x,t))\nabla y(x,t)) &= q(x,t) & \text{on } \Omega \times [0,T], \\
\lambda(y(x,t))\nabla y(x,t) &= b(x,t) & \text{on } \partial\Omega \times [0,T], \\
y(x,0) &= y_0(x) & \text{on } \Omega.
\end{aligned}
\tag{26}
$$

The specific heat capacity C and the heat conduction λ are real valued functions which both depend on the temperature y. q is a source term and b is the heat flux at the boundary. For $\Omega = [0,1]$ we obtain

$$
\begin{aligned}
b(0,t) &= g[y(0,t) - u(t)], \\
b(1,t) &= 0.
\end{aligned}
\tag{27}
$$

Let $x = 1$ denote the interior where the heat flux is zero and let the boundary term at $x = 0$ be given by the difference of the temperatures of the probe $y(0,t)$ and of the surrounding medium $u(t)$ with a constant g. The objective function is to chose u in such a way that

$$
\int_0^T [(y(1,t) - p(t))^2 + \alpha u^2(t)]\, dt
\tag{28}
$$

becomes as small as possible.

We apply a finite element discretization to this problem (linear in x) and obtain the following discretized problem. For details we refer to [17]. In terms of our original problem (1) let $X = \mathbb{R}^{M(N+2)}$ and $Y = \mathbb{R}^{M(N+1)}$ and

$$
\begin{aligned}
u &= (u^2,...,u^{M+1})^T \in \mathbb{R}^M, \\
y &= ((y^2)^T,...,(y^{M+1})^T)^T \in \mathbb{R}^{M(N+1)}, \\
f(u,y) &= \tau \sum_{j=2}^{M+1} [(y_{N+1}^j - p^j)^2 + \alpha(u^j)^2], \\
h^1(u,y) &= A\Gamma(y^2) + D\beta(y^2) + g(y_1^2 - u^2)e^1 - (q^1 + A\Gamma(y^1)), \\
h^j(u,y) &= A(\Gamma(y^{j+1}) - \Gamma(y^j)) + D\beta(y^{j+1}) + g(y_1^{j+1} - u^{j+1})e^1 - q^j.
\end{aligned}
$$

where e^1 is the first unit vector in $I\!\!R^{N+1}$ and

$$
\begin{aligned}
A &= (1/\tau)((b_i, b_j)), \\
D &= ((b_{i\,x}, b_{j\,x})), \\
\beta(y) &= \int_0^y \lambda(s)\,ds, \quad \Gamma(y) = \int_0^y C(s)\,ds, \quad y \in I\!\!R, \\
q_i^j &= \frac{1}{\tau} \int_{t^j}^{t^{j+1}} (q(\cdot, t), b_i)\,dt, \\
p^{j+1} &= \frac{1}{\tau} \int_{t^j}^{t^{j+1}} p(t)\,dt.
\end{aligned}
$$

In order to solve the problem with a reduced SQP method we need to specify the null space representation T. Since the variables separate nicely into u and y, we can use the framework as in (17) and chose $W = I\!\!R^M$. This yields that the secant approximation of the reduced Hessian is of size $M \times M$ only vs. $M(N+2) \times M(N+2)$ for the Hessian of the Lagrangian. In the choice of T according to (17) we have to solve linear equations with coefficient matrix h_y. This is computationally feasible because this matrix is sparse:

$$
h_y(y, u) = \begin{pmatrix}
G(y^2) & 0 & 0 & 0 & 0 \\
E(y^2) & G(y^3) & 0 & 0 & 0 \\
0 & \ddots & \ddots & 0 & 0 \\
0 & 0 & \ddots & G(y^M) & 0 \\
0 & 0 & 0 & E(y^M) & G(y^{M+1})
\end{pmatrix} \in I\!\!R^{M(N+1) \times M(N+1)}
\tag{29}
$$

where $E, G \in I\!\!R^{(N+1) \times (N+1)}$ are also sparse matrices (see [17]). The choice of R and l for right inverse and Lagrange multiplier is made as in the lines below (17).

The numerical results in [17] indicate a 2-step superlinear rate of convergence and one can also observe that the rate is sometimes linear if the initial approximation to the Hessian differs by a noncompact operator in the infinite dimensional counterpart. Since it is a finite dimensional problem, eventually the superlinear convergence takes over but this happens at a very late stage of the iterations depending on the size of the discretization. This is an effect that can be explained nicely with the infinite dimensional theory.

In a recent publication [18] the reduced SQP method was redesigned in the of separable variables to recover 1-step superlinear convergence without any additional updating. The motivation is that in step (5) of Algorithm 1 we obtain x_{k+1} from x_k by a correction using T and R. It is important to note that the correction $R(x_k)h(x_k)$ is independent from the previous steps in the algorithm. Hence it is conceivable that one might try to use this information already at an earlier stage, e.g. in step (2), and use $x_k - R(x_k)h(x_k)$ instead of x_k to improve the performance. This results in the following algorithm.

Algorithm 2

(1) Choose $x_0 \in X, l_0 \in Y^*, B_0 \in \mathcal{L}(W, W)$ self-adjoint and positive definite

(2) Solve $B_k w_k = -\mathcal{T}(x_k - \mathcal{R}(x_k)h(x_k))^* f'(x_k - \mathcal{R}(x_k)h(x_k))$

(3) Set $x_{k+1} = x_k + \mathcal{T}(x_k)w_k - \mathcal{R}(x_k)h(x_k)$

(4) Set $y_k = \mathcal{T}(x_{k+1})^* f'(x_{k+1}) - \mathcal{T}(x_k - \mathcal{R}(x_k)h(x_k))^* f'(x_k - \mathcal{R}(x_k)h(x_k))$

(5) If $w_k = 0$, set $B_{k+1} = B_k$ else update B_{k+1} by a BFGS update

$$B_{k+1} = B_k + \frac{1}{<y_k, w_k>} <y_k, \cdot> y_k - \frac{1}{<w_k, B_k w_k>} <B_k w_k, \cdot> B_k w_k$$

It was shown in [16] that in the case of separable variables Algorithm 2 converges at a 1-step superlinear rate. In [18] this statement could be observed numerically also for the parabolic boundary control problem presented above.

References

[1] J. T. Betts. The application of sparse Broyden updates in the collocation method for optimal control problems. Technical report, Boeing Computer Services, 1988.

[2] H. G. Bock. *Randwertproblemmethoden zur Parameteridentifizierung in Systemen nichtlinearer Differentialgleichungen.* Doktorarbeit, Universität Bonn, 1985.

[3] J. Burger and M. Pogu. Functional and numerical solution of a control problem originating from heat transfer. *J. Optim. Theory Appl.*, 68:49–73, 1991.

[4] R. H. Byrd and J. Nocedal. An analysis of reduced Hessian methods for constrained optimization. *Math. Programming*, 49:285–323, 1991.

[5] R. H. Byrd and R. B. Schnabel. Continuity of the null space basis and constrained optimization. *Math. Programming*, 35:32–41, 1986.

[6] T. F. Coleman and A. R. Conn. On the local convergence of a quasi-Newton method for the nonlinear programming problem. *SIAM J. Numer. Anal.*, 21:755–769, 1984.

[7] T. F. Coleman and D. C. Sorensen. A note on the computation of an orthonormal basis for the null space of a matrix. *Math. Programming*, 29:234–242, 1984.

[8] D. Gabay. Reduced quasi-Newton methods with feasibility improvement for nonlinearly constrained optimization. *Math. Programming Study*, 16:18–44, 1982.

[9] J. C. Gilbert. Une méthode à métrique variable réduite en optimisation avec contraintes d' égalité non linéaires. Technical Report RR-482, INRIA, 1986.

[10] P. E. Gill, W. Murray, and M. A. Saunders. SNOPT:A sparse nonlinear optimizer. In W. W. Hager, D. W. Hearn, and P. M. Pardalos, editors, *Conference on Large Scale Optimization*, 1993.

[11] P. E. Gill, W. Murray, and M. H. Wright. *Practical Optimization*. Academic Press, London - ... - Tokyo, 1981.

[12] C. B. Gurwitz. A two-piece reduced Hessian updating method for solving nonlinear programming problems. Technical report, Department of Computer and Information Science, Brooklyn College, 1991.

[13] C. B. Gurwitz and M. L. Overton. Sequential quadratic programming methods based on approximating a projected Hessian matrix. *SIAM J. Sci. Stat. Comput.*, 10:631–653, 1989.

[14] D. Kraft. Comparing mathematical programming algorithms based on lagrangian functions for solving optimal control problems. In H. E. Rauch, editor, *Control Applications of Nonlinear Programming*. Pergamon Press, 1979.

[15] K. Kunisch and E. W. Sachs. Reduced SQP methods for parameter identification problems. *SIAM J. Numer. Anal.*, 29:1793–1820, 1992.

[16] F.-S. Kupfer. *Reduced Successive Quadratic Programming in Hilbert space with applications to optimal control.* doctoral thesis, Universität Trier, 1992.

[17] F.-S. Kupfer and E. W. Sachs. Numerical solution of a nonlinear parabolic control problem by a reduced SQP method. *Computational Optimization and Applications*, 1:113–135, 1992.

[18] F.-S. Kupfer and E. W. Sachs. Reduced SQP methods for nonlinear heat conduction problems. In R. Bulirsch, A. Miele, J. Stoer, and K. H. Well, editors, *Optimal Control, Oberwolfach 1991*, volume 111 of *Int. Series Num. Math.*, pages 145–160. Birkhäuser, 1993.

[19] J. Nocedal L. T. Biegler and C. Schmid. A reduced Hessian method for large scale constrained optimization. Technical report, Carnegie-Mellon University, 1993. Preprint.

[20] D. C. Luenberger. *Optimization by Vector Space Methods*. Wiley, New York, 1969.

[21] K. C. P. Machielsen. *Numerical solution of optimal control problems with state constraints by sequential quadratic programming in function space.* CWI Tract. Centrum voor Wiskunde en Informatica, Amsterdam, 1988.

[22] J. Nocedal and M. L. Overton. Projected Hessian updating algorithms for nonlinearly constrained optimization. *SIAM J. Numer. Anal.*, 22:821–850, 1985.

[23] C. E. Orozco and O. N. Ghattas. Massively parallel aerodynamic shape optimization. In *Proceedings of the Symposium on High Performance Computing for Flight Vehicles*, 1991. to appear.

[24] C. Schmid and L. T. Biegler. Acceleration of reduced Hessian methods for large-scale nonlinear programming. *Computers chem. Engng.*, 17:451–463, 1993.

[25] S. Vasantharajan, J. Viswanathan, and L. T. Biegler. Reduced successive quadratic programming implementation for large-scale optimization problems with smaller degrees of freedom. *Computers chem. Engng.*, 14:907–915, 1990.

International Series of Numerical Mathematics, Vol. 115, © 1994 Birkhäuser Verlag Basel

Time Optimal Control of Mechanical Systems

W. Schenker*   and   H.P. Geering*

Abstract

We treat the problem of time-optimal control of dynamical systems with the help of differential geometry.

If the problem of time-optimal control is to be solved, a system of first-order differential equations, the so-called adjoint system, has to be integrated. In most cases this cannot be done, neither analytically nor numerically. To reduce the complexity of the adjoint system we suggest to reduce its dimension by using First Integrals. In order to find those we formulate our problem in the language of differential geometry and apply the tools of the theory of dynamical systems.

We state a rule for obtaining First Integrals for several classes of systems. Our procedure and the results we gained with it, mean a step towards optimal control of nonlinear systems.

Introduction

We consider the problem of time-optimal control of holonomic mechanical systems. Let the system equations be given by

$$\dot{x}^i = f^i(\boldsymbol{x}, \boldsymbol{u}) \qquad i = 1, \ldots, n$$

where locally $\boldsymbol{x} \in \boldsymbol{R}^n$ are the states and locally $\boldsymbol{u} \in \boldsymbol{R}^m$ the controls of the system. Applying the methods of calculus of variation we end up with a boundary value problem, the so-called *adjoint* system

$$
\begin{aligned}
\dot{x}^i &= \frac{\partial H^o}{\partial \lambda_i}(x^1, \ldots, x^n, \lambda) \\
\dot{\lambda}_i &= -\frac{\partial H^o}{\partial x^i}(x^1, \ldots, x^n, \lambda)
\end{aligned}
\qquad\qquad H^o = \min_{\boldsymbol{u}}(1 + \lambda_i f^i)
$$

with boundary conditions

$$\boldsymbol{x}(t_0) = \boldsymbol{x}_0 \qquad \text{and} \qquad \lambda(t_f) = \cdots .$$

---

*Measurement and Control Laboratory, Swiss Federal Institute of Technology (ETH), CH-8092 Zurich, Switzerland

We have to solve this boundary value problem but in most cases this is not possible. A numerical treatment fails due to its high dimension and the amount of time thus needed to arrive at a solution. One possibility, however, to get somewhat closer to a solution is to reduce the dimension of the adjoint system by using a First Integral (symmetry) of it. With the help of a First Integral we are able to find a symplectic coordinate transformation $\Phi : \boldsymbol{x} \mapsto \bar{\boldsymbol{x}}, \lambda \mapsto \bar{\lambda}$ such that the transformed system is of the form

$$
\begin{aligned}
\dot{\bar{x}}^i &= \frac{\partial \bar{H}^o}{\partial \bar{\lambda}_i}(\bar{x}^2, \ldots, \bar{x}^n, \bar{\lambda}) \\
\dot{\bar{\lambda}}_1 &= 0 \\
\dot{\bar{\lambda}}_j &= -\frac{\partial \bar{H}^o}{\partial \bar{x}^j}(\bar{x}^2, \ldots, \bar{x}^n, \bar{\lambda}) \, .
\end{aligned}
\qquad
\bar{H}^o = \min_{\bar{\boldsymbol{u}}}(1 + \bar{\lambda}_i \bar{f}^i)
$$

where $j = 2, \ldots, n$. The question now is how to find a First Integral of the adjoint system. One way consists in finding a First Integral of the uncontrolled system $(\dot{\boldsymbol{x}} = \boldsymbol{f}(\boldsymbol{x}, \boldsymbol{u} \equiv 0))$ and then investigating what happens with it if the system becomes controlled.

Setting

Our goal is to apply tools from differential geometry and the theory of dynamical systems to get information about the interior structure of our optimally controlled mechanical system.

Let M be a smooth manifold of dimension n. An atlas gives a coordinate system $(\boldsymbol{x} = (x^1, \ldots, x^n))$ on M. Out of this atlas we are able to construct a configuration as described in Figure (1). There E is a fibre bundle $(E, \pi, T^*(TM))$, with typical fibre B. Let $\{U_i, \phi_i\}$ be a family of local trivializations of E, i.e., we have diffeomorphisms

$$
\phi_i : \pi^{-1}(U_i) \to U_i \times B
$$

for an open covering $\{U_j \mid j \in J\}$ of $T^*(TM)$. On this fibre bundle we have fibre coordinates

$$
(\boldsymbol{x} = (x^1, \ldots, x^n), \boldsymbol{v} = (v^1, \ldots, v^n), \lambda = (\lambda_1, \ldots, \lambda_n), \mu = (\mu_1, \ldots, \mu_n),
$$

$$
\boldsymbol{u} = (u^1, \ldots, u^m)) \, ,
$$

where $(\boldsymbol{u} = (u^1, \ldots, u^m))$ is a coordinate system of B (over a preferred U_i). Furthermore, we consider a submanifold K of E, which is defined by a section $\boldsymbol{u}^o : T^*(TM) \to E$ of the bundle E, i.e., $K = \boldsymbol{u}^o(T^*(TM))$. This arrangement of manifolds and bundles will in the further discussion be called *setting*.

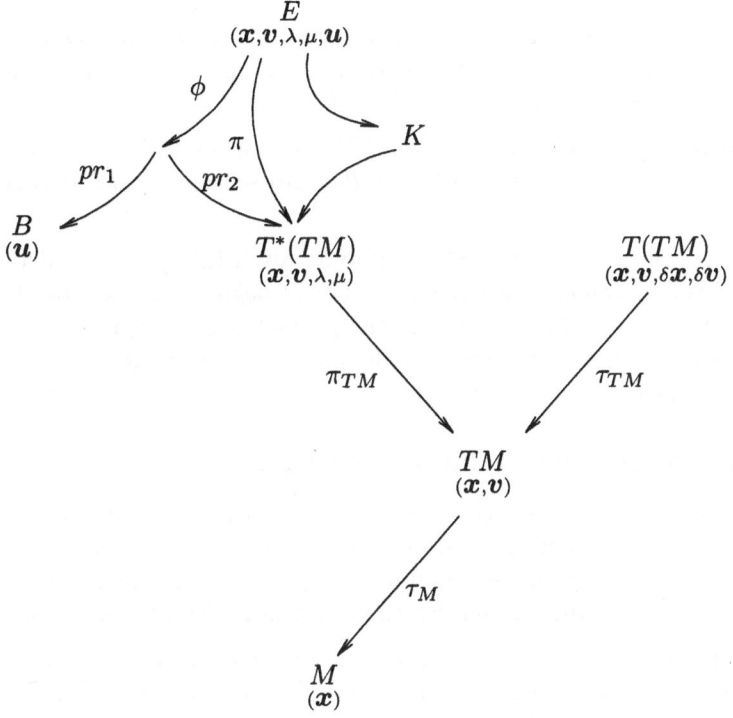

Figure 1: Setting

Setting and Semisprays

We have to optimize a mechanical system that may be described by a second-order differential equation. For this we assume the manifold M (in our setting) to be the configuration manifold of the system. The second-order differential equation is represented by a vector field on TM, a so-called semispray. Now, since M is considered to be the configuration manifold of the system, the structure of $TTM :=T(TM)$ can be recovered by the semispray. In fact, we will see that a semispray determines a connection on TM, from which we can deduce the structure of TTM. So it is interesting for us to consider the relations of this connection and the semispray.

Definition: [1], [4] *A semispray on M (or a second-order differential equation) is a vector field ξ on TM, such that $T\tau_M \circ \xi$ is the identity on TM, whereas $T\tau_M$ is the differential of τ_M.*

Locally ξ is given by

$$\xi = v^i \frac{\partial}{\partial x^i} + \xi^i \frac{\partial}{\partial v^i} \ .$$

Definition: [4] *Let ξ be a semispray on M and C the Liouville vector field that is represented on our coordinate system on M by $C = v^i(\partial/\partial v^i)$. The vector field $\xi^* = [C, \xi] - \xi$ is called the* deviation *of ξ.*

Definition: [4] *A semispray ξ is said to be a* spray *if $\xi^* = 0$ and ξ is C^1 on the zero section. Moreover, if ξ is C^2 on the zero section, then ξ is called a* quadratic spray.

Let ξ be a semispray. Since $\xi^* = 0$ if and only if $[C, \xi] = \xi$, we deduce that ξ is a spray if the functions ξ^i are homogeneous of degree 2 and C^1 on the zero section. Therefore, we see that a spray is the geometrical interpretation of a system of second-order differential equations homogeneous of degree 2 with respect to the first derivatives.

The Canonical Almost Tangent Structure on TM

In this section we introduce a geometric structure which is essential in the Lagrangian formulation of classical mechanics. We will see that on the tangent bundle (TM, τ_M, M) of any smooth manifold M (i.e., on our configuration manifold), we have a canonical almost tangent structure (this is also an explanation of the name).

Definition: [4] *An* almost tangent structure S *on TM is a tensor field S of type $(1,1)$ on TM with constant rank n $(n = \dim M)$ that satisfies $S^2 = 0$. The pair (TM, S) is called an* almost tangent manifold.

Let $(\boldsymbol{x} = (x^1, \ldots, x^n), \boldsymbol{v} = (v^1, \ldots, v^n))$ be a local coordinate system of TM. Then, we can define a $(1,1)$ tensor field S on TM by

$$\boxed{\; S = \frac{\partial}{\partial v^i} \otimes dx^i \;}.$$

Distribution on TTM

Consider now the two *tensor fields* $\boldsymbol{h}$ and $\boldsymbol{v}$ on TM given by

$$\boxed{\; \boldsymbol{h} = (1/2)(\mathrm{Id}_{TTM} - \mathcal{L}_{\xi}S) \;}, \qquad \boxed{\; \boldsymbol{v} = (1/2)(\mathrm{Id}_{TTM} + \mathcal{L}_{\xi}S) \;}.$$

From the fact that $(\mathcal{L}_{\xi}S)^2 = \mathrm{Id}_{TTM}$ follows that $\boldsymbol{h}$ and $\boldsymbol{v}$ have the properties

$$\boldsymbol{h}^2 = \boldsymbol{h}, \qquad\qquad \boldsymbol{v}^2 = \boldsymbol{v},$$
$$\boldsymbol{h} \circ \boldsymbol{v} = \boldsymbol{v} \circ \boldsymbol{h} = 0, \qquad \boldsymbol{h} + \boldsymbol{v} = \mathrm{Id}_{TTM}.$$

These tensor fields are projection operators corresponding to the direct sum decomposition of tangent space, i.e., $T_pTM = \mathrm{im}\,\boldsymbol{h}_p \oplus \mathrm{im}\,\boldsymbol{v}_p$ at each point p of TM. The kernel of $\boldsymbol{h}$ coincides with the image of $\boldsymbol{v}$, hence the kernels of $\boldsymbol{h}$ and $\boldsymbol{v}$ are complementary subspaces. Let

$$\ker \boldsymbol{h} = V, \qquad \ker \boldsymbol{v} = H,$$

then $TTM = V \oplus H$ which is the desired distribution. This construction works for any semispray or second-order differential equation. Note that (from the above construction) the vertical subspace V_p for $p \in TM$ is in fact ker $T_p \tau_M$.

Connections on (TM, S)

We define a connection in the almost tangent manifold (TM, S), where S is the canonical almost tangent structure. This definition was proposed by [3].

Definition: *A* connection Γ *in (TM, S) is a smooth tensor field of type $(1, 1)$ on TM, such that*

$$S\Gamma = S, \qquad \Gamma S = -S,$$

where S is the canonical almost tangent structure on TM.

Proposition:*Let S be the canonical almost tangent structure on TM. Then for any semispray ξ on TM,*

$$\Gamma = -\mathcal{L}_\xi S, \tag{1}$$

is a connection in (TM, S).

Proof: The proof is done with the help of the Nijenhuis tensor.

Now let $TTM = H \oplus V$ be our distribution and let h and v be the horizontal and vertical projections, respectively. Then we can define a covariant derivative for vector fields X and Y on M and $p \in TM$ as

$$\nabla_X^\Gamma Y := J_p^{-1}(Y_*(X_{\tau_M(p)})) \in T_{\tau_M(p)} M. \tag{2}$$

All this now shows us how to get a covariant derivative ∇^Γ out of our connection $\Gamma = -\mathcal{L}_\xi S$ on (TM, S), namely by first passing from Γ to its associated distribution of $TTM = H \oplus V$ and then going from this distribution to the covariant derivative ∇^Γ. Locally, (2) yields

$$\boxed{\nabla_X^\Gamma Y = \left(X^i \frac{\partial Y^j}{\partial x^i} + X^i \Gamma_i^j(x, v) \right) \frac{\partial}{\partial x^j}.}$$

Systems on the Setting

We consider the manifold M of our setting to be the configuration manifold of the system and we learned how to represent a second-order differential equation on M. We will see how the equations of the adjoint Hamiltonian system to a second-order differential equation will be related to our setting. Moreover, we are interested in the representation of the uncontrolled mechanical system in Hamiltonian and Lagrangian formalism on our setting. We will then be able to recover how the different representations of our mechanical system (Hamiltonian and Lagrangian) are related to each other.

Lagrange Systems on the Setting

The system that we consider is a Lagrangian system on the manifold M (with coordinate system $(\boldsymbol{x} = (x^1, \ldots, x^n))$) with Lagrange function $L : TM \to \boldsymbol{R}$. Then, in our setting ω_L is given as

$$\omega_L = \frac{\partial^2 L}{\partial x^j \partial v^i} dx^i \wedge dx^j + \frac{\partial^2 L}{\partial v^i \partial v^j} dx^i \wedge dv^j$$

where

$$\xi_L \lrcorner \omega_L = dE_L \qquad \text{and} \qquad E_L = \mathcal{L}_C - L$$

and the semispray as

$$\xi_L = v^i \frac{\partial}{\partial x^i} + \xi_L^i \frac{\partial}{\partial v^i} .$$

This system represents the uncontrolled mechanical system without forces.

Hamiltonian Systems on the Setting

System on $T^*(TM)$

Out of the semispray which describes the Lagrangian system on M we can form a Hamiltonian system on $T^*(TM)$ with Hamiltonian function $S : T^*(TM) \to \boldsymbol{R}$. We will show how this can be done. The symplectic two-form of the system corresponding to the local coordinates on $T^*(TM)$ is

$$\omega_S = d\lambda \wedge d\boldsymbol{x} + d\mu \wedge d\boldsymbol{v} .$$

The Hamiltonian vector field lives on $T^*(TM)$ with the components

$$\boldsymbol{X}_S = \frac{\partial S}{\partial \lambda_i} \frac{\partial}{\partial x^i} + \frac{\partial S}{\partial \mu_i} \frac{\partial}{\partial v^i} - \frac{\partial S}{\partial x^i} \frac{\partial}{\partial \lambda_i} - \frac{\partial S}{\partial v^i} \frac{\partial}{\partial \mu_i} .$$

The Hamiltonian system is computed out of the semispray $\xi_L = v^i(\partial/\partial x^i) + \xi_L^i(\partial/\partial v^i)$ of our system. Under the consideration of the forces f^i we have

$$\begin{aligned} \dot{x}^i &= v^i \\ \dot{v}^i &= \xi_L^i + f^i . \end{aligned}$$

Now, by adjoining the dual variables (λ, μ) to $(\boldsymbol{x}, \boldsymbol{v})$, we can construct the Hamiltonian function

$$S = v^i \lambda_i + \xi_L^i \mu_i + f^i \mu_i .$$

We obtain the system vector field $\boldsymbol{X}_S$ as

$$\boldsymbol{X}_S = \operatorname{sgrad} S = v^i \frac{\partial}{\partial x^i} + (\xi_L^i + f^i) \frac{\partial}{\partial v^i} - \frac{\partial \xi_L^l}{\partial x^i} \mu_l \frac{\partial}{\partial \lambda_i} - \left(\frac{\partial \xi_L^l}{\partial v^i} \mu_l + \lambda_i \right) \frac{\partial}{\partial \mu_i}$$

which yields

$$\boldsymbol{X}_S \lrcorner \omega_S = -dS .$$

System on E

Let us introduce the control u. We assume the control u to be a section of the bundle E. Furthermore, we suppose the control system to be of the form

$$\ddot{x}^i = \xi_L^i + f^i + u^i .$$

System on K

Out of the system on E we will then construct the Hamiltonian adjoint system.
We first introduce the function $H' : E \to R$ by setting

$$H' = L_{perf} + H .$$

Since we perform time-optimal control it must be that $L_{perf} = 1$. According to Pontryagin we have to minimize H'. We set

$$H^o = \min_{u} H' .$$

The section u of the bundle E that minimizes H' is called the optimal u and is denoted by u^o. Since u^o is a section of bundle E and thus the submanifold K (Chapter) of E can be defined. We recognize that K is a symplectic manifold, and

$$H^o : K \to R .$$

The Hamiltonian system on K is thus given by the vector field

$$X_{H^o} = \text{sgrad} H^o .$$

Symmetries on the Setting

Spray and Symmetries of the Lagrangian system on TM

We are trying to find symmetries of our system in Lagrangian form on our setting. In a first step we will look for symmetries of the uncontrolled system without forces. We assume this system to be described by a homogeneous Lagrange function $L : TM \to R$. It is therefore described by a spray ξ_L. We look for symmetries of this spray.

A vector field W is an *infinitesimal symmetry* of the spray ξ_L iff

$$\mathcal{L}_W \xi_L = [W, \xi_L] = 0 .$$

Proposition: Y^C, *the complete lift of a vector field Y on M, is an infinitesimal symmetry of the spray ξ_L if and only if Y is an infinitesimal affine transformation of the corresponding affine connection.*

Proof: We fix a Riemannian metric g on M and a connection Γ which is compatible with g. With $\nabla = \nabla^\Gamma$ we denote the associated covariant derivative. The Proof had been formulated in terms of lifts.

Hamiltonian System on $T^*(TM)$ and its Symmetries

In what follows we assume the Lagrange function of the system to be given by $(1/2)g_{ij}v^iv^j$, where g_{ij} is a Riemannian metric on M. We see that the spray which stands for the Lagrange equation with Lagrangian $L = (1/2)g_{ij}v^iv^j$ is given by

$$\xi_L = v^i\frac{\partial}{\partial x^i} - v^jv^k\Gamma^i_{jk}\frac{\partial}{\partial v^i} \ .$$

In [2] it is shown that the Lagrange equation of a holonomic system with homogeneous Lagrange function can be written

$$\dot{v}^i = f^i - \Gamma^i_{jk}v^jv^k$$

where $f^i : M \to \mathbf{R}$ are the forces. We are able to construct the Hamiltonian system with Hamiltonian function $S : T^*TM \to \mathbf{R}$,

$$S = v^i\lambda_i + f^i\mu_i - \Gamma^i_{jk}v^jv^k\mu_i \ .$$

Let $\mathbf{Y} = Y^i(\partial/\partial x^i)$ be a vector field in TM that generates an affine transformation on M.

Proposition: *The vector field $\mathbf{Y}_S = \mathrm{sgrad}\, F_S$, with $F_S = Y^i\lambda_i + (\partial Y^i/\partial x^j)v^j\mu_i$, a smooth function from $T^*(TM)$ to M, is symmetric to $\mathbf{X}_S$ if and only if*

$$[\mathbf{Y}, \mathbf{f}] = 0$$

where $f = (f^1, \ldots, f^n)$ are the forces.

Proof: For F_S to be a First Integral we have to verify that

$$\{F_S, S\} = \frac{\partial F_S}{\partial \lambda_s}\frac{\partial S}{\partial x^s} + \frac{\partial F_S}{\partial \mu_s}\frac{\partial S}{\partial v^s} - \frac{\partial F_S}{\partial x^s}\frac{\partial S}{\partial \lambda_s} - \frac{\partial F_S}{\partial v^s}\frac{\partial S}{\partial \mu_s} = 0 \ .$$

Hamiltonian Systems and Control

As we introduce the control forces, our system becomes a controlled system. Let us separate $\mathbf{u}$ into a unit vector α that gives the direction of $\mathbf{u}$ and a scalar $s : K \to \mathbf{R}$ that performs a stretching of α. We set

$$\mathbf{u} = s\alpha$$

with $0 \le s \le s_{max}$ where we have taken the constraints for $\mathbf{u}$ into account. Then H becomes

$$H = v^i\lambda_i + f^i\mu_i - \Gamma^i_{jk}v^jv^k\mu_i + s\alpha^i\mu_i \ .$$

Optimal Control and Symmetries

According to Pontryagin's minimum principle [5], we have to construct a Hamiltonian function $H' : E \to \mathbf{R}$ by

$$H' = L_{perf} + v^i \lambda_i + f^i \mu_i - \Gamma^i_{jk} v^j v^k \mu_i + s \alpha^i \mu_i .$$

Then we have to minimize H' with respect to the control $\mathbf{u}$ to arrive at

$$H^o = \min_{\mathbf{u}} H' .$$

Clearly, the Hamiltonian H' will become minimal if the vector α runs the opposite direction of the vector μ and $s = s_{max}$. From the definition of the Riemannian metric on M, (i.e., $< \cdot , \cdot >$)

$$< \eta, \xi > = g_{ij} \eta^i \xi^j$$

we obtain for α

$$\alpha^i = -\frac{\mu^i}{|\mu|} = -\frac{g^{il} \mu_l}{\sqrt{g^{kl} \mu_k \mu_l}} .$$

For the Hamiltonian we obtain

$$H^o = 1 + v^i \lambda_i + f^i \mu_i - \Gamma^i_{jk} v^j v^k \mu_i - s_{max} \sqrt{g^{kj} \mu_k \mu_j} .$$

Let $\mathbf{Y} = Y^i (\partial / \partial x^i)$ be a vector field on M.

Proposition $\mathbf{Y}_{H^o} = \operatorname{sgrad} F_{H^o}$ *with* $F_{H^o} = Y^i \lambda_i + (\partial Y^i / \partial x^j) v^j \mu_i$ *is a symmetry of the system with Hamiltonian* H^o *if and only if* $\mathcal{L}_{\mathbf{Y}} \mathbf{f} = \mathcal{L}_{\mathbf{Y}} g^{ij} = 0$.

Proof: For F_{H^o} to be a First Integral we have to verify that

$$\{F_{H^o}, H^o\} = \frac{\partial F_{H^o}}{\partial \lambda_s} \frac{\partial H^o}{\partial x^s} + \frac{\partial F_{H^o}}{\partial \mu_s} \frac{\partial H^o}{\partial v^s} - \frac{\partial F_{H^o}}{\partial x^s} \frac{\partial H^o}{\partial \lambda_s} - \frac{\partial F_{H^o}}{\partial v^s} \frac{\partial H^o}{\partial \mu_s} = 0 .$$

We therefore are able to find symmetries of the optimally controlled system. All the proofs and a description of the reduction process are given in the Thesis of W. Schenker.

References

[1] R. Abraham, J.E. Marsden, T. Ratiu: Manifolds, Tensor Analysis, and Applications, Second Edition, Springer New York Berlin Heidelberg London Paris Tokyo, 1988.

[2] B.A. Dubrovin, A.T. Fomenko, S.P. Novikov: Modern Geometry, Methods and Applications, Part 1., Springer-Verlag, New York Berlin Heidelberg Tokyo, 1984.

[3] J. Grifone: Estructure presque tangente et connexions I,II, Ann. Inst. Fourier, Grenoble, 22, 3 (1972), 287–334 and 22, 4 (1972), 291–338.

[4] M. de Leon: Methods of Differential Geometry in Analytical Mechanics, North-Holland, 1989.

[5] L.S. Pontryagin, V.B. Boltjanskij, R.V. Gramkrelidze, E.F. Mischenko: Mathematische Theorie optimaler Prozesse, Oldenburg München, Wien, 1964.

3 Algorithms for Optimal Control Calculations

International Series of Numerical Mathematics, Vol. 115, © 1994 Birkhäuser Verlag Basel

Second Order Algorithm for Time Optimal Control of a Linear System

Mark D. Ardema*   and   Han-Chang Chou*

Abstract

In previous papers, zero-order solutions for time optimal control of singularly perturbed third-order systems have been obtained by the method of matched asymptotic expansions (MAE). The resulting open-loop control laws were founded to give good results, provided the singular perturbation parameter is small. In this paper, we use the MAE method to derive a second-order open-loop controller for a representative third-order system. Numerical simulations show that the second-order controller gives significantly better performance than the zero-order controller.

Introduction

In many engineering systems it is desired to move a mass from position to position in minimum time. One such engineering system is the actuator head in a diskdrive information storage system [1]. Although the theory for minimum time control of linear systems is well established, the implementation of the resulting optimal control laws is very difficult, particularly for high-order systems [2,3].

An attractive method for dealing with prohibitively high-order systems is to use singular perturbation theory. This method separates a dynamic system into reduced-order subsystems on different time scales, and provides a way to synthesize a near-optimal controller by combining the two reduced system controllers (see, for example, [4] and [5]). This approach has been investigated for linear time-optimal systems in [5] and [6]. These references show that the optimal control of the full system is much like the optimal control of the reduced system (perturbation parameter set to zero) except that there are rapid control switches near the end of the process which bring to rest the dynamics neglected by the reduced solution.

In previous papers [7,8], we have derived control laws to zero-order in the small parameter for a third-order linear system and applied the results to control of a disk-drive actuator. Simulations showed that the control laws give a high degree of accuracy except for very short moves on the disk and large values of the parameter. In this paper, we use the method of matched asymptotic expansions

*Department of Mechanical Engineering, Santa Clara University, Santa Clara, CA 95053, U.S.A.

(MAE) to derive second-order corrections to the control law for the third-order system studied in [7].

Finally, we simulate the performance of the singular perturbation open-loop controller for typical values of the system parameters. As measured by errors in the final values of the state variables, the second-order controller gives much greater accuracy than the zero-order controller.

Problem Formulation

The problem we will consider is the following third-order linear system:

$$
\begin{aligned}
\dot{x} &= y, \\
\dot{y} &= z, \\
\varepsilon\dot{z} &= -z + u.
\end{aligned}
\tag{1}
$$

We wish to transfer the system from one rest position to another, the latter taken to be the state space origin,

$$
\begin{array}{llll}
x(0) &= x_0, & x(t_f) &= 0, \\
y(0) &= 0, & y(t_f) &= 0, \\
z(0) &= 0, & z(t_f) &= 0,
\end{array}
\tag{2}
$$

while minimizing the transfer time:

$$
J = \int_0^{t_f} dt.
\tag{3}
$$

The control u is bounded, i.e., $|u| \leq 1$.

The necessary conditions for optimal control are easily stated. For this purpose, form the Hamiltonian function H such that

$$
H = -1 + y\lambda_x + z\lambda_y + (-z + u)\lambda_z,
\tag{4}
$$

and the adjoint equations

$$
\begin{aligned}
\dot{\lambda}_x &= 0, \\
\dot{\lambda}_y &= -\lambda_x, \\
\varepsilon\dot{\lambda}_z &= -\lambda_y + \lambda_z.
\end{aligned}
\tag{5}
$$

The Maximum Principle [9] then gives the optimal control u as

$$
u = sgn\,\lambda_z.
\tag{6}
$$

Lemmas 14.1 and 14.2 of [9] establish that extremal control is bang-bang for this problem and that there are at most three control switches, including the one at the final time. The second switch time should occur at a time of the order of ε

before the end of the process. From equation (6) these switches occur at the zeros of λ_z. The adjoint equations (5) may be integrated to give

$$
\begin{aligned}
\lambda_x &= C_1, \\
\lambda_y &= -C_1 t + C_2, \\
\lambda_z &= -C_1 t + C_2 - \varepsilon C_1 + C_3 e^{t/\varepsilon}.
\end{aligned}
\tag{7}
$$

Thus the control switch times depend on the constants C_1, C_2, and C_3, but determining these constants is nontrivial.

Asymptotic Analysis and Control Law

Instead of solving for the switching function to get the optimal control switch times, we use the fact that the optimal control is bang-bang with at most two switches. The unique solution of the state equations that meets all boundary conditions with two control switches must be the optimal solution, and this fact can be used to solve for the optimal switch times, as first observed in [10]. Thus solutions of the state equations are needed, and our approach is to obtain these solutions approximately by the method of matched asymptotic expansions.

The asymptotic analysis proceeds by dividing the motion into five segments as follows: (1) an initial boundary layer in which the z state variable rapidly and asymptotically approaches its equilibrium value; (2) an outer region ending at the first switch time; (3) an interior boundary layer beginning at the first switch time in which z approaches its new equilibrium value; (4) an outer region ending at the second switch time; and (5) a terminal boundary layer.

Assuming that $x_0 < 0$, there are two switches : at $t = t_{s1}$ from $u = +1$ to $u = -1$ and at $t = t_{s2}$ from $u = -1$ to $u = +1$. At $t = t_f$ (the end of the process) the control switches from $u = +1$ to $u = 0$. In a previous paper [7], we have found the switch times to zero-order in the small parameter ε to be

$$
\begin{aligned}
t_{s1} &= \sqrt{-x_0}, \\
t_{s2} &= 2\sqrt{-x_0}, \\
t_f &= 2\sqrt{-x_0} + \varepsilon \ln 2.
\end{aligned}
\tag{8}
$$

The next logical step to improving the estimates of switch times is to obtain the first-order corrections. When the MAE method was used to find these corrections for the problem defined by (1), however, it was found that the first-order terms in the composite expansions are zero. To verify this result, the method of [6] was applied; the first-order corrections were again found to be zero.

We now turn to the second-order solution, using MAE ([6] does not give a method for second and higher order solutions). By substituting

$$
\begin{aligned}
x &= x_0 + x_1 \varepsilon + x_2 \varepsilon^2, \\
y &= y_0 + y_1 \varepsilon + y_2 \varepsilon^2, \\
z &= z_0 + z_1 \varepsilon + z_2 \varepsilon^2,
\end{aligned}
\tag{9}
$$

into system (1), the solution for the first outer region $(t \in (0, t_{s1}], u = +1)$ is obtained as

$$
\begin{aligned}
x^o &= (t^2/2 + C_4 t + C_5) + \varepsilon(C_6 t + C_7) + \varepsilon^2(C_8 t + C_9), \\
y^o &= (C_4 + t) + \varepsilon C_6 + \varepsilon^2 C_8, \\
z^o &= 1.
\end{aligned}
\tag{10}
$$

By introducing a new independent variable τ such that

$$
\tau = t/\varepsilon, \qquad (\cdot)\prime = d(\cdot)/d\tau,
\tag{11}
$$

system (1) is transformed to

$$
\begin{aligned}
\dot{x}\prime &= \varepsilon y, \\
\dot{y}\prime &= \varepsilon z, \\
\dot{z}\prime &= -z + 1.
\end{aligned}
\tag{12}
$$

Applying boundary conditions (2), the solution for this initial boundary layer $(t \in (0, t_{s1}], u = +1)$ is

$$
\begin{aligned}
x^i &= (t^2/2 + x_0) - \varepsilon t + \varepsilon^2(1 - e^{-\tau}), \\
y^i &= t + \varepsilon(e^{-\tau} - 1), \\
z^i &= 1 - e^{-\tau}.
\end{aligned}
\tag{13}
$$

Next, we match the first outer and initial boundary layer solutions as $\varepsilon \to 0, t \to 0, \varepsilon/t \to 0$ to solve for C_4–C_9, and then form the uniform solution (additive composition) for $t \in [0, t_{s1}]$; the result is

$$
\begin{aligned}
x &= (t^2/2 + x_0) - \varepsilon t + \varepsilon^2(1 - e^{-\varepsilon/t}), \\
y &= t + \varepsilon(e^{-\varepsilon/t} - 1), \\
z &= 1 - e^{-\varepsilon/t}.
\end{aligned}
\tag{14}
$$

Which is identical to the first boundary layer solution.

In order to solve for the second outer region and interior boundary layer, we expand the first and second switch times (t_{s1} and t_{s2}) and final time (t_f) to second order in ε:

$$
\begin{aligned}
t_{s1} &= t_{10} + t_{11}\varepsilon + t_{12}\varepsilon^2, \\
t_{s2} &= t_{20} + t_{21}\varepsilon + t_{22}\varepsilon^2, \\
t_f &= t_{f0} + t_{f1}\varepsilon + t_{f2}\varepsilon^2.
\end{aligned}
\tag{15}
$$

Using a similar procedure as before, yields the uniform solution for $(t \in [t_{s1}, t_{s2}], u = -1)$:

$$
\begin{aligned}
x &= (x_0 - t^2/2 + 2t_{10}t - t_{10}^2) + \varepsilon(-2t_{10}t_{11} - 2t_{10} + 2t_{11}t + t) \\
&\quad + \varepsilon^2(-2t_{11} - 1 + 2e^{-\tau\prime} + t_{12}t - t_{10}t_{12} - t_{11}^2), \\
y &= (-t + 2t_{10}) + \varepsilon(1 + 2t_{11} - 2e^{-\tau\prime}) + 2t_{12}\varepsilon^2, \\
z &= -1 + 2e^{-\tau\prime}.
\end{aligned}
\tag{16}
$$

where

$$\tau\prime = (t - t_{s1})/\varepsilon. \tag{17}$$

The terminal boundary layer is obtained by stretching time-to-go $(t_f - t)$, by introducing a new independent variable

$$\alpha = (t_f - t)/\varepsilon, \qquad (\cdot)\prime = d(\cdot)/d\alpha, \tag{18}$$

in system (1). This results in

$$
\begin{aligned}
\dot{x}\prime &= \varepsilon y, \\
\dot{y}\prime &= \varepsilon z, \\
\dot{z}\prime &= z - 1,
\end{aligned}
\tag{19}
$$

with

$$
\begin{aligned}
x_b(0) &= 0, \\
y_b(0) &= 0, \\
z_b(0) &= 0.
\end{aligned}
\tag{20}
$$

The solutions for this terminal boundary layer $(t \in [t_{s2}, t_f], u = +1)$ are

$$
\begin{aligned}
x &= \varepsilon^2(-e^\alpha + \alpha^2/2 + \alpha + 1), \\
y &= \varepsilon(-1 - \alpha + e^\alpha), \\
z &= 1 - e^\alpha.
\end{aligned}
\tag{21}
$$

Patching equations (16) and equations (21) at t_{s2}, the switch times are obtained as:

$$
\begin{aligned}
t_{s1} &= \sqrt{-x_0} + \varepsilon^2(\ln 2)^2/2\sqrt{-x_0}, \\
t_{s2} &= 2\sqrt{-x_0} + \varepsilon \ln 2 + \varepsilon^2(\ln 2)^2/\sqrt{-x_0}, \\
t_f &= 2\sqrt{-x_0} + 2\varepsilon \ln 2 + \varepsilon^2(\ln 2)^2/\sqrt{-x_0},
\end{aligned}
\tag{22}
$$

A second-order open loop control algorithm is now easily stated. First, precompute t_{s1}, t_{s2}, and t_f from equation (22). Then begin the process with control $u = +1$. Switch the control to $u = -1$ when $t = t_{s1}$, then switch back to $u = +1$ when $t = t_{s2}$. Finally, end the process at $t = t_f$.

Numerical Examples

In this section, we use the open-loop, second-order control algorithm in a simulation program that numerically integrates system (1) subject to the initial conditions (2).

Figure 1 shows the time-histories of the state variables x, y, and z for the case of a change in displacement of 5 units ($x_0 = -5$) with $\varepsilon = 1$. It is seen that all

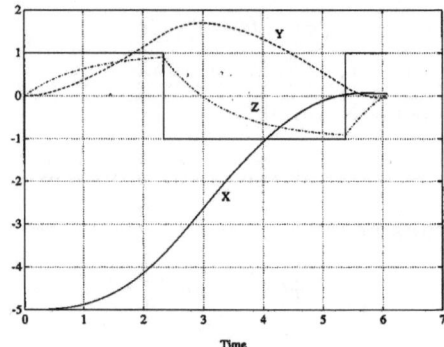

Figure 1: Time Histories of State Variables for $x_0 = -5$ and $\varepsilon = 1$.

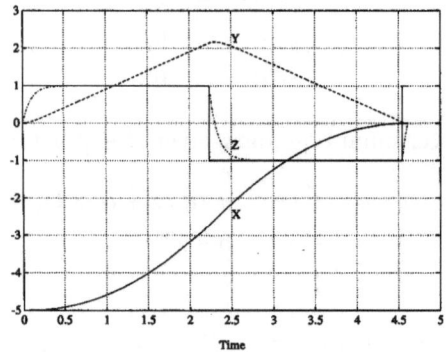

Figure 2: Time Histories of State Variables for $x_0 = -5$ and $\varepsilon = 0.1$.

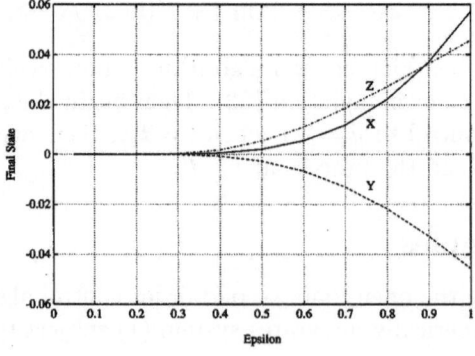

Figure 3: Effect of ε for $x_0 = -5$

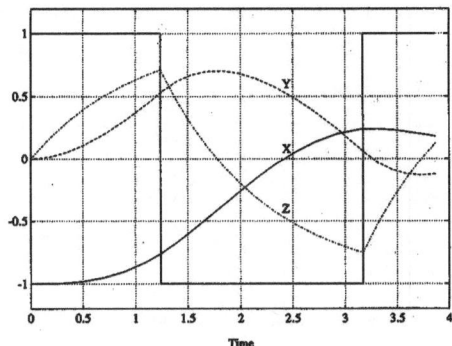

Figure 4: Time Histories of State Variables for $x_0 = -1$ and $\varepsilon = 1$.

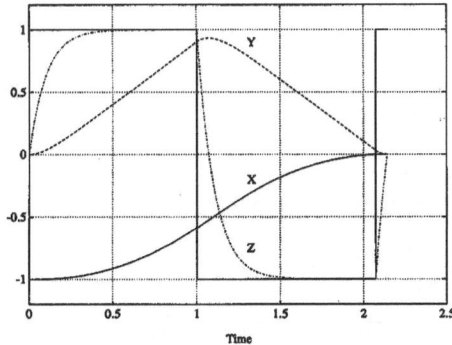

Figure 5: Time Histories of State Variables for $x_0 = -1$ and $\varepsilon = 0.1$.

states are near zero at the final time. Figure 2 shows the case $\varepsilon = 0.1$ for $x0 = -5$. As expected this case shows an extremely high degree of accuracy. The final values of the state variables as a function of ε are shown by figure 3 for the case $x_0 = -5$. It is seen that all variables approach zero as ε approaches zero. This confirms the asymptotic validity of the algorithm.

The case of a change in displacement of 1 unit (Figs. 4, 5, and 6) is considered next. Figure 4 shows that the error in the states at the final time is greater than for a move of 5 units. Figure 5 shows that for $\varepsilon = 0.1$, the accuracy is again quite good. Figure 6 shows the error in the final values of the three states.

In Figure 7, the error in final position, $x(t_f)$, is shown as a function of ε for the two displacement changes. The second-order controller makes final errors in the state extremely small, even when ε is "not so small".

In Figures 8 and 9, a comparison of the error in the final position, $x(t_f)$, as a function of ε, is made between the zero-order and the second-order controller

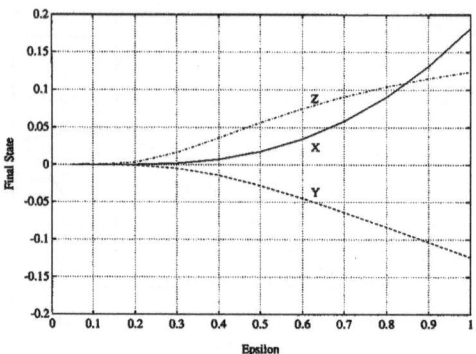

Figure 6: Effect of ε for $x_0 = -1$

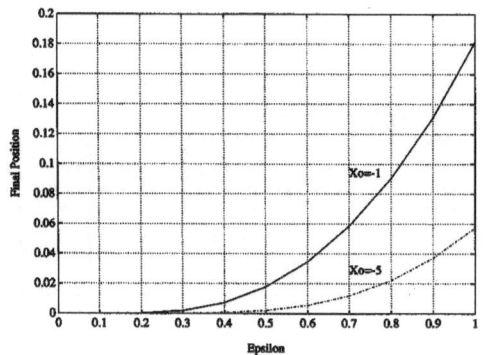

Figure 7: Error in final position.

for the two displacement changes. As shown in Figure 8 ($x_0 = -5$), the errors in final position for the second-order controller are significantly smaller than for the zero-order. For the case of $x_0 = -1$, Figure 9, although the errors are about the same for these two controllers for $\varepsilon = 1$, for ε less than about 0.7 the second-order corrections give significant improvement.

Concluding Remarks

We have obtained asymptotically valid second-order estimates of the optimal switching times of a third-order linear system (the problem studied does not have firstorder corrections). The method used to derive the control law was the method of matched asymptotic expansions. Numerical results show that the second-order controller gives results that are significantly better than for zero-order controllers, as measured by errors in the final values of the state variables. This indicates that

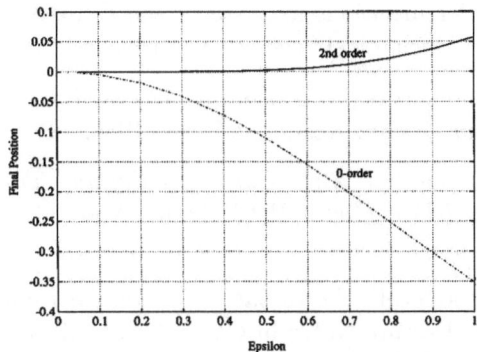

Figure 8: Error in final position for $x_0 = -5$.

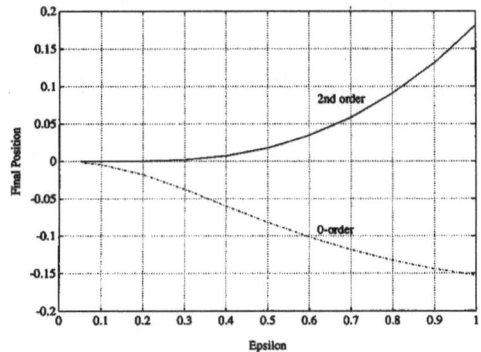

Figure 9: Error in final position for $x_0 = -1$.

the MAE method is useful for constructing higher-order near-optimal controllers for linear systems.

References

[1] Cooper, E., *Minimizing Power Dissipation in a Disk File Actuator*, IEEE Trans. on Magnetics, Vol. 24, No. 3, May 1988.

[2] Yastreboff, M., *Synthesis of Time-Optimal Control by Time Interval Adjustment*, IEEE Trans. Auto. Control, Dec. 1969, pp. 707.

[3] Kassam, S.A., Thomas, J.B., and McCrumm, J.D., *Implementation of Sub-Optimal Control for a Third-Order System*. Comput. Elect. Engng., Vol. 2 1975, pp. 307.

[4] Ardema, M.D., *An Introduction to Singular Perturbations in Nonlinear Opti-*

mal Control, Singular Perturbations in Systems and Control, M.D. Ardema, ed., International Centre for Mechanical Sciences, Courses and Lectures No. 280, 1983.

[5] Kokotovic, P.V., Khalil, H.K., and O'Reilly, J.,*Singular Perturbation Methods in Control: Analysis and Design*, Academic Press, 1986.

[6] Kokotovic, P.V. and Haddad, A.H., *Controllability and Time-Optimal Control of Systems with Slow and Fast Modes*, IEEE Trans. Auto. Control, Feb. 1975, pp. 111.

[7] Ardema, M.D. and Cooper, E., *Perturbation Method for Improved Time-Optimal Control of Disk Drives*, Lecture Notes in Control and Information Sciences, Vol. 151, J.M. Skowronski et. al. (eds), Springer-Verlag, 1991, pp. 37.

[8] Ardema, M.D. and Cooper, E., *Singular Perturbation Time-Optimal Controller For Disk Drives,* Meeting on Optimal Control, Oberwolfach, Germany, May 1991.

[9] Leitmann, G., *The Calculus of Variations and Optimal Control*, Plenum, 1981.

[10] L. S. Pontryagin, V. G. Boltyanskii, R. V. Gamkrelidze, and E. F. Mishchenko, *The Mathematical Theory of Optimal Process*. New York: Interscience, 1962.

International Series of Numerical Mathematics, Vol. 115, © 1994 Birkhäuser Verlag Basel

An SQP-type Solution Method for Constrained Discrete-Time Optimal Control Problems

E. Arnold*   and   H. Puta*

Keywords. Constrained optimal control, discrete-time systems, structured nonlinear programming, hydroelectric power-station systems

Introduction

The considered nonlinear, constrained discrete-time optimal control problem is stated as follows:

$$J = F(x^N) + \sum_{k=0}^{N-1} f_0^k(x^k, u^k) \longrightarrow \text{Min}$$

subject to the *state equation*:
$$x^{k+1} = f^k(x^k, u^k), \quad k = 0, \ldots, N-1, \tag{1}$$
and *inequality constraints*:
$$c^k(x^k, u^k) \leq 0, \qquad k = 0, \ldots, N-1,$$
$$c^N(x^N) \leq 0,$$
$$f^k : \mathbf{R}^n \times \mathbf{R}^m \to \mathbf{R}^n, \; c^k : \mathbf{R}^n \times \mathbf{R}^m \to \mathbf{R}^{r^k}$$

with sufficiently smooth functions F, f_0^k, f^k, c^k. The constraints include fixed initial or final states as well as bounds for state and control variables or more general constraints.

Usual numerical solution methods for (1) are based on Dynamic Programming, in particular Differential Dynamic Programming [MY84], the Maximum Principle, or hierarchical optimization algorithms [ATW94]. Here (1) is treated as a large structured nonlinear programming problem in the state *and* control variables.

Sequential Quadratic Programming (SQP) methods are now standard in nonlinear programming ([Pow78],[Fle81]). The special structure of the problem under consideration can be used during an essential step of the SQP algorithm, the solution of a quadratic programming (QP) problem with linear equality constraints.

*Institut für Automatisierungs- und Systemtechnik, Technische Universität Ilmenau, D-98684 Ilmenau, Germany

This gives a new extension of the well-known Riccati solution procedure for unconstrained linear-quadratic control problems to the equality constrained case. Hence we obtain an efficient, robust and precise algorithm for this special optimal control problem.

The second part of the paper shows an application to optimal release control of connected reservoirs with hydroelectric power-stations. A system consisting of four reservoirs and the river segments between them is described by a deterministic, discrete-time model with a discretization step of one month. The main objective is to maximize electrical energy production over a fixed horizon by a suitable release strategy. This work was done at Institute of Automatic Control of Warsaw University of Technology as a part of the IIASA project 'Water Resources' ([ATW94]).

The main algorithm

The constrained discrete-time optimal control problem (1) is considered as a nonlinear programming problem.

$$\min_{y} \left\{ J(y) \mid h(y) = 0, \quad c(y) \leq 0 \right\} \tag{2}$$

The state and control variables of each time instant are handled as independent decision variables y. Therefore, the state equations become additional equality constraints, whereas the inequality constraints remain as before.

$$y = \begin{bmatrix} x^0 \\ u^0 \\ \vdots \\ x^N \end{bmatrix}, \; h(y) = \begin{bmatrix} f^0(x^0, u^0) - x^1 \\ f^1(x^1, u^1) - x^2 \\ \vdots \\ f^{N-1}(x^{N-1}, u^{N-1}) - x^N \end{bmatrix}, \; c(y) = \begin{bmatrix} c^0(x^0, u^0) \\ c^1(x^1, u^1) \\ \vdots \\ c^N(x^N) \end{bmatrix}$$

A sequential quadratic programming (SQP) method is adapted to solve this high dimensional and structured nonlinear programming problem.

In each SQP iteration step the solution of a linear-quadratic approximation of problem (2) gives a correction of the optimization variables [Pow78]. This approximation is gained by a quadratic cost function which includes second order information of the original cost function and constraints and a linearization of the equality and inequality constraints. An SQP-algorithm consists of the following principal steps:

(1) Start with an approximation y to the solution vector, Lagrange multipliers p and λ, and a positive definite approximation H to the Hessian $\nabla_{yy}^2 L$ of the Lagrangian

$$L(y, p, \lambda) = J(y) + p^T h(y) + \lambda^T c(y) \tag{3}$$

(2) Solve the quadratic programming problem

$$\min_{s} \left\{ \frac{1}{2} s^T H s + J_y(y)^T s \mid h_y(y)s + h(y) = 0, c_y(y)s + c(y) \leq 0 \right\}$$

for s^+, p^+, λ^+.

(3) Use s^+ as a search direction

$$y^+ = y + \alpha s^+$$

and an appropriate line search cost function to determine α.

(4) Update H by a quasi-Newton-formula using

$$y^+ - y \quad \text{and} \quad L_y(y^+, p^+, \lambda^+) - L_y(y, p^+, \lambda^+)$$

sustaining the block-diagonal structure of $\nabla_{yy} L$ in H (see below).   □

The linear-quadratic (quadratic programming) problem of step (2) is solved in a finite number of steps by an active set method. This approach generates a sequence of linear equality constrained subproblems which differ in the selected (active) inequalities ([Ber82],[Fle81]).

The Hessian of the Lagrangian shows the following block structure:

$$
\nabla_{yy}^2 L =
\begin{bmatrix}
\square & \square & & & & \\
\square & \square & & & & \\
& & \square & \square & & \\
& & \square & \square & & \\
& & & & \ddots & \\
& & & & & \square
\end{bmatrix}
\tag{4}
$$

If this structure is preserved during the update in step (4), then a sequence of linear equality constrained linear-quadratic discrete-time optimal control problems is to be solved in step (2). The new solution algorithm given in the next section is a generalization of the known Riccati solution procedure for unconstrained LQ problems [GJ84] to the constrained case.

Solution of the equality constrained LQ problem

In this section the following constrained LQ-problem is considered

$$
\bar{J} = \frac{1}{2} \left(x^N\right)^T H_{xx}^N x^N + \left(f_{0x}^N\right)^T x^N +
$$

$$
+ \sum_{k=0}^{N-1} \left\{ \frac{1}{2} \begin{bmatrix} x^k \\ u^k \end{bmatrix}^T \begin{bmatrix} H_{xx}^k & H_{xu}^k \\ (H_{xu}^k)^T & H_{uu}^k \end{bmatrix} \begin{bmatrix} x^k \\ u^k \end{bmatrix} + \begin{bmatrix} f_{0x}^k \\ f_{0u}^k \end{bmatrix}^T \begin{bmatrix} x^k \\ u^k \end{bmatrix} \right\}
$$

with the *linearized state equation*:   (5)

$$x^{k+1} = f_x^k x^k + f_u^k u^k + f^k, \quad k = 0, \ldots, N-1$$

and *linearized equality constraints* (active inequalities):

$$g_x^k x^k + g_u^k u^k + g^k = 0, \quad k = 0, \ldots, N-1$$
$$g_x^N x^N + g^N = 0$$

where x^k, u^k are subvectors of the search direction s, matrices $H^k_{..}$ submatrices of the approximated Hessian H of the Lagrangian of the original problem, $g^k(\cdot, \cdot)$ active components of $c^k(\cdot, \cdot)$, and $f^k_{0.}$, $f^k_.$, $g^k_.$ the corresponding partial derivatives.

Using the accompanying *Lagrangian*

$$\bar{L} = \bar{J} + \left(\lambda^N\right)^T \left(g^N_x x^N + g^N\right) + \sum_{k=0}^{N-1} \Big\{ \left(p^k\right)^T \left(f^k_x x^k + f^k_u u^k + f^k - x^{k+1}\right) + \\ + \left(\lambda^k\right)^T \left(g^k_x x^k + g^k_u u^k + g^k\right) \Big\} \tag{6}$$

with $p^k \in \mathbf{R}^n$, $k = 0, \ldots, N-1$, $\lambda^k \in \mathbf{R}^{r^k}$, $k = 0, \ldots, N$, the necessary optimality conditions – a structured system of linear equations of order $o(N)$ – can be formulated.

This linear system is solved by a *modified Riccati solution* procedure: during a 'backward run' (backward in time), starting from the end of the optimization horizon, the control variables are expressed as (linear) functions of the (unknown) states. A general solution of the underdetermined and possibly rank deficient linear system of the constraint equations is used here in the constrained case. These 'control laws' together with the state equations are used to calculate step by step the optimal state and control variables, and Lagrange multipliers in the 'forward run'.

(1) Backward run: establish 'control laws':

Set $\quad V^N_{xx} = H^N_{xx}, \quad V^N_x = f^N_{0x},$
$\quad\quad\ \bar{g}^N_x = g^N_x, \quad\ \ \bar{g}^N = g^N$

for $k = N-1, \ldots, 0$:
Transformation of the constraints

- Transformation of the state constraints $\bar{g}^{k+1}_x x^{k+1} + \bar{g}^{k+1} = 0$ of step $k+1$ into constraints for x^k, u^k by means of the state equation,

- Decomposition of all constraints of step k into pure state constraints and state dependent control constraints using a $QR\bar{Q}$ factorization (algorithm HFTI, see [LH74]):

$$B = \left[\begin{array}{c} g^k_u \\ \bar{g}^{k+1}_x f^k_u \end{array} \right]^T = [Q^k_1 \ Q^k_2] \left[\begin{array}{cc} R^k_1 & 0 \\ 0 & 0 \end{array} \right] \bar{Q}^k \left(P^k\right)^T \tag{7}$$

with B $-$ (m, r_0^k) matrix, rank r_1^k,
                                $0 \le r_1^k \le \min(m, r_0^k)$, $r_0^k = r^k + \dim(\bar{g}^{k+1})$,
$[Q_1^k \, Q_2^k]$ $-$ orthogonal (m, m) matrix;
                                first r_1^k columns forming Q_1^k,
R_1^k $-$ regular (r_1^k, r_1^k) upper triangular matrix,
$\bar{Q}^k$ $-$ orthogonal (r_0^k, r_0^k) matrix,
P^k $-$ orthogonal (r_0^k, r_0^k) matrix,
                                containing column change information of B.

$$\begin{bmatrix} \tilde{g}_x^k \\ \bar{g}_x^k \end{bmatrix} = \bar{Q}^k \left(P^k\right)^T \begin{bmatrix} g_x^k \\ \bar{g}_x^{k+1} f_x^k \end{bmatrix}, \quad \begin{bmatrix} \tilde{g}^k \\ \bar{g}^k \end{bmatrix} = \bar{Q}^k \left(P^k\right)^T \begin{bmatrix} g^k \\ \bar{g}^k + \bar{g}_x^{k+1} f^k \end{bmatrix}$$

Control laws and recursion formulas for V_{xx}^k, V_x^k:
Express u^k, p^k, λ^k as linear functions of the (unknown) x^k and $\bar{\lambda}^k$.

$$R_{ux}^k = Z^k \left(\left(Z^k\right)^T G_{uu}^k Z^k\right)^{-1} \left(Z^k\right)^T \left(\left(G_{xu}^k\right)^T - G_{uu}^k S^k \tilde{g}_x^k\right) + S^k \tilde{g}_x^k$$

$$R_u^k = Z^k \left(\left(Z^k\right)^T G_{uu}^k Z^k\right)^{-1} \left(Z^k\right)^T \left(G_u^k - G_{uu}^k S^k \tilde{g}^k\right) + S^k \tilde{g}^k$$

$$R_{\lambda x}^k = \left(S^k\right)^T \left(\left(G_{xu}^k\right)^T - G_{uu}^k R_{ux}^k\right) \tag{8}$$

$$R_\lambda^k = \left(S^k\right)^T \left(G_u^k - G_{uu}^k R_u^k\right)$$

$$V_{xx}^k = G_{xx}^k - G_{xu}^k R_{ux}^k - \left(\tilde{g}_x^k\right)^T R_{\lambda x}^k$$

$$V_x^k = G_x^k - G_{xu}^k R_u^k - \left(\tilde{g}_x^k\right)^T R_\lambda^k$$

using

$$G_{xx}^k = H_{xx}^k + \left(f_x^k\right)^T V_{xx}^{k+1} f_x^k, \quad G_x^k = f_{0x}^k + \left(f_x^k\right)^T \left(V_{xx}^{k+1} f^k + V_x^{k+1}\right),$$

$$G_{xu}^k = H_{xu}^k + \left(f_x^k\right)^T V_{xx}^{k+1} f_u^k, \quad G_u^k = f_{0u}^k + \left(f_u^k\right)^T \left(V_{xx}^{k+1} f^k + V_x^{k+1}\right),$$

$$G_{uu}^k = H_{uu}^k + \left(f_u^k\right)^T V_{xx}^{k+1} f_u^k, \quad S^k = Q_1^k \left(R_1^k\right)^{-T}, \quad Z^k = Q_2^k$$

(2) **Calculate x^0, $\bar{\lambda}^0$ by generalized elimination** (using a QR factorization):

$$\left(\bar{g}_x^0\right)^T = Q \begin{bmatrix} R \\ 0 \end{bmatrix} = [Q_1 \, Q_2] \begin{bmatrix} R \\ 0 \end{bmatrix} = Q_1 R$$

Q – orthogonal, R – regular upper triangular ,

$$S = Q_1 R^{-T}, \quad Z = Q_2 \tag{9}$$

$$x^0 = Z \left(Z^T V_{xx}^0 Z\right)^{-1} Z^T \left(V_{xx}^0 S \bar{g}^0 - V_x^0\right) - S \bar{g}^0$$

$$\bar{\lambda}^0 = -S^T \left(V_{xx}^0 x^0 + V_x^0\right)$$

(3) **Forward run: calculate the optimal solution**
for $k = 0, \ldots, N - 1$:

$$
\begin{aligned}
u^k &= -R_{ux}^k x^k - R_u^k && \text{(control law)} \\
x^{k+1} &= f_x^k x^k + f_u^k u^k + f^k && \text{(state equation)} && (10) \\
\begin{bmatrix} \lambda^k \\ \bar{\lambda}^{k+1} \end{bmatrix} &= P^k \left(\bar{Q}^k \right)^T \begin{bmatrix} -R_{\lambda x}^k x^k - R_\lambda^k \\ \bar{\lambda}^k \end{bmatrix} \\
p^k &= V_{xx}^{k+1} x^{k+1} + V_x^{k+1} + \left(\bar{g}_x^{k+1} \right)^T \bar{\lambda}^{k+1}
\end{aligned}
$$

Set $\lambda^N = \bar{\lambda}^N$. $\square$

This algorithm for equality constrained LQ problems becomes identical to the standard Riccati solution procedure for vanishing constraints. It gives the solution with $o(N)$ operations (instead of $o(N^3)$ as would to be expected without taking into account the special structure of the linear system). All steps are well defined, if second order sufficient optimality conditions are fulfilled.

Optimal control of a reservoir system

The method described above was applied to find out optimal release strategies for a system of reservoirs with hydroelectric power-stations on the Zambezi river in southern Africa [ATW94]. This work was done at Institute of Automatic Control of Warsaw University of Technology as a part of the IIASA project 'Water Resources' ([ATW94]).

The structure of the reservoir system is given in Figure 1.

Its discrete-time model has a discretization step of one month and a control horizon ranging from one year up to five years. State equations are derived from volume balance equations of the reservoirs with the volumes of stored water as state variables. Control variables are the energetic outflow (outflow through the turbines) and the spillage.

Balance equation $(i = 1, \ldots, 4, \quad k = 0, \ldots, N - 1)$:

$$
V_i^{k+1} = V_i^k + Z_i^k + \sum_{j \in U_i} \left(Q_j^{k-\Theta_j} + F_j^{k-\Theta_j} \right) - Q_i^k - F_i^k - E_i^k \cdot f_{Ai}(V_i^k)
$$

$$
\begin{aligned}
V_i^k & \quad - \text{reservoir volume; state variable} \\
Z_i^k & \quad - \text{natural inflow (deterministic)} \\
Q_i^k & \quad - \text{energetic outflow; control variable} \\
F_i^k & \quad - \text{spillage; control variable} \\
U_i & \quad - \text{upstream reservoirs} \\
\Theta_j & \quad - \text{time delay} \\
E_i^k \cdot f_{Ai} & \quad - \text{evaporation}
\end{aligned}
$$

Since a deterministic control problem is considered, natural inflows are supposed to be known scenarios from historical data. Each of the balance equations

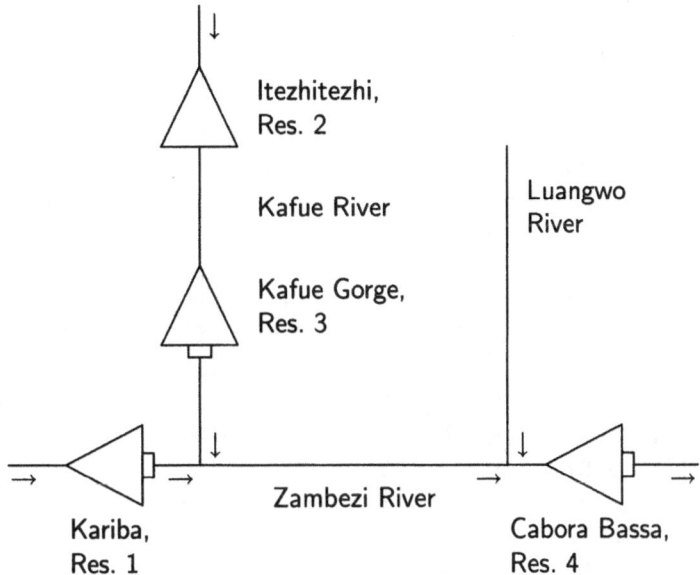

Figure 1: Structure of the Zambezi river system.

contains a nonlinear evaporation term $E_i^k \cdot f_{Ai}(V_i^k)$. River dynamics between reservoirs (time delay Θ_j) are modeled by linear difference equations with additional state variables. Thus, the four reservoir system has six state variables and seven controls.

State and control bounds (or box constraints) arise from physical restrictions and additional control demands, e. g. flood control and ecological demands.

$$
\begin{aligned}
V_{i\,min}^k &\leq V_i^k \leq V_{i\,max}^k && \text{(reservoir capacity)} \\
Q_{i\,min}^k &\leq Q_i^k \leq Q_{i\,max}^k && \text{(turbine flow bounds)} \\
F_{i\,min}^k &\leq F_i^k \leq F_{i\,max}^k && \text{(flood control etc.)}
\end{aligned}
$$

The main control goal is the maximization of total electrical energy production. Therefore, the cost function includes a nonlinear mixed state-control term (outflow multiplied by state dependent water height level). Other terms of the cost function result from the desired small deviations in time of the energy production and the final water storage demands, respectively.

$$
J = \sum_{i=1}^{4} \left\{ \phi \left(V_i^N - V_{i\,ref} \right)^2 + \sum_{k=0}^{N-1} \left\{ -\alpha Q_i^k f_{Hi}(V_i^k, V_i^{k+1}) + \psi \left(Q_i^k - Q_{i\,ref}^k \right)^2 \right\} \right\}
$$

f_{Hi} – average water–level

$V_{i\,ref}$ – final reservoir volume reference value

$Q_{i\,ref}^k$ – reference trajectory for energetic outflow

This optimization problem fits into the class of optimal control problems defined in the previous section, if only the volume V_i^{k+1} in the water-level term of the cost function is replaced by the right-hand side of the balance equation.

The SQP-based algorithm solved satisfactorily all variations of the problem with different time horizon, varying inflow scenarios and cost function parameters. Solutions of high accuracy were found with a quite small number of SQP-iterations. A typical iteration course of the SQP-method is given below:

iter.	objective	gradient LF	infeasibility	AS iter.
0	-103843.80	6538.8655	0.6826716	73
1	-110438.56	385.05873	0.1390460	18
2	-111558.44	165.85204	0.0053325	7
3	-111607.95	63.661547	0.0000246	6
4	-111612.57	30.723903	9.003E-06	2
5	-111613.62	13.776013	0.0000366	7
6	-111613.92	8.3063499	0.0000251	1
7	-111614.00	7.3195985	8.246E-06	1
8	-111614.03	7.1489734	2.417E-06	1
9	-111614.04	7.1326521	6.243E-07	

Iteration terminated.

Examples of optimal trajectories are shown in Figure 2.

A principal drawback which limits somewhat the efficiency of the algorithm (in particular for high dimensional problems with many active constraints) is the necessary number of active set steps to find out the correct set of binding inequality constraints at the solution. Optimization results, together with a short comparison of the working behaviour of the algorithm with an up-to-date version of the augmented price method are reported in [ATW94].

Conclusions

An SQP method for constrained discrete-time optimal control problems is proposed here with the same advantages as SQP for nonlinear programming problems, for instance: robustness, solutions of high accuracy, and convergence within a small number of major iterations (and cost function and gradient evaluations).

The algorithm is similar to the Differential Dynamic Programming method in some aspects (see [MY84], for instance), but here the state equations are treated as equality constraints and are violated during the iteration process in general.

Some further development is needed with respect to the active set strategy and a structure-preserving quasi-Newton update formula.

It should be mentioned that the method is also suitable for solution of constrained continuous-time optimal control problems via discretization.

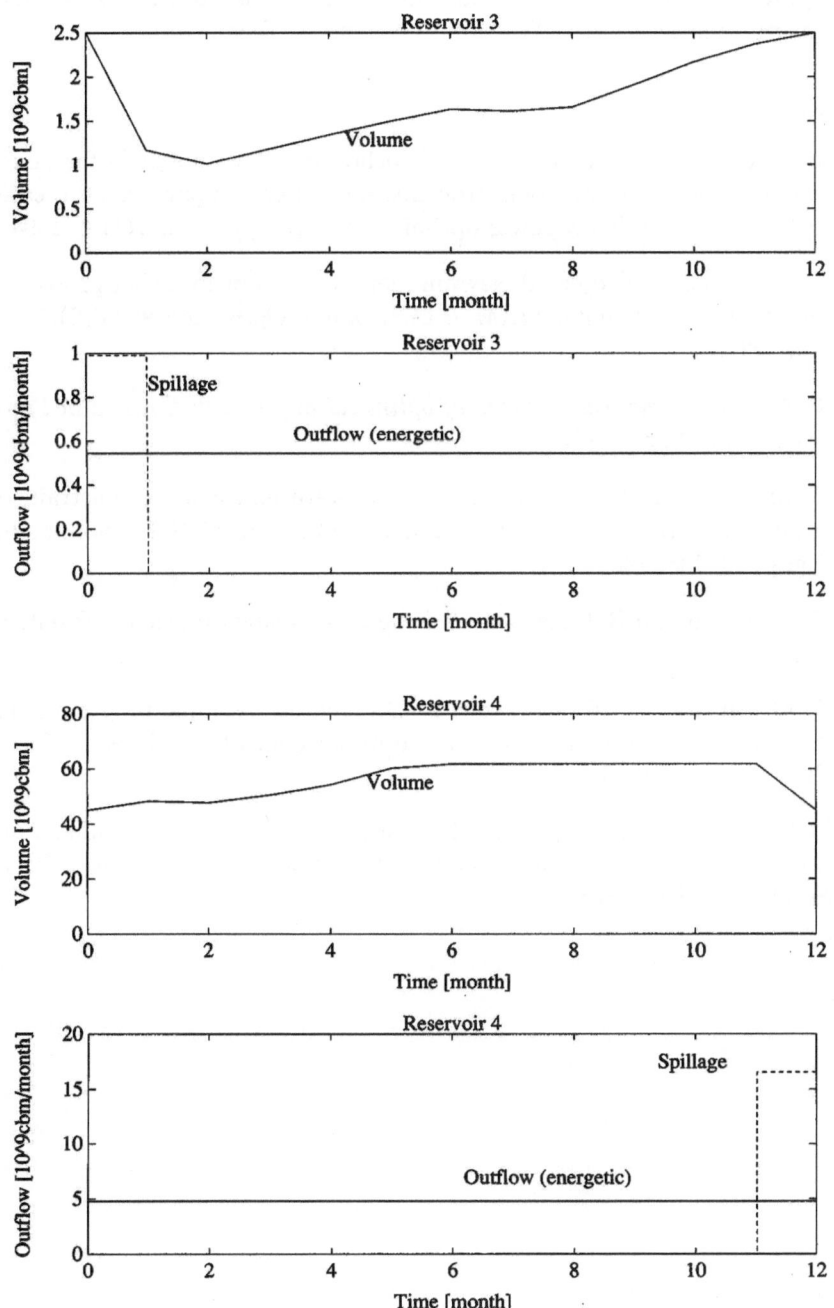

Figure 2: Optimal trajectories. Reservoirs 3 and 4, time horizon 12 months (1960).

Acknowledgement. Part of this work was done during a visit of the first author at Institute of Automatic Control, Warsaw University of Technology.

References

[ATW94] E. Arnold, P. Tatjewski, and P. Wołochowicz. Two methods for large scale nonlinear optimization problems and their comparison on a case study example of hydropower optimization. *To appear in JOTA*, 1994.

[Ber82] D. P. Bertsekas. Projected Newton methods for optimization problems with simple constraints. *SIAM J. Control and Optimization*, 20(2):221–246, 1982.

[Fle81] R. Fletcher. *Practical methods of optimization, Vol. 2: Constrained optimization.* Wiley, 1981.

[GJ84] T. Glad and H. Jonson. A method for state and control constrained linear quadratic problems. In *9th World Congress of IFAC*, volume 9, pages 229–233, 1984.

[LH74] C.L. Lawson and R.J. Hanson. *Solving least squares problems.* Prentice Hall, 1974.

[MY84] D. M. Murray and S. J. Yakowitz. Differential dynamic programming and Newton's method for discrete optimal control problems. *JOTA*, 43(3):395–414, 1984.

[Pow78] M.J.D. Powell. A fast algorithm for nonlinearly constrained optimization calculations. In G.A. Watson, editor, *Numerical analysis, Dundee 1977.* Springer Verlag, 1978.

International Series of Numerical Mathematics, Vol. 115, © 1994 Birkhäuser Verlag Basel

Numerical Methods for Solving Differential Games, Prospective Applications to Technical Problems

N.D.Botkin*   V.M.Kein†   V.S.Patsko*

V.L.Turova*   and   M.A.Zarkh*

Keywords. Differential Games Numerical Methods, Backward Procedure, Solvability Set, Optimization-Based Control System Design

Introduction

The majority of numerical methods for solving differential games is based on the backward procedure for computing the solvability sets [1-3]. The solvability set comprises all initial positions of a game such that one player can bring the state vector to a terminal target set for any possible actions of the opposite player. In the case of linear differential games, each step of the backward procedure consists of two operations with convex polyhedra: finding both an algebraic sum and a geometric difference of polyhedra. For two- or three-dimensional state space, these operations were implemented by the authors in terms of computing the convex hull of some piecewise-linear "almost" convex function. For this low-dimensional case, especially effective algorithms, which use information about local violation of convexity, have been created [4]. These algorithms were applied to various problems of an aircraft guidance in the presence of wind disturbances [5-7]. The way used for solving the above mentioned problems includes such steps as linearization of the corresponding nonlinear dynamic system with respect to some nominal trajectory, solving an auxiliary linear differential game, computing optimal strategies, and simulating the nonlinear dynamic system, using the strategies obtained from the auxiliary linear differential game.

This paper outlines a way for extending the backward procedure to the case of relatively high dimension of the state space. An algorithm for implementation of the operations of finding the algebraic sum and geometric difference based on the methods of [8,9] is proposed. An application of the numerical methods to the problem of active protection of a building from oscillations caused by earthquake excitations is considered. This problem is solved with the use of the linearization as it was done for the problems of aircraft guidance.

---

*Institute of Mathematics and Mechanics, S.Kovalevskaya Str. 16, Ekaterinburg 620219, Russia

†Civil Aviation Academy, Pilotov Str. 38, S.-Petersburg, 196210, Russia

Numerical algorithm

The dynamics of a game is described by the following equation

$$\dot{x} = A(t)x + B(t)u + C(t)v, \; x \in R^n. \tag{1}$$

Here u and v are control parameters of the first and the second player, respectively. We consider parameters u and v to be restricted as follows:

$$u \in P, \quad v \in Q,$$

where P and Q are convex compacts in the space of an appropriate dimension. The first player seeks to lead the first m coordinates $(m \leq n)$ of the state vector to a target set $M \subset R^m$ at a given time moment T. That is, he seeks to satisfy the inclusion

$$\{x_1(T), x_2(T), ..., x_m(T)\} \in M. \tag{2}$$

The objective of the second player is opposite. It is assumed that the players use feedback strategies [1].

Introduce the solvability set as a set of all initial positions (t_0, x_0) such that the first player can satisfy the inclusion (2) for any possible actions of the opposite player when the system (1) starts from the initial position (t_0, x_0).

It should be noted that the Isaaks' condition is fulfilled in the case considered. So, we have the following alternative: if the initial position does not belong to the solvability set, then there exists a strategy of the second player which brings all trajectories out of the target set.

As we consider differential games with fixed terminal time, it is convenient to make the following variables transformation

$$y(t) = X_m(T, t)x(t),$$

where $X(T, t)$ is the fundamental matrix of the homogeneous part of (1). The index m means that we use only the first m rows of $X(T, t)$. Making this transformation, we obtain an equivalent differential game with the dynamics

$$\dot{y} = D(t)u + E(t)v, \; y \in R^m, \tag{3}$$

where the matrices $D(t)$ and $E(t)$ have the following form

$$\begin{aligned} D(t) &= X_m(T, t)B(t), \\ E(t) &= X_m(T, t)C(t). \end{aligned}$$

Note that the solvability set of the equivalent differential game lies in the space $[0, T] \times R^m$.

Denote by W the solvability set of the equivalent differential game. Let $t_1, t_2, ..., t_N$ be a partition of the interval $[0, T]$ and

$$W(t_i) = \{x \in R^m, \; (t, x) \in W\}.$$

To compute approximations of the sets $W(t_i)$, we apply the backward procedure proposed by L.S.Pontryagin [2]. Following this method, we set

$$Z_N = \Upsilon(M)$$
$$Z_{i-1} = (Z_i + P_i) \underset{*}{} Q_i.$$

Here, P_i and Q_i are defined by formulas

$$
\begin{aligned}
P_i &= -(t_i - t_{i-1})X_m(T, t_i)D(t_i)\Upsilon(P), \\
Q_i &= (t_i - t_{i-1})X_m(T, t_i)E(t_i)\Upsilon(Q).
\end{aligned}
$$

The symbol $\Upsilon(\cdot)$ denotes a polyhedral approximation of a convex set. The sign "+" indicates the operation of finding an algebraic sum of two sets and the sign "$*$" denotes the operation of finding a geometric difference of two sets. We remind that

$$A * B = \{x \in R^m : x + B \subset A\}.$$

Each set Z_i approximates the corresponding set $W(t_i)$. Under certain assumptions, the following estimate holds (see [10]):

$$\text{dist}(Z_i, W(t_i)) \leq K \sup_j (t_j - t_{j-1}) + L\varepsilon, \quad \text{for any } i = 1, ..., N.$$

Here, K and L are some constants, ε is a measure of the precision of the operation $\Upsilon(\cdot)$. So, we have the backward procedure for the approximate computing the solvability set W. Each step of this procedure requires the following operations:

1. Approximation of the convex sets M, P, Q by polyhedra,

2. Finding an algebraic sum of two polyhedra,

3. Finding a geometric difference of two polyhedra.

In addition, we have to supply our algorithm with an operation for the approximation of complex polyhedra by more simple polyhedra because the algebraic sum of two polyhedra is a polyhedron with a large number of vertices and faces. Unless doing this approximation, the complexity of Z_i would exponentially increase from step to step.

For the state space dimension equal 2 and 3, effective algorithms for implementation of operations 1) - 3) have been proposed [4]. These operations are implemented in terms of computing the convex hull of the following piecewise-linear positive-homogeneous function

$$\eta_{i-1}(l) = \max_{y \in Z_i}\langle l, y \rangle + \max_{u \in P_i}\langle l, D(t_i)u \rangle - \max_{v \in Q_i}\langle l, E(t_i)v \rangle, \quad l \in R^m,$$

which is the support function of Z_{i-1}. The algorithms for constructing the convex hull of the function η_{i-1} use informations about local violation of its convexity.

These algorithms are very specific because they are based on the fact that one can introduce a special order for normals and faces of a convex polyhedron in 2- or 3-dimensional space. Unfortunately, these ideas cannot be extended to the case of higher dimensions. We now describe a way for implementation of these operations in an $m-$ dimensional space.

To find an algebraic sum of two polyhedra, we use the procedure suggested in [8]. We assume that the polyhedra Z_i and P_i are represented by systems of linear inequalities. So, we have

$$Z_i: \ \langle f_j, x \rangle + \mu_j \leq 0, \ j = 1, ..., n_z$$
$$P_i: \ \langle g_k, y \rangle + \nu_k \leq 0, \ k = 1, ..., n_p.$$

Then the algebraic sum of Z_i and P_i is given by the formula

$$Z_i + P_i = \text{Projection}_x \{(x, y):$$

$$-\langle f_j, y \rangle + \langle f_j, x \rangle + \mu_j \leq 0, \ \langle g_k, y \rangle + \nu_k \leq 0,$$

$$j = 1, ..., n_z; \ k = 1, ..., n_p\}.$$

So, if we eliminate y from the above system of linear inequalities, we obtain a system which defines $Z_i + P_i$.

A method for eliminating variables from systems of linear inequalities had been proposed by Fourier many years ago. According to that method, one has to combine certain inequalities with some appropriate positive coefficients. This method was advanced in [9], where the way to avoid from considering a lot of combinations which generate inactive inequalities was suggested.

Thus, applying the methods of [9], we obtain the representation of $Z_i + P_i$ via a system of linear inequalities

$$Z_i + P_i: \ \langle l_j, x \rangle + b_j \leq 0, \ j = 1, ..., n_s. \tag{4}$$

To implement the operation of finding the geometric difference, we assume that the polyhedron Q_i is given as the convex hull of a finite point set, that is

$$Q_i = \text{co}\{q_1, q_2, ..., q_r\}.$$

Then $(Z_i + P_i) * Q_i$ is represented by the following system of linear inequalities:

$$(Z_i + P_i) * Q_i: \ \langle l_j, x \rangle + b_j + \max_k \langle l_j, q_k \rangle \leq 0, \ j = 1, ..., n_s. \tag{5}$$

The systems (4) and (5) may contain a lot of inactive inequalities. To eliminate these inequalities, we use a procedure which is based on the simplex method in linear programming. This procedure also provides the elimination of such inequalities which give, in a certain sense, small faces to the resulting polyhedron. Applying this procedure, we obtain the polyhedron Z_{i-1} with a reasonable number of faces.

The above considered backward procedure for constructing the sequence $\{W(t_i)\}$ was implemented on a computer with the efficiency of 7 megaflops. The running times for various space dimensions are represented by the following table.

m	2	3	4	5	6	7
time	30 s	3 min	15 min	40 min	2 hours	4 hours

It is assumed that 150 steps of the backward procedure were done and the maximal number of faces was limited by 500.

In this paper, we do not consider the questions of finding optimal strategies of the players. We would only like to emphasize that these strategies can be found on the basis of the sets $W(t_i)$, using the extremal aiming procedure [1] or more special procedure proposed in [11]. The latter way gives optimal strategies of the players in the form of switch surfaces which are convenient for practical employments.

Protection from earthquake

As an illustration of our numerical methods, we consider the problem of actively reducing the harmful oscillations of buildings caused by an earthquake. We use a nonlinear building's model, which may seem to be artificial but even so describes many features of reality. The approach we use here for solving this nonlinear control problem with unknown disturbances is analogous to the one which had been employed for aircraft control problems [5-7, 11].

The model of a building is assumed to be a planar construction consisting of rods connected by means of elastic hinges (Fig. 1). It is assumed that each hinge produces a turning back moment and viscosity friction. So, the moment M_i applied to the ith rod is

$$M_1 = -k_1 \cdot \varphi_1 - h_1 \cdot \dot{\varphi}_1$$
$$M_i = -k_i \cdot (\varphi_i - \varphi_{i-1}) - h_i \cdot (\dot{\varphi}_i - \dot{\varphi}_{i-1}), \ i = 2, 3, 4.$$

This choice of M_i provides the asymptotic stability of the considered construction in the absence of disturbances. Disturbances are horizontal and vertical accelerations (v_1, v_2) of the ground caused by the earthquake excitation. The bounds on v_1 and v_2 are assumed to be known. The stabilizing control u is associated with a force produced by the water pumping with an acceleration from the reservoir A to the reservoir B and back.

The above described model leads to a nonlinear system of the Lagrange equations:

$$\sum_{j=1}^{4} R_{ij}(\varphi_1, ..., \varphi_4)\ddot{\varphi}_j = Q_i(\varphi_1, ..., \varphi_4, \dot{\varphi}_1, ..., \dot{\varphi}_4, u, v_1, v_2), \ i = 1, ..., 4.$$

Here R_{ij}, Q_{ij} are certain functions. For example, the first equation looks as follows

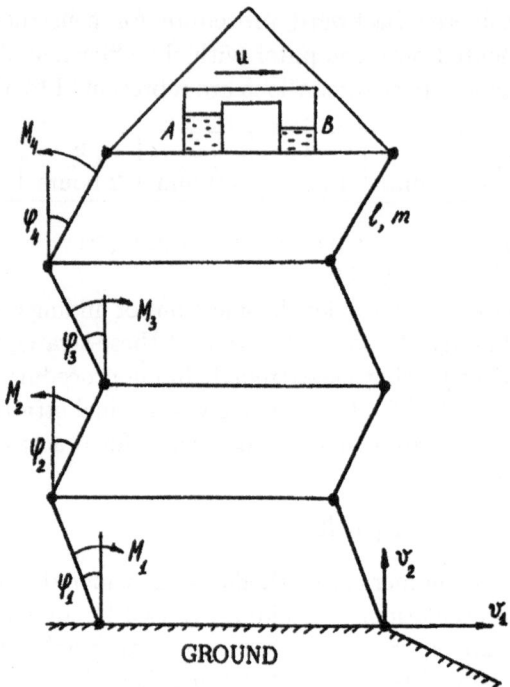

Figure 1: Sketch of the Considered Structure

$$\frac{32}{3}l\ddot{\varphi}_1 + 8l\ddot{\varphi}_2 \cos(\varphi_2 - \varphi_1) + 5l\ddot{\varphi}_3 \cos(\varphi_3 - \varphi_1) + 2l\ddot{\varphi}_4 =$$

$$8l\sin(\varphi_2 - \varphi_1)\dot{\varphi}_2^2 + 5l\sin(\varphi_3 - \varphi_1)\dot{\varphi}_3^2 + 2l\sin(\varphi_4 - \varphi_1)\dot{\varphi}_4^2 -$$

$$\frac{2(k_1 + k_2)}{ml}\varphi_1 - \frac{2(h_1 + h_2)}{ml}\dot{\varphi}_1 - \frac{2k_2}{ml}\varphi_2 - \frac{2h_2}{ml}\dot{\varphi}_2 + 2g\sin\varphi_1 + \cos\varphi_1 v_1.$$

Let us choose the following values of the parameters involved into the model:

$$k_1 = 45 \cdot 10^3 \, \text{kNm}, \ k_2 = 30 \cdot 10^3 \, \text{kNm}, \ k_3 = k_4 = 25 \cdot 10^3 \, \text{kNm},$$

$$h_i = 10^4 \, \text{kNms}, \ i = 1, ..., 4, \ l = 4 \, \text{m}, \ m = 10^4 \, \text{kg}.$$

Linearization of the nonlinear system with respect to the equilibrium state $\varphi_i = 0$, $\dot{\varphi}_i = 0$ gives the following linear system:

$$\ddot{\varphi}_i = \sum_{j=1}^{4}(a_{ij}\varphi_j + a'_{ij}\dot{\varphi}_j) + b_i u + c_{i1}v_1 + c_{i2}v_2, \quad i = 1, ..., 4, \tag{6}$$

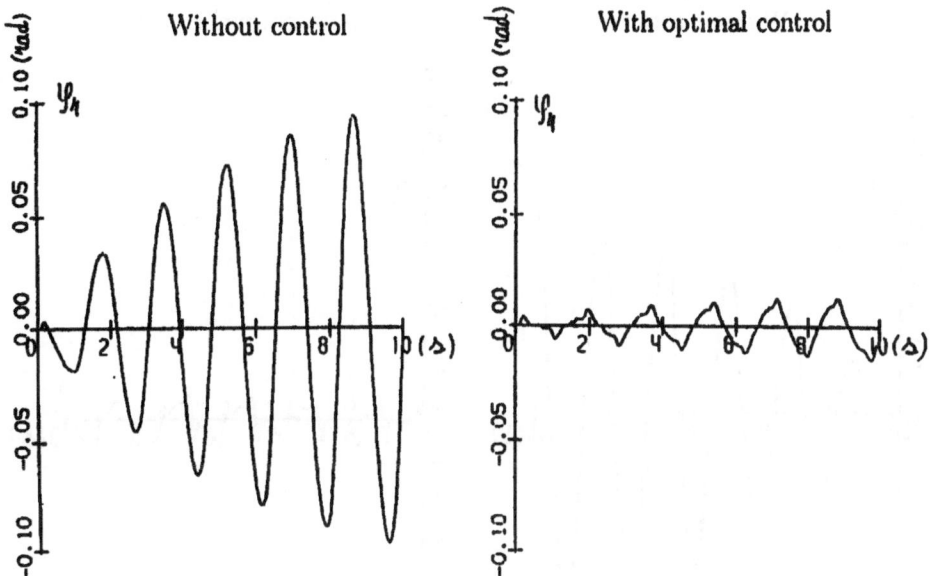

Figure 2: Angle φ_4 without and with Control

$$(a_{ij}) = \begin{pmatrix} -2764.147 & 1568.023 & -334.285 & 42.743 \\ 1175.74 & -1232.62 & 733.928 & -213.716 \\ -336.063 & 782.83 & -1030.713 & 515.767 \\ 51.702 & -264.665 & 811.071 & -548.156 \end{pmatrix}$$

$$(a'_{ij}) = \begin{pmatrix} -74.202 & 53.799 & -13.336 & 1.717 \\ 33.511 & -43.997 & 29.179 & -8.549 \\ -10.372 & 29.179 & -40.919 & 20.631 \\ 1.596 & -10.258 & 32.257 & -23.366 \end{pmatrix}$$

$$(b_i) = \begin{pmatrix} 0 \\ -0.001 \\ 0.004 \\ -0.008 \end{pmatrix}, \quad (c_{i1}) = (c_{i2}) = \begin{pmatrix} 0.927 \\ -0.135 \\ 0.02 \\ -0.003 \end{pmatrix}.$$

We have set $(c_{i2}) = (c_{i1})$ instead of $(c_{i2}) = 0$ in order to increase abilities of the disturbance in the linearized model (6). We consider the process on the time interval $[0, T]$, where $T = 10s$. The payoff functional is defined as follows

$$g(x(\cdot)) = \max\{|\varphi_3(T)|, |\varphi_4(T)|, |\dot{\varphi}_3(T)|, |\dot{\varphi}_4(T)|\}. \tag{7}$$

Let the bounds on controls and disturbances be

$$|u| \le 100\,\text{kN}, \quad |v_1| \le 1\,\text{m/s}^2, \quad |v_2| \le 1\,\text{m/s}^2. \tag{8}$$

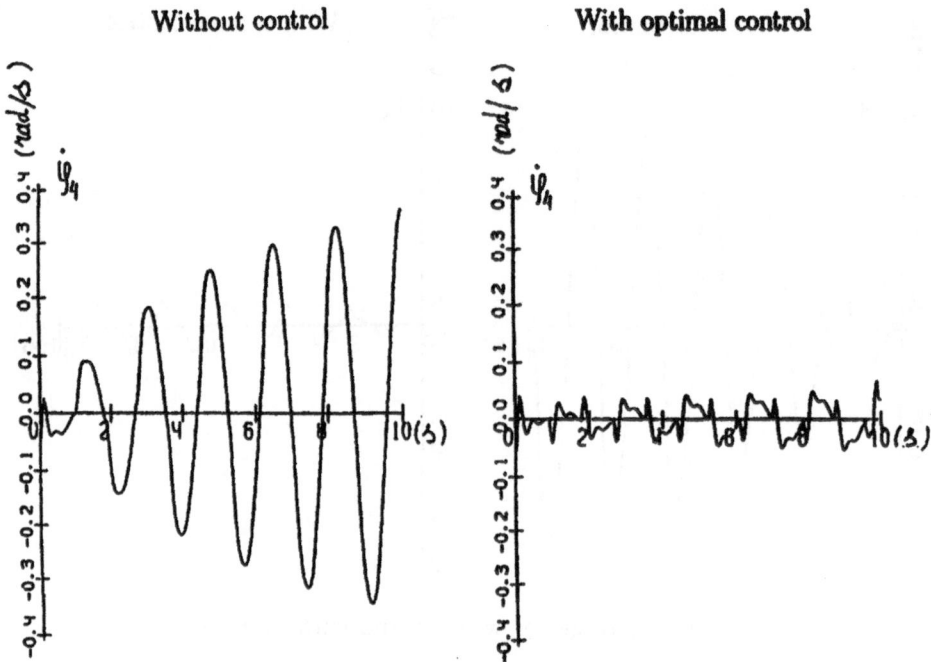

Figure 3: Angular Velocity $\dot\varphi_4$ without and with Control

Note, that the vector (b_i) is not collinear to the vectors (c_{i1}) and (c_{i2}). From the viewpoint of the robust control methods [12], this means that so-called matching conditions are far from being fulfilled. Therefore, it might be difficult to construct a stabilizing control in the way used in [13], where some linear model assuming the matching conditions was considered.

Being applied to the differential game (6)-(8), the numerical methods give us optimal strategies of the players: A feedback strategy $U(t, x)$ for stabilizing control and optimal feedback strategies $V_1(t, x)$ and $V_2(t, x)$ for the components of the disturbance. These strategies are used when simulating the nonlinear dynamic system.

Figures 2 and 3 show the simulation results.

References

[1] Krasovskii, N.N. and Subbotin, A.I.: Game-Theoretical Control Problems. New York: Springer-Verlag, 1988.

[2] Pontryagin, L.S.: Linear differential games. 2, Soviet Math. Dokl., Vol. 8, pp. 910–912, 1967.

[3] Pshenichnii, B.N. and Sagaidak, M.I.: Differential games of prescribed dura-

tion. Cybernetics, Vol. 6, No. 2, pp. 72–83, 1970 (in Russian).

[4] Subbotin, A.I. and Patsko, V.S. (Editors): Algorithms and programs for solving linear differential games. Ural Scientific Center, Academy of Sciences of the USSR, Sverdlovsk, 1984 (in Russian).

[5] Botkin, N.D., Kein, V.M., Krasov, A.I. and Patsko, V.S.: Control of aircraft lateral motion during landing in the presence of wind disturbances. Institute of Mathematics and Mechanics, Sverdlovsk, Civil Aviation Academy, Leningrad, Report No.81104592/02830078880, VNTI Center, 1983 (in Russian).

[6] Botkin, N.D., Kein, V.M., Patsko, V.S. and Turova, V.L.: Aircraft landing control in the presence of windshear. Problems of Control and Information Theory, Vol. 18, No. 4, pp. 223–235, 1989.

[7] Botkin, N.D., Zarkh, M.A., Kein, V.M., Patsko, V.S. and Turova, V.L.: Differential games and aircraft control problems in the presence of wind disturbances. Izvestia Akademii Nauk. Tekhnicheskaya Kibernetika, No. 1, pp. 68–76, 1993 (in Russian).

[8] Bushenkov, V.A. and Lotov, A.V.: Methods and algorithms for linear system analysis on the basis of constructing the generalized attainable sets. J. of Computational Mathematics and Mathematical Physics, Vol. 20, No. 5, pp. 1130–1141, 1980 (in Russian).

[9] Chernikov, S.N.: Lineare Ungleichungen. Verlag Wissenschaft, 1971 (translated from Russian).

[10] Botkin, N.D.: Evaluation of numerical construction error in differential game with fixed terminal time. Problems of Control and Information Theory, Vol. 11, no. 4, pp. 283–295, 1982.

[11] Botkin, N.D. and Patsko ,V.S.: Positional control in a linear differential game. Engineering Cybernetics, Vol. 21, No.4, pp.69–76, 1983.

[12] Chen, Y.H. and Leitmann, G.: Robustness of uncertain systems in the absence of matching assumptions. Int. J. Control, Vol. 45, No. 5, pp. 1527–1542, 1987.

[13] Kelly, J.M., Leitmann, G. and Soldatos, A.G.: Robust control of Based-Isolated Structures under Earthquake Excitation. J. of Optimization Theory and Applications, Vol. 53, No. 2, pp. 159–180, 1987.

International Series of Numerical Mathematics, Vol. 115, © 1994 Birkhäuser Verlag Basel

Construction of the Optimal Feedback Controller for Constrained Optimal Control Problems with Unknown Disturbances

Michael H. Breitner*

Abstract

Many optimal control problems can be solved today efficiently and conveniently by means of direct and indirect optimization methods. Thereby, unpredictable or unmeasureable influences are modelled as deterministic functions. Therefore, algorithms for the computation of feedback control laws are necessary in order to apply optimal solutions to real processes especially in the presence of uncertainties. A new approach for the construction of the optimal feedback controller is outlined, which guarantees a maximal value of the minimum performance index against all disturbances. Numerical results are presented for the optimal reentry maneuver of a hypersonic glider in the presence of uncertain air density.

Keywords. Optimal control problems, uncertainties, differential games, feedback controller, space-shuttle reentry.

Introduction

Direct optimization methods, e.g. the direct collocation method [SB92], enable the solution of difficult optimal control problems in quite different areas, such as aeronautics, astronautics, robotics and economics. However, the precalculated optimal open-loop controls cannot be applied for a realistic problem without having a procedure for the computation of feedback controls. This is due to the fact, that important circumstances cannot be included in the mathematical model used for the optimization. The use of less accurate approximations for these circumstances leads generally to large deviations from the precalculated optimal trajectories. The way to overcome this difficulty is to consider interval functions for the unpredictable disturbances in the model. Consider the dynamical system with the initial and terminal conditions

$$\dot{z}(t) = f(z(t), u(t), w(t)), \qquad z(0) = z_0, \qquad \|B(z(t_f))\|_p \leq \varepsilon \qquad (1)$$

and state variable inequality constraints of the type

$$C(z(t), u(t), w(t)) \leq 0 \qquad \forall t \in [0, t_f]. \qquad (2)$$

---

*Technische Universität München, Mathematisches Institut, D-80290 München, Germany

Here the dot denotes the derivative w. r. t. the independent variable t. $t \in [0, t_f]$ is the time or the normalized time, 0 is the initial time, t_f the free or fixed final time. Furthermore, $z(t)$ denotes the state vector, $u(t) \in U(t)$ the control vector and $w(t) \in W(t)$ the unknown disturbance vector within the convex set $U(t)$, resp. $W(t)$. The initial state vector is given by z_0 and the vector of prescribed terminal conditions for the state by $B(z(t_f))$. $\|.\|_p, p \in 2 * \mathbb{N}$ is a differentiable norm and ε is the prescribed accuracy for the final conditions. C is a vector valued function for the constraints. The goal is to minimize a performance index $I(z(t_f))$ subject to the constraints (1) and (2) against all possible disturbances $w(t)$. Therefore, two primary goals for the optimization must be guaranteed: the terminal conditions must be reached within the prescribed accuracy and the inequality constraints must be satisfied, too. If and only if the primary goals are reached, the secondary goal of minimizing $I(z(t_f))$ can be considered additionally. An appropriate global worst-case analysis for the unknown disturbances $w(t)$ must be taken into account for these three different goals. Since the occurrence of the worst case $w^*(t)$ is extremely unlikely, it is necessary to approximate the optimal controls by a feedback control law properly. Here a highly accurate approximation procedure is presented, which can be used for quite general optimal control problems. Due to the large amount of computation needed, this method is applicable in real-time only for slow systems, e.g. macroeconomics. Nevertheless, the high accuracy enables a comparison of those feedback controllers, which are designed for real-time computations, see e.g. [BR94]. Various concepts of real-time feedback control laws working also for fast systems can be found in [PE89], [KP90], [BB91], [BS91] and [LE89].

Model of the Hypersonic Glider

Assuming no planet rotation and no oblateness, no winds and a stationary atmosphere, a point mass vehicle and a constant drag polar, the equations of motion for a US Space-shuttle vehicle in a flight path oriented coordinate system can be written as

$$\dot{h} = v \sin \gamma \,, \tag{3}$$

$$\dot{\Lambda} = \frac{v}{R+h} \cos \gamma \sin \chi \,, \tag{4}$$

$$\dot{\Theta} = \frac{v}{R+h} \frac{\cos \gamma \cos \chi}{\cos \Lambda} \,, \tag{5}$$

$$\dot{v} = -C_L{}^n \frac{S v^2 \rho}{2\,m} \sigma - C_{D_0} \frac{S v^2 \rho}{2m} \sigma - g \sin \gamma \,, \tag{6}$$

$$\dot{\chi} = C_L \sin \mu \frac{S v \rho}{2\,m \cos \gamma} \sigma - \frac{v}{R+h} \cos \gamma \cos \chi \tan \Lambda \,, \tag{7}$$

$$\dot{\gamma} \;=\; C_L \cos\mu \, \frac{S\,v\,\rho}{2\,m}\,\sigma + \left(\frac{v}{R+h} - \frac{g}{v}\right)\cos\gamma\;. \qquad (8)$$

Here h denotes the altitude above the Earth's surface, Λ the cross-range angle, Θ the down-range angle, v the velocity, χ the heading angle, and γ the flight path angle (see Fig. 1). Besides these six state variables there are two control variables,

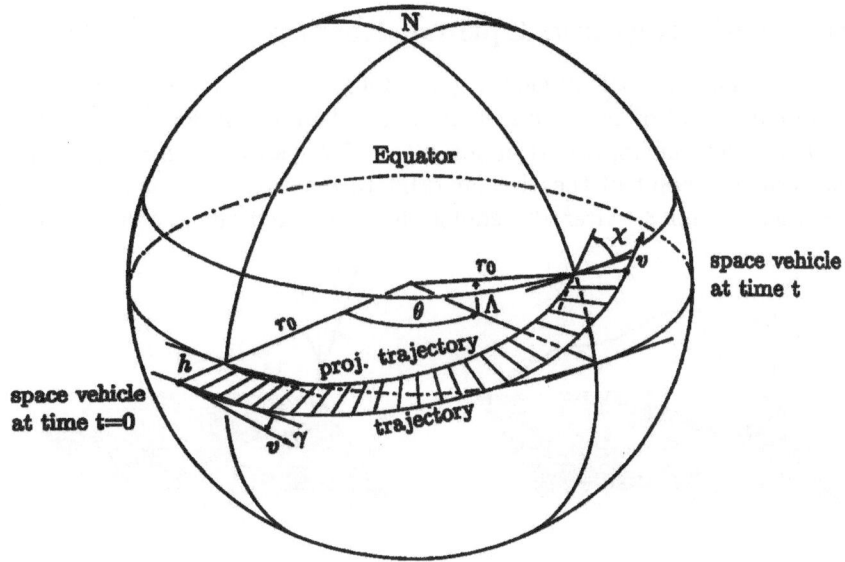

Fig. 1: Position of the shuttle (h, Λ, Θ), velocity vector (v, χ, γ), both in spherical coordinates, the trajectory and its projection.

the lift coefficient C_L and the bank angle μ. Moreover, R is the Earth's radius and $g(h)$ the gravity acceleration. S, m, C_{D_0}, and n are the aerodynamic reference area, the mass, the zero-drag coefficient, and the exponent of C_L in the induced drag of the shuttle, respectively. The air density $\rho\sigma$ is based on a prescribed air density model $\rho(h)$, e.g. $\rho = \rho_0 \exp(-\beta h)$. The deviation factor σ describes the density fluctuation. Five control and state variable inequality constraints are introduced for technical reasons: the dynamic pressure constraint $q(h, v, \sigma) \leq q_{max}$, the heating constraint for the stagnation point $\vartheta(C_L, h, v, \sigma) \leq \vartheta_{max}$, the load factor constraint $n(C_L, h, v, \sigma) \leq n_{max}$, the flight path angle constraint $\gamma \leq \gamma_{max}$, and the bank angle constraint $|\mu| \leq \mu_{max}$. For the US Space-shuttle, realistic limits are $q_{max} = 40\,\text{kN/m}^2$, $\vartheta_{max} = 1500\,^\circ\text{C}$, $n_{max} = 2$, $\gamma_{max} = 1/2$ deg and $\mu_{max} = 60$ deg. The first four constraints are transformed into $C_L \geq C_{L,min,q}$, $C_L \leq C_{L,max,\vartheta}$, $C_L \leq C_{L,max,n}$ and $C_L \leq C_{L,max,\gamma}$. Since $C_L \geq C_{L,min,q}$ was never active for all calculated solutions, the other constraints for the lift coefficient are combined to

$$C_L \leq \left(C_{L,\max,\vartheta}{}^{-4} + C_{L,\max,n}{}^{-4} + C_{L,\max,\gamma}{}^{-4} \right)^{-\frac{1}{4}} \tag{9}$$

which guarantees that $C_L \leq \min \{ C_{L,\max,\vartheta}, C_{L,\max,n}, C_{L,\max,\gamma} \}$. In addition, the constraint

$$|\mu| \leq \mu_{\max} \tag{10}$$

must be taken into account separately.

Deorbit Trajectory and Optimal Reentry Trajectories

A typical US Space-shuttle mission, e.g. a satellite start or a repair, takes place in a stable circular orbit of an altitude of 242 km and an inclination of 28.5 deg. After the mission, a deorbit impulse is imposed ($v = 7.758$ km/s $\rightarrow v = 7.673$ km/s) at the southernmost point of the circular orbit to deform the orbit into an elliptic one, see Figs. 2 and 3. After the shuttle has reentered the Earth's atmosphere

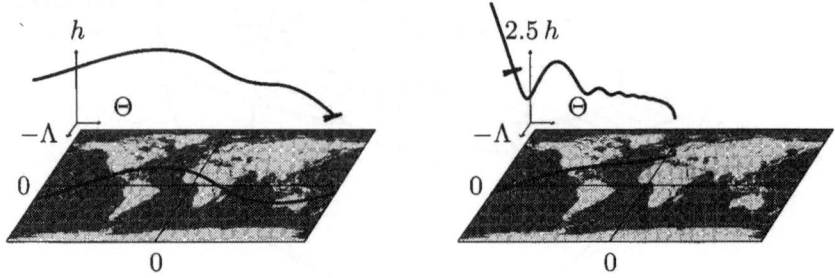

Fig. 2 and Fig. 3: Stable circular mission orbit of the shuttle ($h = 242$ km), deorbit impuls for an elliptic orbit at the southernmost point, reentry into the Earth's atmosphere ($h = 95$ km) and optimal reentry trajectory (constraints not considered).

($t = 0$ s : $h = 95$ km, $\Lambda = 0$ deg, $\Theta = -158.7$ deg, $v = 7.85$ km/s, $\chi = 28.5$ deg, $\gamma = -1.25$ deg) , the lift force vector can be directed by the controls C_L and μ. For the optimal reentry trajectory the performance index

$$I := -(1-\alpha) * \Lambda(t_f) - \alpha * \Theta(t_f) , \quad \alpha = 0.38 \tag{11}$$

is to be minimized with the controls $C_L(t)$ and $\mu(t)$ subject to the constraints (9) and (10). The unprescribed terminal time t_f is fixed by the terminal condition

$$F := \left\| \left(\frac{h(t_f) - h_f}{h_f}, \frac{v(t_f) - v_f}{v_f}, \frac{\chi(t_f) - \chi_f}{\chi_f}, \frac{\gamma(t_f) - \gamma_f}{\gamma_f} \right)^\top \right\|_2 \leq \varepsilon \tag{12}$$

with $h_f = 30$ km, $v_f = 1.116$ km/s, $\chi_f = 45$ deg, $\gamma_f = -2.7$ deg, and the desired

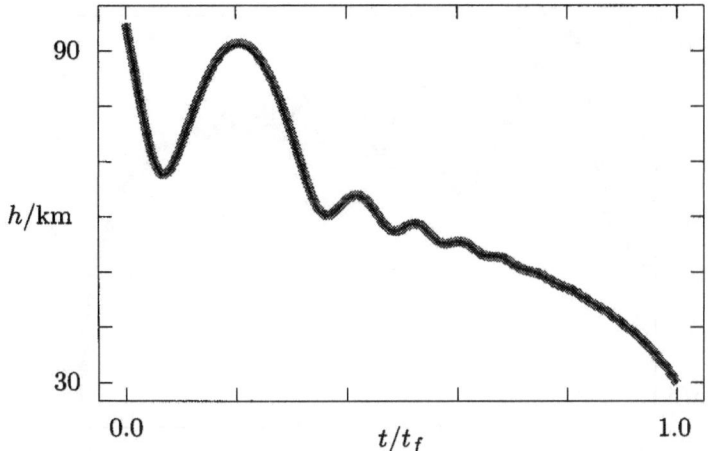

Fig. 4: Altitude h versus time t : comparison between collocation method (thick gray curve) and multiple shooting method (thin curve).

accuracy $\varepsilon = 0.01$. The final conditions enable a quasi stationary glide to a landing site everywhere in Germany. The calculation of the optimal trajectory $z^*(t)$, $z := (h, \Lambda, \Theta, v, \chi, \gamma)^\top$, and of the optimal open-loop controls $C_L{}^*(t)$ and $\mu^*(t)$ for $t \in [0, t_f]$ can be easily done with a new direct collocation method, see [SB92]. To simplify the further procedure, the constraints (9) and (10) are not included firstly, although the collocation method can also solve the constrained optimization problem very conveniently. This collocation method not only computes the optimal trajectory $z^*(t)$, but also yields an accurate estimate of the adjoint vector $\lambda^*(t)$, $\lambda := (\lambda_h, \lambda_\Lambda, \lambda_\Theta, \lambda_v, \lambda_\chi, \lambda_\gamma)^\top$. The adjoint vector λ^* can be interpreted as the gradient of the optimal performance index I^* w.r.t. the state vector z. All these estimates are very useful to start an highly accurate indirect optimization method such as the multiple shooting method ([BU71], [SB93], [HI93]). As usual, the Hamiltonian

$$H(z, \lambda, C_L, \mu, \sigma) := \lambda^\top f(z, C_L, \mu, \sigma) = C_1(z)\, A(z, \lambda, C_L, \mu)\, \sigma + C_2(z, \lambda) \quad (13)$$

is defined and the optimal open-loop controls are obtained by the minimum principle ($\sigma \equiv 1$ for the optimal control problem)

$$(C_L{}^*, \mu^*) := \arg \min_{C_L, \mu} H(z, \lambda, C_L, \mu, \sigma), \quad (14)$$

as well as the nonlinear multipoint boundary-value problem (MPBVP) for $t \in [0, t_f]$

$$\dot{z} = f(z, C_L{}^*(z, \lambda), \mu^*(z, \lambda), \sigma) \quad (15)$$

$$\dot{\lambda} = -\frac{\partial}{\partial z} H(z, \lambda, C_L{}^*(z, \lambda), \mu^*(z, \lambda), \sigma)^\top \quad (16)$$

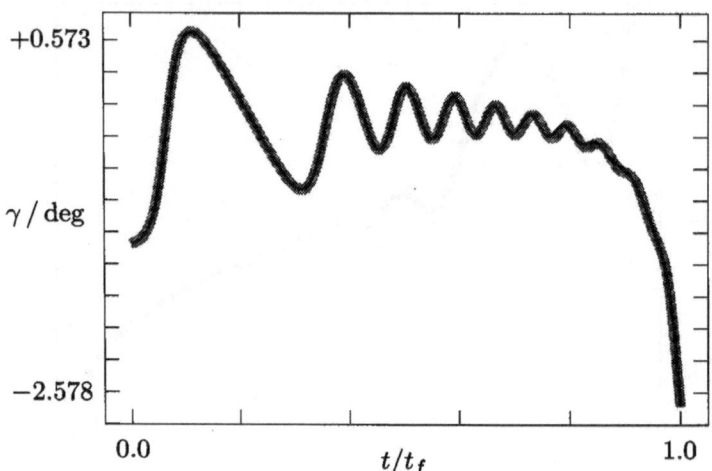

Fig. 5: Flight path angle γ versus time t : comparison between collocation method (thick gray curve) and multiple shooting method (thin curve).

with the boundary conditions

$$z(0) = z_0 \quad \text{and} \quad F\left(z\left(t_f\right)\right) - \varepsilon = 0 \tag{17}$$

and the transversality conditions

$$\lambda(t_f) - I_z\left(z(t_f)\right)^\top + \frac{I_z\left(z(t_f)\right)\, f(t_f)}{F_z\left(z(t_f)\right)\, f(t_f)} F_z\left(z(t_f)\right)^\top = 0 . \tag{18}$$

For details see, e.g., [BRHO75]. The MPBVP (15) – (18) is solved with the multiple shooting method of [HI93]. For a comparison between the numerical results of the direct collocation method and the indirect multiple shooting method, see Figs. 3 – 6. The constraints (9) and (10) are not taken into account. The history of the optimal controls $C_L{}^*$ and μ^* can be seen in Figs. 10 and 11.

Air-Density Worst-case Analysis as a Differential Game

An air density model such as $\rho = \rho_0 \exp(-\beta h)$ is based on many measurements in various altitudes, at various locations on the Earth, and for different seasons. With this vast amount of data, the nominal air density ρ and a related fluctuation tube can be approximated by a data fit. The fluctuation σ, the boundaries $\sigma_{\min}(h)$ and $\sigma_{\max}(h)$ (including measurement errors) and some realistic measurements are depicted in Fig. 7. Note that the maximum fluctuation of the air density ($\pm 15\,\%$) is obtained at an altitude of 75 km. The unknown air density $\sigma(t)\ \rho(h(t))$, in particular $\sigma(t)$ with

$$\sigma_{\min}(h(t)) \ \leq\ \sigma(t) \ \leq\ \sigma_{\max}(h(t)) \qquad \forall t \in [0, t_f] \tag{19}$$

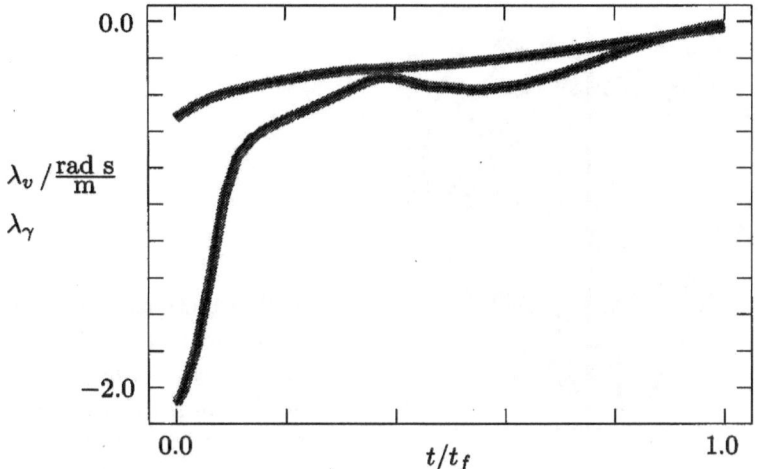

Fig. 6: Adjoint variables λ_γ (lower curves) and λ_v (upper curves) versus time t : comparison between collocation method (thick gray curves) and multiple shooting method (thin curves).

will be considered in the sequel as the control of an antagonistic player. Thereby, the global worst case for air density disturbances during the whole reentry maneuver will be investigated. It is assumed that the shuttle controller has no information about the air density in the future, except the fluctuation tube. This model leads to a two-player differential game: the first player O, the optimizer, controls C_L and μ whereas the second player D, the disturber, controls σ, see also [BP93]. In general the differential game is of nonzero-sum type. However, the example presented here leads to a zero-sum differential game. The theory of differential games can be found, e.g., in [IS65], [BO82] or [BPG93] and will not be presented here in detail. In the following in contrast to control variables, a tilde indicates strategies of the two players. Let Σ be the set of all admissible strategies $\tilde{\sigma}(z)$ for D, where (19) is included, and let Ξ_u and Ξ_c be the sets of all admissible strategies $(\tilde{C}_L, \tilde{\mu})(z)$ for O, where either the constraints (9) and (10) are not included or included. First, we investigate the worst-case analysis for $(\tilde{C}_L, \tilde{\mu})(z) \in \Xi_u$, i.e. the unconstrained case. O's primary goal is to ensure reaching the terminal set defined by (12) from the initial position z_0 against all disturbances (controllability precondition), i.e.,

$$\max_{\tilde{\sigma} \in \Sigma} \quad \min_{(\tilde{C}_L, \tilde{\mu}) \in \Xi_u} \quad \min_{t \in [0, \infty[} F\left(z\left(t\right)\right) \leq \varepsilon . \tag{20}$$

Here, it is tacitly assumed that the min and max operations commute. Note that both players O and D have to use feedback controls. Due to the maneuvrability of the shuttle, O can enforce (20) for all states z which occurred in the numerical calculations. Therefore O and D both turn to the secondary goal of minimizing,

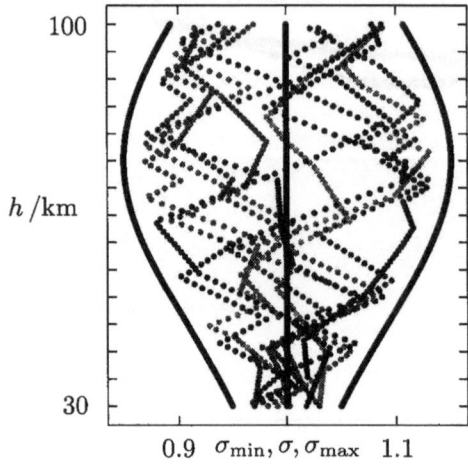

Fig. 7: Fluctuation tube for the air density model $\sigma\rho$: $\sigma_{\min}(h)$ (left curve), $\sigma \equiv 1$ (middle line), $\sigma_{\max}(h)$ (right curve) and realistic measurements (gray curves).

resp. maximizing, the performance index $I\left(z\left(t_f\right)\right)$ at any state z and any time t

$$\max_{\tilde{\sigma} \in \Sigma} \min_{(\tilde{C}_L, \tilde{\mu}) \in \Xi_u} I\left(z\left(t_f\right)\right) \quad \text{with} \quad t_f := \min\{t \in [0, \infty[: F\left(z\left(t\right)\right) = \varepsilon\} \ . \quad (21)$$

Again, it is assumed that the min and max operations commute. Similar to optimal control problems, the numerical calculation of the optimal performance index $I^*(z)$ and the calculation of the optimal feedback controllers $\tilde{\sigma}^*(z)$ and $(\tilde{C}_L^*, \tilde{\mu}^*)(z)$ is impossible for complex problems. However, by means of the indirect multiple shooting method, an open-loop representation of the optimal feedback controllers along optimal trajectories, i.e. both players O and D play optimally at any time, can be obtained. The major difference to optimal control problems is that the minimum principle has to be replaced by the more general minimax principle

$$(\tilde{C}_L^*, \tilde{\mu}^*)(t) := \arg \max_{\sigma \in \Sigma} \min_{(C_L, \mu) \in \Xi_u} H\left(z^*(t), \lambda^*(t), C_L, \mu, \sigma\right) , \quad (22)$$

$$\tilde{\sigma}^*(t) := \arg \min_{(C_L, \mu) \in \Xi_u} \max_{\sigma \in \Sigma} H\left(z^*(t), \lambda^*(t), C_L, \mu, \sigma\right) . \quad (23)$$

The assumption that the min and max operators commute, is necessary for the differential game to be of zero-sum type. Non-zero sum differential games are more difficult and are described, e.g., in [BO82]. Along the optimal trajectory $z^*(t)$ the direct collocation method estimates the functions $\lambda^*(t)$, $C_L^*(t)$ and $\mu^*(t)$. With the help of these estimates, an estimate of the switch function A, see (13), can be calculated. Therefore, in the neighborhood of an optimal trajectory ($\sigma_{\max} \approx 1$ and $\sigma_{\min} \approx 1$) it is possible to estimate $\tilde{\sigma}^*(z^*(t))$ by means of (24), see Fig. 8. Next, the solution of the unconstrained optimal control problem is determined highly accurately by the indirect multiple shooting method, where the estimates of both the state variables $z^*(t)$ and the adjoint variables $\lambda^*(t)$ are obtained from

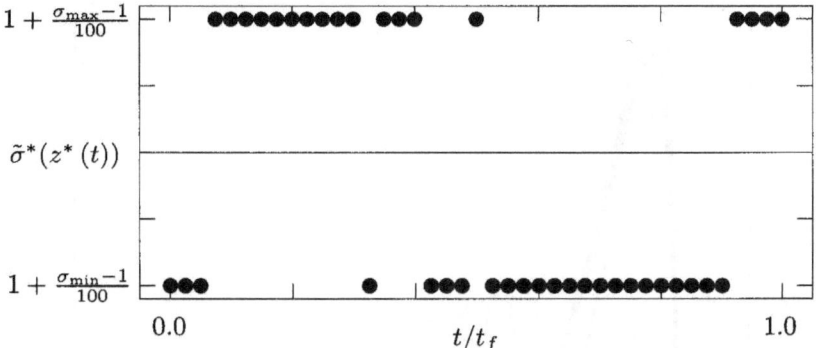

Fig. 8: Worst-case estimate of the air density fluctuation $\tilde{\sigma}^* \left(z^*(t) \right)$ in the neighborhood of the solution of the optimal control problem (calculated by the direct collocation method).

the collocation method (this so-called hybrid approach is due to [SB92]). Thereby, the optimal controls $C_L{}^*$ and μ^* are calculated via (14). Since σ appears linearly in the Hamiltonian H, see (13), and since $C_1(z) > 0$ the worst-case air-density fluctuation $\tilde{\sigma}^* \left(z^*(t) \right)$ is determined via

$$
\tilde{\sigma}^* \left(z^*(t) \right) = \begin{cases} \sigma_{\min} & : \quad A \left(z^*(t), \lambda^*(t), C_L{}^*(t), \mu^*(t) \right) < 0 \\ \sigma_{\max} & : \quad A \left(z^*(t), \lambda^*(t), C_L{}^*(t), \mu^*(t) \right) \geq 0 \end{cases} , \tag{24}
$$

if $\tilde{\sigma}^* \left(z^*(t) \right)$ is of bang-bang type. After the switching function $A(t)$ is obtained from the solution of the optimal control problem, the boundaries $\sigma_{\min}(h)$ and $\sigma_{\max}(h)$ for $\tilde{\sigma}^* \left(z^*(t) \right)$ can now be unchained from 1. Thereby, a continuation is performed from the optimal control problem to the differential game. The homotopy to the realistic values of $\sigma_{\min}(h)$ and $\sigma_{\max}(h)$ given in Fig. 7 is depicted in Figs. 9 – 12. The bang-bang type strategy $\tilde{\sigma}^* \left(z^*(t) \right)$ induces stronger oscillations in the optimal controls $C_L{}^*(t)$ and $\mu^*(t)$ and also in the reentry trajectory. Due to the worst case of the air density, the performance index $I \left(z \left(t_f \right) \right)$ is increased by 0.17 %. Beside the primary goal to meet the terminal conditions, the optimizer O has to obey the constraints (9) and (10) against all possible air density fluctuations. Note, that the lift coefficient limit $C_{L,\max}(h, v, \gamma, \sigma)$ is a function of the unknown σ, i.e., O has to guarantee $C_L \leq C_{L,\max}$ at any time $t \in [0, t_f]$ and for all $\sigma \in [\sigma_{\min}(h), \sigma_{\max}(h)]$. Therefore the constraint (9) must be transformed to the much more severe constraint for all $t \in [0, t_f]$

$$
C_L \leq \min_{\sigma} C_{L,\max} (h, v, \gamma, \sigma) \quad \text{with} \quad \sigma \in [\sigma_{\min}(h), \sigma_{\max}(h)] . \tag{25}
$$

The functions $A(t)$, $C_L{}^*(t)$ and $\mu^*(t)$ for the resulting reentry trajectory are depicted in the Figs. 9 – 11. The trajectory for the worst-case air-density fluctuation

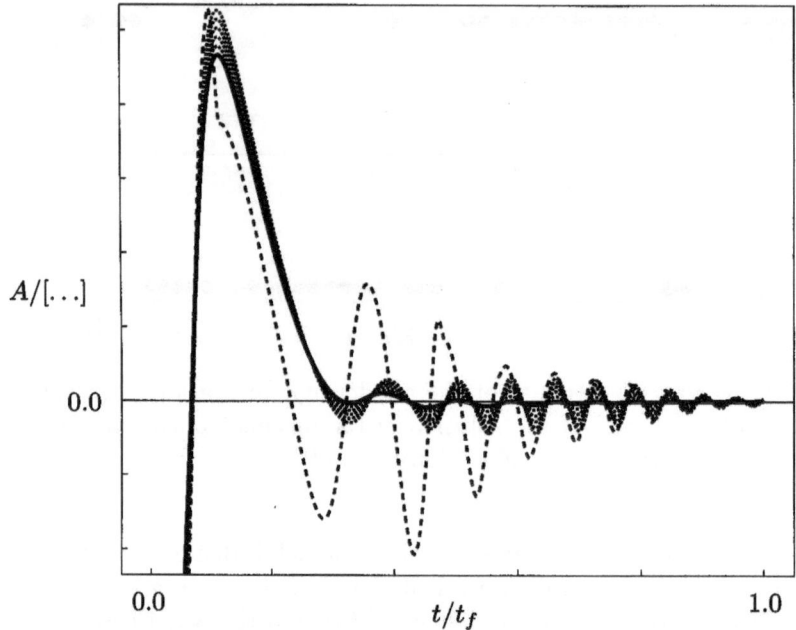

Fig. 9: Switching function $A(t)$ for the worst-case air-density fluctuation $\tilde{\sigma}^*(z^*(t))$ (thick solid curve for $\sigma \equiv 1$ and 5 dotted curves, constraints not considered) with increasing oscillations for an increasing fluctuation tube. Switching function $A(t)$ for the realistic fluctuation tube with all constraints obeyed (dashed curve).

$\tilde{\sigma}^*(z^*(t))$ itself is compared to the unconstrained case in Fig. 13. Due to the constraint (25), the performance index $I^*(z(t_f))$ is increased by an additional amount of 0.27 % . Since the reentry trajectory is the solution of a zero-sum differential game for the optimizer O and the disturber D, the associated performance index $I^*(z)$ has a saddlepoint property. There exists an optimal feedback controller $(\tilde{C}_L^*, \tilde{\mu}^*)(z)$ for the optimizer O which guarantees for every possible $\sigma(t)$ that the resulting performance index $I(z)$ is less or equal $I^*(z)$. The other way round, there exists an optimal feedback controller $\tilde{\sigma}^*(z)$ for the disturber D which prevents a reducing of $I(z)$ below $I^*(z)$ by the optimizer O. However, the optimal feedback controller of interest $(\tilde{C}_L^*, \tilde{\mu}^*)(z)$ can only be obtained along very unlikely trajectories $z^*(t)$, $t \in [0, t_f]$ with $\sigma \equiv \tilde{\sigma}^*(z^*(t))$ and not for realistic $\sigma(t)$.

Construction of the Optimal Feedback Controller

The reentry of the shuttle into the Earth's atmosphere is an extremely difficult maneuver from the viewpoint of control theory. Usually a feedback controller tracking a precalculated trajectory is used which has the disadvantage that the optimal va-

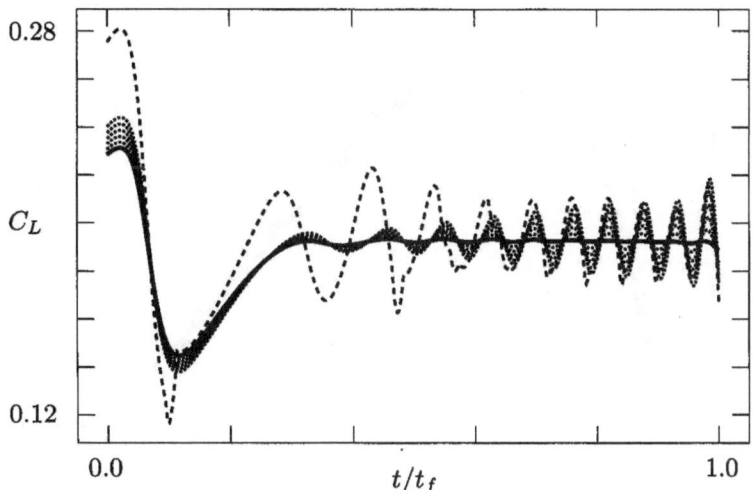

Fig. 10: Optimal lift coefficient $C_L^*(t)$ for the worst-case air-density fluctuation $\tilde{\sigma}^*(z^*(t))$ (thick solid curve for $\sigma \equiv 1$ and 5 dotted curves, constraints not considered) with increasing oscillations for an increasing fluctuation tube. Optimal lift coefficient $C_L^*(t)$ for the realistic fluctuation tube with all constraints obeyed (dashed curve).

lue I^* of the performance index often cannot be achieved at $t = t_f$. Even feedback controllers using neighboring optimal trajectories cannot guarantee a minimum of the performance index I against all possible fluctuations of the air density, see Fig. 15. The procedure to calculate the optimal feedback controller $(\tilde{C}_L^*, \tilde{\mu}^*)(z(t))$ for an arbitrary air-density fluctuation $\sigma(t)$ is sketched in the sequel:

1. Initialization:

   Choose the parameter $\Delta t_{\max}$ and η for the step size control (here: $\Delta t_{\max} = 30\,\text{s}$ and $\eta = 0.01$). Set initial time $t_s := 0$ and initial state $z_s := z_0$. Solve the MPBVP for the related differential game (see Section 4), store the guaranteed performance index $I_s^* = I^*(t_s, z_s)$ and store $z^*(t)$ and $\lambda^*(t)$ for all $t \in [0, \Delta t_{\max}]$. Set $\Delta t := \Delta t_{\max}$.

2. Integration:

   Compute $z_r(t_s + \Delta t)$ by numerical integration of the differential equations (3) – (8) using $C_L(t) = \tilde{C}_L^*(z_r(t), \lambda^*(t), \sigma(t))$ and $\mu(t) = \tilde{\mu}^*(z_r(t), \lambda^*(t), \sigma(t))$ from (22) and the air density fluctuation $\sigma(t)$.

3. First step size control for Δt:

   If $\| z^*(t_s + \Delta t) - z_r(t_s + \Delta t) \|_2 \,/\, \| z_r(t_s + \Delta t) \|_2 > \eta$ then set $\Delta t := \Delta t/2$ and goto 2.

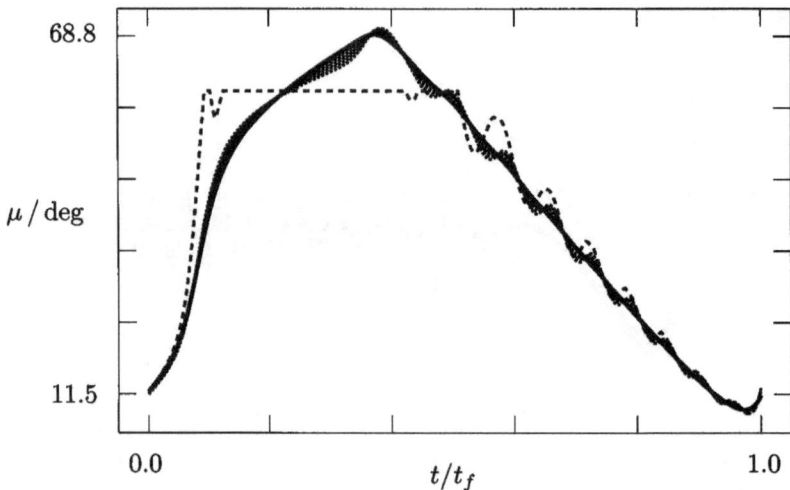

Fig. 11: Optimal bank angle $\mu^*(t)$ for the worst-case air-density fluctuation $\tilde{\sigma}^*(z^*(t))$ (thick solid curve for $\sigma \equiv 1$ and 5 dotted curves, constraints not considered) with increasing oscillations for an increasing fluctuation tube. Optimal bank angle $\mu^*(t)$ for the realistic fluctuation tube with all constraints obeyed (dashed curve).

4. Second step size control for $\triangle t$:

   Solve the MPBVP for the related differential game with the new initial time $t_s + \triangle t$ and the new initial state $z_r(t_s + \triangle t)$. If the new $I^*(t_s + \triangle t, z_s + \triangle t)$ is greater than $I^*(t_s, z_s)$, set $\triangle t := \triangle t/2$ and goto 2.

5. Stop condition:

   If $v(t_s + \triangle t) \leq v_f$ then stop the iteration, do the final approach to the terminal conditions with a special feedback controller and store all numerical results.

6. Iteration:

   Set $t_s := t_s + \triangle t$, $z_s := z_r(t_s + \triangle t)$ and $\triangle t := \triangle t * 4$. Solve the MPBVP for the related differential game, store the guaranteed performance index $I_s^* = I^*(t_s, z_s)$, store $z^*(t)$ and $\lambda^*(t)$ for all $t \in [t_s, t_s + \triangle t_{\max}]$ and goto 2.

The solution of the previous MPBVP can always serve as an initial guess to solve the updated MPBVP with the indirect multiple shooting method in Steps 4 and 6. Note that the calculation of the solution of the MPBVPs requires an automatic adaption of the switching structure. Both step size controls combined with the stop condition cannot produce an infinite iteration loop. For the numerical calculation of

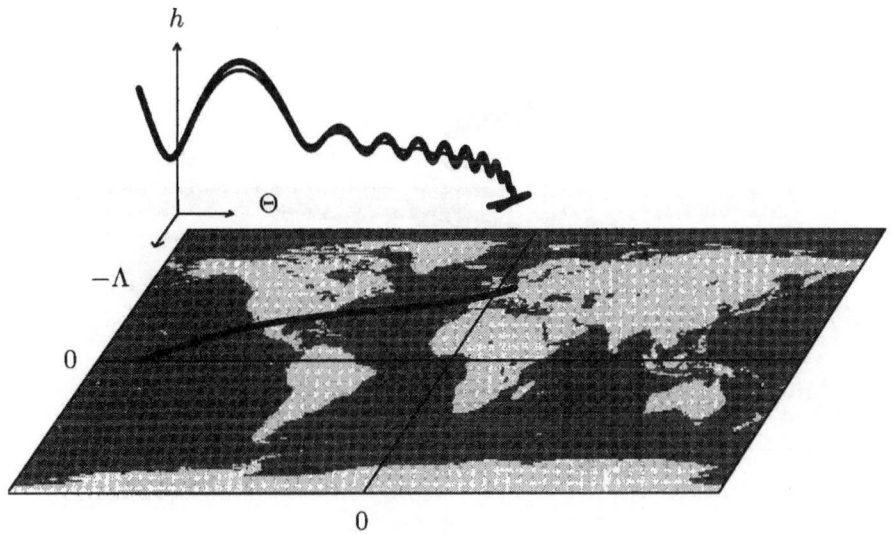

Fig. 12: Optimal reentry trajectories (constraints not considered): solution of the optimal control problem without air density fluctuation $\sigma \equiv 1$ (thin curve) and solution of the differential game (worst-case analysis) $\sigma \equiv \tilde{\sigma}^* (z^* (t))$ (thick curve).

the resulting reentry trajectory (constraints not considered) with $\sigma \equiv \sigma_{\max}(h)$, see Fig. 14, 756 MPBVPs have been solved on a SUN Sparc 2 Workstation in two and a half day. Although a real-time application of the optimal feedback controller is not possible today, the resulting trajectories and optimal control histories enable a comparison and validation of real time applicable feedback controllers. As an example serves an implementation of a real time applicable feedback controller which follows the optimal trajectories in the neighborhood of an a priori calculated optimal reentry path. The optimal feedback controller $(\tilde{C}_L^*, \tilde{\mu}^*)$ calculated via the outlined iteration procedure guarantees that the maximal performance index I^* cannot increase in contrast to the real time applicable feedback controller, see Fig. 15.

Acknowledgement. This Research has been supported by Deutsche Forschungsgemeinschaft within the project "Anwendungsbezogene Optimierung und Steuerung". The author would like to thank Oskar von Stryk for providing him with his collocation method DIRCOL (version 04).

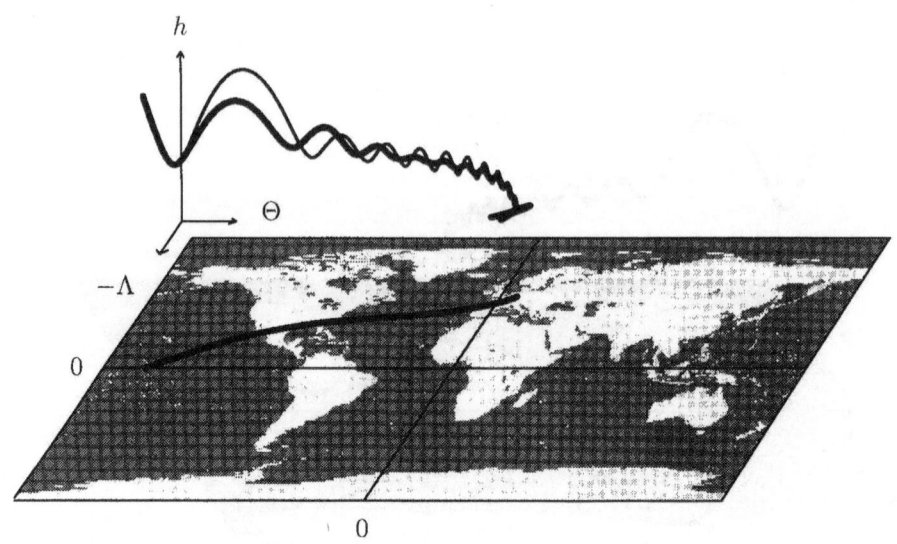

Fig. 13: Optimal trajectories for the worst-case air-density fluctuation $\tilde{\sigma}^*\left(z^*\left(t\right)\right)$: constraints not considered (thin curve) and all constraints considered (thick curve).

References

[BB91] Başar, T. and Bernhard, P. : H^{∞}–Optimal Control and Related Minimax Design Problems. Birkhäuser, Boston, 1991.

[BO82] Başar, T. and Olsder, G. J. : Dynamic Noncooperative Game Theory. Academic Press, London, Great Britain, 1982.

[BS91] Bardi, M. and Sartori, C. : Differential Games and totally risk-averse optimal control of systems with small disturbances. Lecture Notes in Control and Information Sciences 156, Springer , Berlin, 1991.

[BP93] Breitner, M. H. and Pesch H. J. : Reentry Trajectory Optimization under Atmospheric Uncertainty as a Differential Game. In : Annals of the ISDG, Vol. 1, Advances in Dynamic Games and Applications, ed. by Başar, T. and Haurie, A. . Birkäuser Boston, Basel, Berlin, 1993.

[BPG93] Breitner, M. H. , Pesch, H. J. and Grimm, W. : Complex Differential Games of Pursuit-Evasion Type with State Constraints, Part 1: Necessary Conditions for Optimal Open-Loop Strategies, Part 2: Numerical Computation of Optimal Open-Loop Strategies. To appear in JOTA, autumn 1993.

[BR94] Breitner, M. H. : Real-Time Applicable Feedback Controller for Differential Games. In preparation for : Proceedings of the Sixth International Symposium on

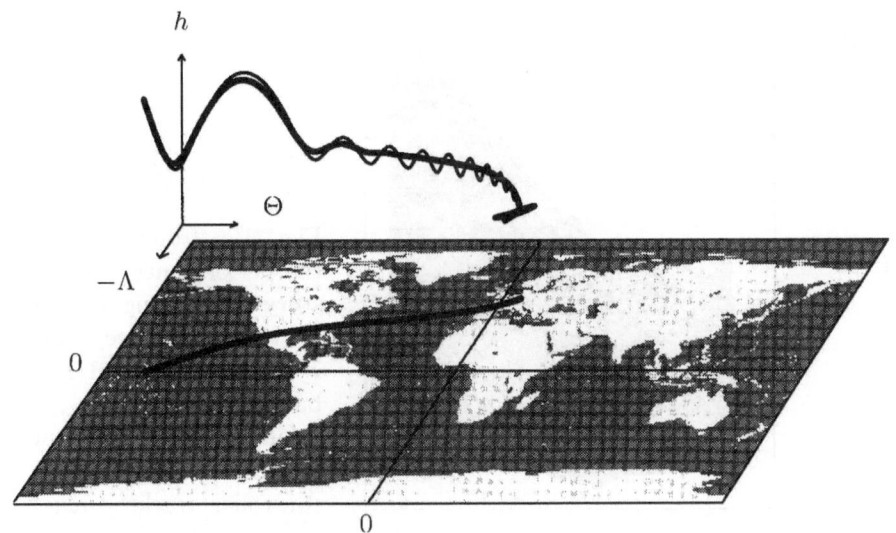

Fig. 14: Comparison of the reentry trajectories for different air-density fluctuations (constraints not considered): Worst case $\sigma(t) \equiv \tilde{\sigma}^*(z^*(t))$ (thin curve) and simulated fluctuations $\sigma(t) \equiv \sigma_{\max}(h(t))$ (thick curve).

Dynamic Games and Applications, St-Jovite, Québec, Canada, July 13–15, 1994.

[BU71] Bulirsch, R. : *Die Mehrzielmethode zur numerischen Lösung von nichtlinearen Randwertproblemen und Aufgaben der optimalen Steuerung.* Report of the Carl-Cranz Gesellschaft, Oberpfaffenhofen, Germany, 1971. Reprint: Department of Mathematics, Munich University of Technology, Germany, 1993.

[BH75] Bryson, A. E. and Ho, Y.-C. : *Applied Optimal Control.* Hemisphere Publ. Corp., New York, 1975.

[HI93] Hiltmann, P. : *Numerische Lösung von Mehrpunkt-Randwertproblemen und Aufgaben der optimalen Steuerung mit Steuerfunktionen über endlichdimensionalen Räumen.* Doctoral Thesis, Report No. 448 of the "Schwerpunkt Anwendungsbezogene Optimierung und Steuerung" of the Deutsche Forschungsgemeinschaft, München, Germany, 1993.

[IS65] Isaacs, R. : *Differential Games.* Wiley, New York, 1965.

[KP90] Kugelmann, B. and Pesch, H. J. : *A New General Guidance Method in Constrained Optimal Control, Part 1: The Numerical Method, Part 2: Application to Space Shuttle Guidance.* JOTA, Vol. 67 (3), 1990.

[LE89] Leitmann, G. : *Deterministic Control of Uncertain Systems Via a Construc-*

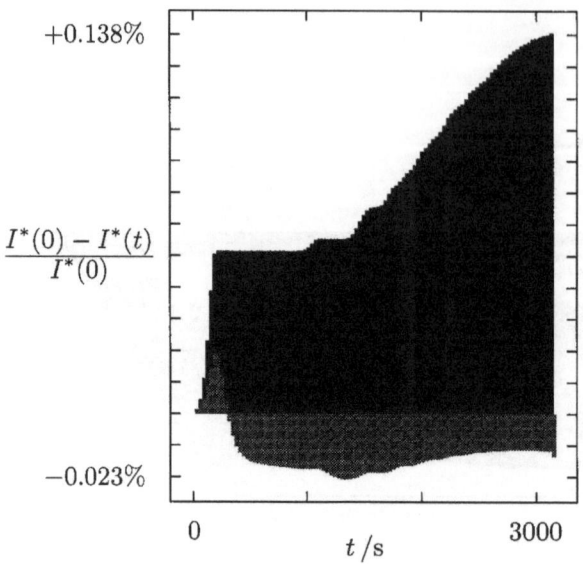

$$+0.138\%$$

$$\frac{I^*(0) - I^*(t)}{I^*(0)}$$

$$-0.023\%$$

0   3000

$t\,/\mathrm{s}$

Fig. 15: Comparison of the performance index history: Decreasing, guaranteed performance index I^* for the optimal feedback-controller $(\tilde{C}_L{}^*, \tilde{\mu}^*)$ (black) and performance index for the feedback-controller tracking neighboring optimal trajectories (gray).

tive Use of Lyapunov Stability Theory. Lecture Notes in Control and Information Sciences 143, Springer, Berlin, 1989.

[PE89] Pesch, H. J. : *Real-time Computation of Feedback Controls for Constrained Optimal Control Problems, Part 1: Neighbouring Extremals, Part 2: A Correction Method Based on Multiple Shooting.* OCAM, Vol. 10, pp. 129–145, 1989.

[SB93] Stoer, J. and Bulirsch, R. : *Introduction to Numerical Analysis.* Springer, New York, Berlin, Heidelberg, 1993[3].

[SB92] Stryk, O. von and Bulirsch, R. : *Direct and Indirect Methods for Trajectory Optimization.* Annals of Operations Research, Vol. 37, pp. 357–373, 1992.

International Series of Numerical Mathematics, Vol. 115, © 1994 Birkhäuser Verlag Basel

Repetitive Optimization for Predictive Control of Dynamic Systems under Uncertainty

Krzysztof Malinowski*

Abstract

Control of dynamic systems by repetitive on-line use of optimization is considered. A basic predictive control scheme and a modified scheme based on two-stage optimization are presented. The paper focuses on important issues related to the one-stage and two-stage controllers. Application is presented for flood control of large reservoirs together with a brief summary of extensive simulation experiments. Those experiments consisted of the simulation of over fifty control sessions for each version of the control mechanism.

Introduction

Control of dynamic systems by repetitive on-line use of optimization is now a well established area of applications. Those applications range from predictive control algorithms for regulation — based on linear models (e.g. de Kayser et al. 1988) — to more advanced control schemes for nonlinear systems involving optimization under the constraints on control and, possibly, state variables. Predictive control is used also in hierarchical control structures (Findeisen et al. 1980). Those algorithms are, however, largely based on the open-loop feedback approach and involve optimization of control trajectory for a given prediction (forecast) of external uncontrolled inputs, if such inputs are explicitly considered. Algorithms of the considered class can be found in production control, in control of transportation networks, hydropower generation, control of water storage reservoirs etc. More advanced approaches involve the use of repetitive two-stage optimization.

In order to make the application of repetitive predictive control successful, it is also necessary to provide very fast routines for on-line dynamic optimization. The modern computer technology provides extremely powerful microprocessors to do the job, but still a lot of skill and experience is required to implement both *fast* and *robust* computational algorithms. Finally, it also very important to remember that formal theoretical analysis of the properties of control by repetitive optimization (predictive control) is possible only in very special cases and that those advanced computer based control techniques can be fully evaluated and assessed only through extensive simulation. The simulation experiments have to

---

*Institute of Automatic Control, Warsaw University of Technology, 00-665 Warszawa, Poland

be well organized and supported to make the task of assessment through simulation practically manageable.

The paper presents the above concepts and focuses on important issues related to the two-stage scheme for on-line control of nonlinear dynamic systems. Application is presented for flood control of large reservoirs together with a brief summary of extensive simulation experiments. Those experiments consisted of the simulation of over fifty control sessions for each version of the control mechanism.

Fixed-Form Control Rules and Predictive Control

Consider the following dynamic system,

$$\dot{x}(t) = f(x(t), m(t), z(t)) \tag{1}$$

where $x(t) \in R^n$ is the state vector at time t, $m(t) \in R^r$ is the control input and $z(t) \in R^p$ is the external uncertain uncontrolled input. The observations (measurements) available are given by

$$y(t) = h(x(t), m(t), z(t)) \tag{2}$$

where $y(t) \in R^s$ is the measurement vector. Measurement errors will not be explicitly considered since those errors are not most important in control problems discussed and since, if needed, those errors can be represented as the components of $z(t)$. Observations can be made available to the control station at times $t \in T_s$, where T_s is the set of scanning instants.

Assume that the process is controlled after starting at time t_0, where $t_0 \in T_s$, and that the control process (control session) terminates at time t_f. It is possible to consider the case in which t_f is infinite. Assume further that any decisions regarding control inputs can be made at times t_i, $i = 0, \ldots$, where $t_i \in T_s$, and denote by m_i, $m_i \in R^w$ the vector of parameters by which the values of $m(t)$ for $t \in (t_i, t_{i+1}]$ are specified; in the simplest case of zero-hold control we have $m_i = m(t)$, $t \in (t_i, t_{i+1}]$.

As far as constraints on the process variables are concerned, it is necessary to state that this is a quite important and difficult point. No simple general approach is available to describe properly the constraints on $m(t)$ and/or $x(t)$ under uncertainty concerning external uncontrolled inputs. In this paper it is assumed that the constraints are defined in such way that for given measurement history $I(t_i)$ at time t_i, $I(t_i) = \{y(t), t \in T_s, t \leq t_i\}$, it is possible to specify constraints on m_i in the form

$$m_i \in M(I(t_i)) \tag{3}$$

and that by satisfying those constraints the control station will be able to generate admissible control inputs for $t \in (t_i, t_{i+1}]$.

The objective of the control algorithm will be then to generate admissible controls satisfying Eq. (3) and, in addition, such that some performance measures will be satisfactory. In particular, assume that one would like to minimize the following performance function

$$J = \int_{t_0}^{t_f} q(x(t), m(t), z(t)) \, dt \, . \tag{4}$$

It is obvious that the value of this performance measure cannot be calculated prior to the actual control session since $z(t)$, $t \in [t_0, t_f]$, is unknown. It is then typical to assume that one wishes to minimize the expected value of (4) or some other a priori representation of this performance function. In order to compute an optimal control law minimizing the expected value of (4) it is necessary to have a stochastic model of the uncertain input $z(t)$, for example in the form of a disturbance generator driven by a white noise process. Optimal solution of the closed-loop stochastic control problem is possible only in very particular cases; essentially it requires a LQG formulation of the control design problem. For a nonlinear case and/or in presence of constraints (3) such solution is practically impossible. It is also worthwhile to mention that in practice it can be very difficult or costly to provide a satisfactory stochastic model of $z(t)$. Quite often one should consider the situation in which the a priori knowledge of a dynamic process lying behind the formation of $z(t)$ (the uncontrolled input generator) is very poor and only either records of trajectories of $z(t)$ from the past sessions are available or some simplified generators can be proposed and verified. Such is the case, for example, as far as meteorological (hydrological) phenomena are concerned or when market behavior is considered.

There are then basically two practical ways by which a control law for the above dynamic system with uncertain inputs can be provided, assuming that one wishes to generate feasible control inputs and to obtain a satisfactory level of performance (4). The first is to introduce a *fixed-form* control law, i.e. the parameter dependent control rule

$$m_i = \gamma_i(I(t_i), \alpha_i) \tag{5}$$

where $\alpha_i \in R^{v_i}$ are the vectors of parameters and the predetermined functions γ_i are such that the constraints given by Eq. (3) are satisfied for $\alpha_i \in A_i \subseteq R^{v_i}$. In most cases $\alpha_i = \alpha$ for every i. The "best" values of α_i (or α) are found prior to the actual control session by using some tuning procedure. Tuning may be done using a set of given explicit rules (e.g. the use of Ziegler-Nichols rules for tuning a PID controller) or by performing parametric optimization with respect to α_i. In the latter case the operation of the dynamic system under control rule (5) is simulated for recorded (in the past) or/and randomly generated trajectories of $z(t)$; simulation may be repeated for several control sessions on the interval $[t_0, t_f]$. In this way the estimated value of (4) can be computed and used to iterate on α_i.

It should be noted that such an approach requires a considerable computing effort to perform off-line optimization, while, on the other hand, the computations required on-line are usually simple and straightforward. It is also worthwhile to observe that the tuning procedure involves a simulated closed-loop operation of the controlled process and therefore it is the *closed-loop* behavior of this process which is, albeit parametrically, optimized. The main disadvantage of using the fixed-form control rule is, apart of often costly off-line optimization, the difficulty to choose good functions γ_i. Nevertheless, applications of this approach are very popular and range from PID controllers to the use of neural nets. Neural nets for control have, in particular, became recently a popular topic due to their proven approximation properties and the reported success in applications. Yet one should remember that greater flexibility with fixed-form rules can be achieved, as in the case of neural controllers, at the expense of the introduction of a large number of parameters, and therefore at the expense of costly and prolonged tuning (learning).

The second practical approach consists of shifting the computational burden to the real time by introducing control schemes with repetitive optimization (predictive control). This is made possible due to largely increased computing power of available microprocessors which can be used for on-line control applications. The basic predictive control scheme (BPC) consists, at each real time "intervention" instant t_i of the following steps:

1) Obtain an estimate (or measurement) $\hat{x}(t_i)$ of the current process state $x(t_i)$. Modify (update) the process model (1) if necessary.

2) Generate a prediction of the external disturbance input in form of a single trajectory $\bar{z}^i(t)$ for $t \in (t_i, t_{fi}]$, where t_{fi} is the end time of the *prediction horizon* $(t_i, t_{fi}]$, where $t_{fi} \leq t_f$.

3) Solve the optimization problem
   BPC-P:

$$\text{minimize} \int_{t_i}^{t_{fi}} q(x(t), m(t), \bar{z}^i(t)) \, dt \tag{6}$$

   with respect to $m_i, m_{i+1}, \ldots, m_{fi-1}$, subject to Eqs. (1), (2) and (3) (with $z(t) = \bar{z}^i(t)$ for $t \in (t_i, t_{fi}]$) and assuming that $x(t_i) = \hat{x}(t_i)$. Additional constraints on the controls m_k, $k \geq i$, can be imposed. For example, it is typical to assume when solving the problem BPC-P that $m_k = m_{k-1}$ for $k \geq k_{i0}$, where $i \leq k_{i0} \leq fi - 1$. This means that the control inputs are assumed to be "frozen" near the end of the prediction horizon $(t_i, t_{fi}]$. Such extra constraints can improve the behavior of the control scheme.

Once the solution of the optimization problem BPC-P is obtained in form of a sequence of predicted *open-loop* optimal control actions $\hat{m}_i, \hat{m}_{i+1}, \ldots, \hat{m}_{fi-1}$, the first component, that is $\hat{m}_i$, is taken to determine the control actions till time t_{i+1} at which instant the whole procedure is repeated.

It may be noted that in the BPC scheme one can choose $t_{fi} = t_i + \Delta t$ (moving horizon version) or $t_{fi} = t_f$ (receding horizon version). The constraints (3) can be observed when solving BPC-P since for a given prediction $\bar{z}^i(\cdot)$ of $z(\cdot)$ the predicted "future" information $I(t_j)$ can be computed for each t_j, $t_i < t_j \leq t_{fi}$.

The BPC scheme is widely implemented by now in a variety of control algorithms (e.g. de Kayser et al. 1988, Balchen et al. 1992). In fact, almost all predictive control algorithms use this framework, varying in choice of the way in which prediction of the uncontrolled input is made, in choice of the performance function q and in such details as choosing the "freezing" instant k_{i0}. DMC and QDMC algorithms can be mentioned here as popular industry standards by now.

The advantages of the BPC scheme over the parametric fixed-form rules are: flexibility with respect to a situation changing during a control session and no need to determine the form of a control law and to tune this law before the actual control session.

The characteristic features are:

- the implementation requires complex fast and reliable on-line tools for forecasting and optimization,

- assessment of the operation through computer simulation is much more difficult when using an explicit control rule (5),

- the BPC controller is an open-loop-feedback controller.

As far as fast on-line optimization is concerned, the present computer technology provides a powerful hardware. Yet the standard optimization procedures often need plenty of time to converge on the solution and, more often than not, they are not very reliable. Therefore, for a successful implementation of the BPC controller it is necessary to develop (or adapt) an optimization problem solver, which would be capable, without failing, of solving the BPC-P in prescribed time and with prescribed accuracy. The development of such a solver should involve taking into account particular features of the controlled process and the performance function. Any analytical elements of the solution are welcome; such preprocessing is definitely possible and is made, for example, in case of linear models (1), (2) and quadratic performance function. Thus most BCP controllers for linear system do not require on-line numerical optimization, some employ reliable linear problem solvers. In case of nonlinear process models the situation is more difficult and the solutions will most likely be specific to a given problem in order to avoid excessive computations. It should also be noted at this point that just "reasonable" accuracy is required when solving the BCP-P.

The general analysis of the BPC scheme is difficult and is restricted so far to the stability analysis (e.g. Chen and Shaw 1982, Mayne and Michalska 1990). An evaluation of the BCP scheme through computer simulation is difficult for obvious reasons: this simulation requires repetitive multiple solution of the forecasting and the optimization problems. In particular, the simulation of the forecasting procedure can be time consuming and also difficult to arrange in case when the forecasting depends on many on-line data. In such situation a forecast dummy, i.e. a procedure producing dummy forecasts for simulation purposes may be used. Such a forecast dummy should produce forecasts with the same error characteristics as the actual forecasting routine. A forecast dummy can be much simpler than the actual procedure and requires less external data since this dummy may use the simulated future "real" trajectory of $z(\cdot)$.

The last characteristic of the BPC controller, i.e. the open-loop character, or more precisely the open-loop-feedback character, is sometimes forgotten or neglected. Yet it is clear that in case when forecasting of the future uncontrolled inputs is inaccurate, the scheme may lead to a performance being much more degraded than when using a fixed-form rule tuned in closed-loop manner and less prone, if any, to errors in forecasting the values of $z(t)$. The two-stage predictive controller described in the next section can be hoped to cope with the uncertainty with respect to the future values of $z(t)$ in a better and more flexible way.

Two-Stage Predictive Controller

The basic weakness of the BPC is, as just noted, the open-loop character of the algorithm. In other words, the algorithm used at a given time to compute control action does not anticipate the future information which will be made available to the control station. The simplest way to modify the BPC controller in a closed-loop manner is first to assume that at a given time t_i the forecasting module will provide, instead of a single forecast, a set of, say, N forecasts. Hence the steps 1) and 2) of the BPC scheme will be now

1a) Obtain an estimate (or measurement) $\hat{x}(t_i)$ of the current process state $x(t_i)$. Modify (update) the process model (1) if necessary.

2a) Generate N predicted realizations of the external disturbance input in form of the set of N trajectories, i.e. $\bar{z}^{i,j}(t)$ for $t \in (t_i, t_{fi}]$, $j = 1, \ldots, N$, where t_{fi} is the end time of the *prediction horizon* $(t_i, t_{fi}]$, and where $t_{fi} \le t_f$.

Now, the optimization mechanism which is to produce $\hat{m}_i$, is based upon the *assumption* that one of the generated scenarios $\bar{z}^{i,j}(\cdot)$ will actually occur and that it will be known at time t_{i+1} which is the right one.

The decision mechanism of the TSPC controller at time t_i is then the following:

3a)

TSPC-P

$$\min_{m_i} \sum_{j=1}^{N} p_j \left[\int_{t_i}^{t_{i+1}} q(x^j(t), m(t), \bar{z}^{i,j}(t)) \, \mathrm{d}t + \right.$$

$$\left. + \min_{m^j_{i+1}, \ldots, m^j_{fi-1}} \int_{t_{i+1}}^{t_{fi}} q(x^j(t), m^j(t), \bar{z}^{i,j}(t)) \, \mathrm{d}t \right] \qquad (7)$$

where

$$\dot{x}^j(t) = f(x^j(t), m^j(t), \bar{z}^{i,j}(t)), \qquad x^j(t_i) = \hat{x}(t_i)$$

$$\text{and } m^j(t) = m(t) \quad \text{for } t \in (t_i, t_{i+1}] \quad \text{' (hence } m_i^j = m_i)$$

subject to the constraints (3) imposed on m_k^j, $k \geq i$; the information sets $I^j(t_k)$ can be evaluated at t_i for each external input scenario $\bar{z}^{i,j}(\cdot)$ and for a given control input trajectory $m^j(\cdot)$ up to time t_k. Again, it might be required that $m_k = m_{k-1}$ for $k \geq k_{i0}$, where $i \leq k_{i0} \leq fi - 1$.

The weights p_j, $p_j \geq 0$, $\sum_j p_j = 1$, represent the relative importance of the uncontrolled input scenarios. By choosing these weights it is possible to tune the TSPC control scheme. In particular, for $p_1 = 1$ and $p_2 = p_3 = \ldots = p_N = 0$, the TSPC becomes the BPC scheme.

A generalization of the TSPC controller to the K-stage scheme is now apparent; it is only necessary to anticipate at time t_i that at every future time t_k, $i \leq k < i + K$, each uncontrolled input scenario will split into several distinct trajectories.

In some applications it can be worthwhile to consider such K-stage schemes. It should be noted, however, that even the TSPC scheme requires much increased computational effort when compared with the BPC. An N-scenario forecast has to be made available and then the internal minimization in (7) has to be repeated N times at each iteration with respect to m_i. Fortunately, it can be hoped that, providing a careful selection of the scenarios $\bar{z}^{i,j}(\cdot)$ is made, N will not have to be large; e.g. one could consider $N = 2$. Further, parallel computing can be employed to perform multiple-forecasting and then multiple internal optimization in (7) in a concurrent manner.

An analysis of the operation and the performance of the TSPC scheme, prior to actual implementation, and tuning this scheme w.r.t. the values of the weights p_j, the "freezing" horizon k_0, and the details of optimization procedure (e.g. accuracy of a solution) can only be made through a computer simulation. It is therefore one of the most important practical research directions now to establish and develop tools for a computer analysis of predictive control.

It may be interesting at this point to observe that the TSPC scheme is in fact based upon solving an optimal closed-loop control design, performed repetitively at times t_i, assuming at each t_i that the model of uncertainty is given by N future scenarios $\bar{z}^{i,j}(\cdot)$, and assuming that enough information will be available at t_{i+1} to indicate the "real" scenario. The weights p_j can be considered as a priori probabilities associated with the scenarios. The uncontrolled input model considered at time t_i can be regarded thus as a largely simplified model of the uncertainty. By extending this idea to other simplified models of the input $z(\cdot)$ one can introduce new control schemes, each of them based on repetitive optimal control design using a simplified model of the uncertainty.

It is also possible to consider the situation in which the repetitive design is performed at less frequent intervals. Finally, repetitive design can be proposed, whereby at times t_{i_k} the parameters of the control rule (5) are retuned according to the current information w.r.t. the future uncontrolled inputs and the process model; in particular by using multiple forecast $\bar{z}^{i,j}(\cdot)$ of $z(\cdot)$ as in the TSPC scheme.

There is a number of important, mutually related, questions associated with a given problem when one would like to implement either the BPC controller or the TSPC or some other predictive control scheme with repetitive optimization/design:

1. What repetitive scheme to choose, e.g. BPC or TSPC?

2. What forecasting mechanism is to be used?

3. What optimization technique should be used?

4. How and by what means should the computer simulation and tuning of the control scheme be performed; in particular what forecast dummy should be used for computer simulation?

Available experience shows that there are not useful general answers to those questions. The decisions will be largely motivated by specific features of a given problem. On-line optimization technique and tools for computer simulation of the scheme offer a considerable challenge. For applications which allow more time for decision making it can be worthwhile to consider the participation of a human decision maker at the optimization stage. This could make less stringent, in particular, the demands with respect to reliability of optimization routines.

Min-Max Performance

Consider now the case when the performance index is, instead of Eq. (4), given by the following formula,

$$J = \max_{k=0,\ldots,f} Q_k(x(t_k), m_k, z_k) \tag{8}$$

where z_k includes the characteristics of $z(t)$, $t \in (t_i, t_{i+1}]$, needed to evaluate the performance value.

With performance (8), applications of BPC and TSPC control schemes require some consideration. First, at a given time t_i one has to decide whether to take into account the realized (past) value of the performance, i.e.

$$\max_{k=0,\ldots,i-1} Q_k(x(t_k), m_k, z_k)$$

or whether to forget this value, even assuming that the worst might have already happened.

The forgetting option seems to be more flexible and still one loses nothing as far as the optimality is concerned by adopting this option.

When using the TSPC scheme, and assuming the forgetting option, one can consider the following versions of the optimization problem within the scheme.

S1:

TSPC-P1:

$$
\min_{m_i} \max \left\{ \sum_{j=1}^{N} p_j \cdot Q_i(x(t_i), m_i, \bar{z}_i^{i,j}), \right.
$$

$$
\left. \sum_{j=1}^{N} p_j \cdot \left[\min_{m_{i+1}^j, \ldots, m_{fi-1}^j} \max_{k=i+1,\ldots,fi-1} Q_k(x^j(t_k), m_k^j, \bar{z}_k^{i,j}) \right] \right\} \qquad (9)
$$

or

S2:

TSPC-P2:

$$
\min_{m_i} \sum_{j=1}^{N} p_j \cdot \left\{ \max \left[Q_i(x(t_i), m_i, \bar{z}_i^{i,j}), \right. \right.
$$

$$
\left. \left. \min_{m_{i+1}^j, \ldots, m_{fi-1}^j} \max_{k=i+1,\ldots,fi-1} Q_k(x^j(t_k), m_k^j, \bar{z}_k^{i,j}) \right] \right\} \qquad (10)
$$

In the first version the expected (w.r.t. the simplified multiple-scenario model of the uncertainty) cost during the immediate next control interval (from t_i to t_{i+1}) is compared with the expected future cost, after time t_{i+1}, and then the larger of the two is minimized w.r.t. m_i. In case of S2 the expected value of the cost in the future, including interval $(t_i, t_{i+1}]$, is minimized.

Solutions of problems TSPC-P1 and TSPC-P2 are different since, in general, $\max(Ex, Ey) \neq E\{\max(x, y)\}$.

Control of a Storage Reservoir During Flood

The TSPC scheme has been applied to control of a storage reservoir during flood. In this case $x(t) \in R$ denotes the volume of water stored at time t, $m(t) \in R$ denotes discharge from the reservoir at t, and $z(t)$ denotes the uncontrolled inflow. Hence

$$
\dot{x}(t) = z(t) - m(t)
$$

It is assumed that $x(t_i)$ can be accessible through the measurements and so $\hat{x}(t_i) = x(t_i)$. The constraints (3) on $m_i = m(t)$, $t \in (t_i, t_{i+1}]$, are given in the form

$$
m_i^{min}(x(t_i), z(t_i)) \leq m_i \leq m_i^{max}(x(t_i), z(t_i))
$$

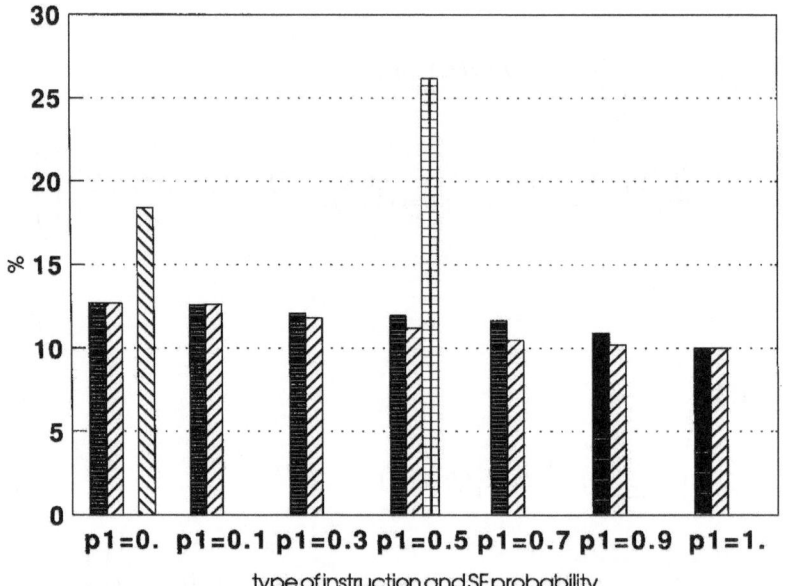

Figure 1: Average reduction of the maximal discharge from the reservoir with respect to unregulated flood wave. Forecast dummy WFM.

The performance index to be minimized is

$$\max_{i=0,\ldots,f} m_i$$

i.e. the objective is to minimize the maximal (peak) discharge.

The forecasting system provides at each time t_i — when the new value of discharge is to be calculated (e.g. at three hour intervals) — the set of two inflow scenarios. One is related to the zero-rainfall option, whereby it is assumed that no future rainfall, after time t_i, will occur. The second scenario anticipates future rainfall as forecasted by the weather service. Thus, there are two inflow scenarios available, generated with the use of complex high-order nonlinear rainfall-runoff model. Obviously, $z^{i,1}(t) \leq z^{1,2}(t)$ for $t \geq t_i$.

The TSPC scheme, with both versions S1 and S2 of the internal optimization problem, has been used to provide control mechanism for the Roznow reservoir on the Dunajec River in the southern part of Poland. Fifty control sessions have been simulated for each choice of the optimization version and of the weights

p_1 and $p_2 = 1 - p_1$. The multiple simulation has been made possible by using the software package FC-ROS developed in the Institute of Automatic Control for simulation of the reservoir operation, including the predictive control. The forecast dummies were provided by the Institute of Meteorology and Water Management. One of these dummies, WFM, produces pseudo-forecasts with similar statistical errors (w.r.t. the considered historical real flood inflows) as the currently used forecasting routine. The second forecast dummy, CFM, represents a forecasting system with a better performance and reflects hopes with respect to the future enhanced forecasting system.

The flood control problem, the forecast dummies and the simulation results are described in more detail in (Karbowski at al. 1992), while the software package FC-ROS for simulation of the reservoir operation is described in (Karbowski 1991b).

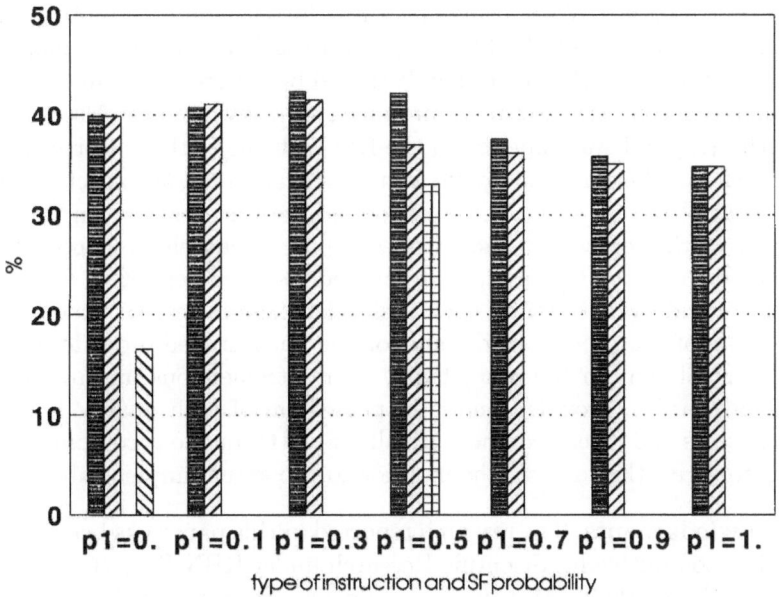

Figure 2: Average reduction of the maximal discharge from the reservoir with respect to unregulated flood wave. Forecast dummy CFM.

The synthetic results of multiple simulations (about three thousand simulation runs) are depicted in Fig.1 for the WFM forecast dummy and in Fig.2 for the

CFM dummy. Both strategies S1 and S2 have been employed. The bar diagrams present the average (calculated from the sample of fifty control sessions) reduction of the maximum flood discharge with respect to to the unregulated flood wave. For the sake of comparison the reduction when using the traditional control rule of type (5), currently in operation, has been also computed. A special solver based upon an analytical solution (Karbowski 1991b) of the BPC-P has been used.

In case of strategy S1, the WFM model and $p_1 = 0.5$ (p_1: "probability" of a small flood SF), the control sessions have also been simulated by the use of additional protection mechanism as proposed and described in (Malinowski and Żelaziński 1990). In fact, only when using this protection mechanism it became possible to out perform the traditional control rule of the fixed-form type. In other cases this rule provided a better performance than the much more sophisticated TSPC scheme. This means that with the actual forecasting system the use of the proposed controller for single reservoir operation during flood will require a careful use of the protection mechanism and will not bring any major improvement with respect to the current control routine.

The situation will be quite different, however, after the quality of the inflow forecasting is improved. While using the CFM forecast dummy, both predictive control schemes (with S1 and S2 two-stage optimization) provide for much better results than the traditional control rule. It should be observed that in the majority of the simulation runs the use of S1 produced the best results. Also the TSPC scheme with $p_1, p_2 \neq 1$ was superior to the BPC scheme (either p_1 or p_2 equal 1).

The basic conclusion to be drawn from this example is, as expected, that the predictive control is very sensitive to model and data accuracy — in this case to the accuracy of the forecasted uncontrolled input. It can easily happen, in case of a poor model or poor forecasts, that this model-based repetitive optimization scheme will provide worse results than a simpler fixed-form rule which does not make much use of the process model and/or the uncontrolled input forecast. The predictive control can, on the other hand, offer a tremendous improvement over classical control rules, provided one is in possession of high quality information about the process and its environment. In all cases a throughout computer analysis is required to verify the control scheme prior to the actual implementation.

Acknowledgments. The research reported in this paper has been supported by the Polish Committee for Scientific Research under KBN Project 3 0219 91 01. I would also like to thank Mrs. E. Niewiadomska-Szynkiewicz for assistance in preparing computational results and in editing.

References

[1] Balchen J.G., D. Ljungquist and S. Strand 1992, State-Space Predictive Control, *J. Chemical Engineering Science*, Vol. 47, No. 4, pp. 787-807.

[2] Chen C.C. and L. Shaw 1982, On Receding Horizon Feedback Control, *Automatica* (GB), Vol. 18, pp. 349-353.

[3] Findeisen W., F.N. Bailey, M. Brdyś, K. Malinowski, P. Tatjewski and A. Woźniak 1980, *Control and Coordination in Hierarchical Systems*, Wiley, London.

[4] Karbowski A. 1991, Optimal Control of Single Retention Reservoir During Flood; Analytical Solution of Deterministic, Continuous-Time Problems, *J. Optim. Theory and Appl.*, 69(1), pp. 55-81.

[5] Karbowski A., K. Malinowski and E. Niewiadomska-Szynkiewicz 1992, Application of Stochastic Inflow Forecast to Real-Time Flood Control in Single Reservoir Systems, manuscript, submitted for publication, Instytut Automatyki, Politechnika Warszawska.

[6] Karbowski A. 1991, Decision Support System for Reservoir Operators During Flood, *Environmental Software*, 6(1), pp. 11-15.

[7] Keyser de R.M., G.A. Van der Velde, F.A. Dumortier 1988, A Comparative Study of Self-Adaptive Long-Range Predictive Control Methods, *Automatica*, 24, pp. 149-163.

[8] Malinowski K. and J. Żelaziński 1990, Reservoir Systems: Operational Flood Control, in *Systems and Control Encyclopedia*, Supplementary Volume 1, Pergamon Press.

[9] Mayne D.Q. and H. Michalska 1990, Receding Horizon Control of Nonlinear Systems, IEEE *Trans. AC*, 35, pp. 814-824.

International Series of Numerical Mathematics, Vol. 115, © 1994 Birkhäuser Verlag Basel

Optimal Control of Multistage Systems Described by High-Index Differential-Algebraic Equations

C.C. Pantelides*   R.W.H. Sargent*   and   V.S. Vassiliadis*

Abstract

An algorithm is described for computing time-varying controls and time-invariant parameters to optimize the performance of a dynamic system whose behaviour is described by differential-algebraic equations (DAEs) of arbitrary index.

The problem formulation deals with a multistage system, in which each stage has its own describing equations. It also accommodates end-point, interior point, and path constraints which can be equalities or inequalities.

The algorithm uses a parameterization of the controls in terms of a given set of basis functions. Path constraints are dealt with via integral constraints. The problem is thus converted into a nonlinear programming problem, for which the objective and constraint functions are evaluated by integration of the system equations, and their gradients with respect to the optimization parameters via the integration of the sensitivity equations.

The special problems arising when the DAE system has index greater than one are explicitly dealt with, using automatic differentiation to generate the necessary additional equations, thus reducing the system index. A constraint stabilization technique for the resulting low index system is proposed.

Introduction

Classical optimal control theory is concerned with the optimal performance of systems whose behaviour is described by a set of ordinary differential equations, and there is now a wide literature on both the underlying theory and numerical algorithms for such problems. However, many systems of practical interest are naturally described in terms of a mixed set of differential and algebraic equations (DAEs) of the form

$$f\left(t,\ \dot{x}(t),\ x(t),\ u(t),\ v\right) = 0 \tag{1}$$

where t is the time, $x(t)$ are the system variables, $u(t)$ are the control variables, v is a set of time invariant variables, and $\dot{x}(t)$ denotes $dx(t)/dt$.

---

*Centre for Process Systems Engineering, Imperial College of Science, Technology and Medicine, London SW7 2BY, United Kingdom

Multistage systems, in which a sequence of processes is used to achieve some overall result, are also frequently of interest. Examples are a multistage rocket designed to achieve a maximum altitude, or the production of chemicals using a sequence of chemical reactions and separations. Each stage is then described by a system like (1), with "junction conditions" linking the successive stages.

Often mixed systems like (1) can be reformulated as systems of ordinary differential equations by a process involving differentiations and algebraic manipulations (Gear and Petzold, 1984, Bachmann *et al.*, 1990). In this paper, we describe a general index reduction method which uses only differentiations and numerical operations. Furthermore, the original system equations appear as a subset of the equations in the resulting low-index system. Thus the solution obtained by integrating the latter using currently available numerical algorithms satisfies all the original equations within the integration tolerance without the need for additional "constraint stabilization" measures (see, e.g., Gear, 1988).

Morison and Sargent (1986) presented a numerical algorithm for the solution of multistage optimal control problems where each stage was described by a set of DAEs of index one, and Gritsis *et al.* (1992) give an algorithm for semi-explicit systems of index two.

Problem Formulation

In this paper, we present a general technique for multistage systems of arbitrary index. Specifically, the algorithm uses the control parameterization method for a multistage system consisting of a sequence of stages, $j = 1, ..., s$, each of which is described by a DAE system of the form (1):

$$f^j \left(t, \dot{x}^j(t), x^j(t), u^j(t), v \right) = 0, \qquad t \in [t_{j-1}, t_j] \tag{2}$$

with $[t_0, t_s] \subset T \subset R$, $x^j(t) \in X^j \subset R^{n_j}$, $u^j(t) \in \bar{U}^j \subset R^{q_j}$, $v \in \bar{V} \subset R^p$, and $f^j : \Omega^j \rightarrow R^{n_j}$, where $\Omega^j = T \times R^{n_j} \times X^j \times \bar{U}^j \times \bar{V}$. The controls and parameters must satisfy the conditions:

$$U_L^j \leq U^j(u^j(t), v) \leq U_U^j, \qquad t \in [t_{j-1}, t_j], \ j = 1, ..., s \tag{3a}$$

and

$$V_L \leq V(v) \leq V_U \tag{3b}$$

where $U^j(.,.)$ and $V(.)$ are appropriate vector functions.

The initial state must satisfy the equations:

$$J^0 \left(\dot{x}^1(t_0), x^1(t_0), u^1(t_0), v \right) = 0 \tag{4a}$$

and there are also junction conditions relating the states in successive stages:

$$J^j \left(t_j, \dot{x}^{j+1}(t_j), x^{j+1}(t_j), u^{j+1}(t_j), \dot{x}^j(t_j), x^j(t_j), u^j(t_j), v \right) = 0 \quad j = 1, ..., s-1 \tag{4b}$$

We assume that equations (2), (4a), (4b) determine unique trajectories $x^j(t)$, $t \in [t_{j-1},\, t_j]$, $j = 1, ..., s$, for given v and $u^j(t)$, $t \in [t_{j-1},\, t_j]$, $j = 1, ..., s$.

We also allow stage end-time constraints of the general form:

$$F_L^j \;\leq\; F^j\left(t_j,\, \dot{x}^j(t_j),\, x^j(t_j),\, u^j(t_j),\, v\right) \;\leq\; F_U^j, \qquad j = 0, ..., s \tag{5a}$$

and path constraints of the form:

$$H_L^j \;\leq\; H^j\left(t,\, \dot{x}^j(t),\, x^j(t),\, u^j(t),\, v\right) \;\leq\; H_U^j, \qquad t \in [t_{j-1},\, t_j], \quad j = 1, ..., s \tag{5b}$$

Each of the path constraints (5b) is converted into an end-point constraint by integrating its violation over the interval and demanding that the integral does not exceed an appropriate tolerance ϵ:

$$\int_{t_{j-1}}^{t_j} \max\left\{0,\, H_L^j - H^j(t),\, H^j(t) - H_U^j\right\} dt \;\leq\; \epsilon \tag{5c}$$

where the max function is applied element-by-element.

We note that we can effectively relax the lower or upper bounds in (3a,b) and (5a,b) by setting the relevant bounds to machine limits, while equality constraints can be imposed by making the corresponding elements of the upper and lower bounds equal.

The initial time, t_0, is taken as fixed, but the stage end-times t_j, $j = 1, ..., s$, may be fixed or free. In the latter case, they may, in fact, be implicitly defined by the junction or stage end-time constraints (4b) or (5a).

The objective function, or performance index, is defined in terms of a function evaluated at the final time t_s:

$$P\left(t_s,\, \dot{x}^s(t_s),\, x^s(t_s),\, u^s(t_s),\, v\right) \tag{6}$$

Control Parameterization

For practical numerical computation, we need a finite parameterization of the controls $u(t)$, and therefore define the class of admissible control functions as follows:

$$u^j(t) \;=\; \theta^j\left(t,\, w^j\right), \qquad t \in [t_{j-1},\, t_j], \quad j = 1, ..., s \tag{7}$$

where the $\theta^j\left(t,\, w^j\right)$ are given basis functions involving a finite set of parameters $w^j \in R^{l_j}$. In fact, we could allow each stage to be subdivided into a fixed number of subintervals, each with its own basis functions, but it is more convenient to treat each such subinterval as a separate "stage"

Thus the solution of (2), (4) is uniquely determined by the set of decision variables

$$\eta \;\equiv\; \{v,\, w^1,\, w^2,\, ...\,,\, w^s,\, t_1,\, t_2,\, ...\,,\, t_s\} \tag{8}$$

yielding, on integration, the values $\dot{x}^j(t)$, $x^j(t)$, $t \in [t_{j-1}, \ t_j]$, $j = 1, ..., s$. The constraint functions (5) and objective function (6) can then be evaluated, thus defining these functions implicitly in terms of the decision variables:

$$\bar{P}(\eta) \ = \ P \ (t_s, \ \dot{x}^s(t_s), \ x^s(t_s), \ u^s(t_s), \ v) \tag{9}$$

and

$$\bar{F}^j(\eta) \ = \ F^j \ (t_j, \ \dot{x}^j(t_j), \ x^j(t_j), \ u^j(t_j), \ v) \qquad j = 0, ..., s \tag{10}$$

assuming that constraints (5c) are included in (5a).

Nonlinear Program

The optimal control problem is thus reduced to the nonlinear program:

$$\min \ \bar{P}(\eta)$$

subject to

$$F_L^j \ \leq \ \bar{F}^j(\eta) \ \leq \ F_U^j, \qquad j = 0, ..., s$$

and

$$U_L^j \ \leq \ U^j(\theta^j(t, \ w^j), \ v) \ \leq \ U_U^j \quad t \in [t_{j-1}, \ t_j], \ j = 1, ..., s, \tag{11a}$$

$$V_L \ \leq \ V(v) \ \leq \ V_U \tag{11b}$$

This is still an infinite dimensional problem since the control constraints (11a) must be satisfied for each $t \in [t_0, \ t_s]$. However, the constraints (3a) are often linear inequalities, and the basis functions can usually be chosen so that verification of the constraints at a finite set of times suffices. Otherwise, we can use the same technique as for the path constraints (5b), converting (3a) to integral constraints. The resulting finite-dimensional nonlinear programme can then be solved by standard methods.

In order to use such a method, we need to be able to integrate the system, and evaluate the objective and constraint functions, together with their derivatives with respect to the decision variables, for any given set of decision variable values. In the next section, we consider properties of the DAE system, and develop the necessary equations for this purpose. We conclude with general comments on our approach.

Properties of the DAE System

In this section we consider a single stage, and hence for simplicity omit the superscripts denoting the stage number. We shall also omit the arguments of f and its partial derivatives, it being understood that they represent the solution at a typical time.

Index Determination

The *index* of the DAE system (2) is defined as the smallest nonnegative integer I such that the system:

$$f\ (t,\ \dot{x}(t),\ x(t),\ u(t),\ v)\ =\ 0 \tag{12a}$$

$$\frac{d^i}{dt^i}\ f\ (t,\ \dot{x}(t),\ x(t),\ u(t),\ v)\ =\ 0, \qquad i = 1, ..., I \tag{12b}$$

defines $\dot{x}(t)$ as a locally unique function of t, $x(t)$, $u(t)$, v. Clearly I can vary as these arguments vary, and strictly (12a,b) defines the *local index*. However, for brevity, we shall refer to this simply as the "index" Of course, the index is well defined only if all the derivatives in (12b) exist. We shall assume not only their existence, but also their continuity with respect to their arguments.

The existence of a finite index is a sufficient condition for the existence of a unique solution trajectory through any point satisfying (12a,b), and this solution can be extended so long as the index remains uniformly bounded.

If $f_{\dot{x}}$ is nonsingular at a point on the solution trajectory, then by the Implicit Function Theorem, (12a) defines $\dot{x}$ as a locally unique function of the remaining arguments, and the index is zero.

Otherwise, we can apply Gaussian elimination with row and column interchanges to (12a) to yield:

$$L_0\ P_0\ f_{\dot{x}}\ Q_0\ =\ \begin{pmatrix} f_{\dot{x}}^1 \\ \cdot \end{pmatrix} \tag{13}$$

where L_0 is nonsingular and lower-triangular, P_0 and Q_0 are permutation matrices, and $f_{\dot{x}}^1$ is of full rank $(r_1 < n)$, with its leading $r_1 \times r_1$ submatrix being unit upper-triangular.

We can then define:

$$\begin{pmatrix} f^1 \\ g^1 \end{pmatrix}\ \equiv\ L_0\ P_0\ f, \qquad \begin{pmatrix} \dot{f}^1 \\ \dot{g}^1 \end{pmatrix}\ \equiv\ L_0\ P_0\ \dot{f} \qquad etc. \tag{14}$$

where it is important to note that $\dot{f}^1$ should *not* be interpreted as df^1/dt since L_0 may itself be a nonlinear function of t, $\dot{x}$, x, u, v.

From the differentiability assumptions, both L_0 and the transformed vectors in (13) and (14) are continuously differentiable. However, since f^1, g^1 represent a partition such that $g_{\dot{x}}^1 = 0$, their dimensions may vary from point to point. Nevertheless, if r_1 is constant over any sub-domain of the arguments, then $g_{\dot{x}}^1$ is zero over this sub-domain, and g^1 is therefore independent of $\dot{x}$. From the continuity assumption, the points at which r_1 changes, if any, form a set of measure zero.

Linearizing (12a,b) about a given point at which r_1 does not change, and premultiplying by $L_0 P_0$, we obtain, with some rearrangement:

$$\begin{pmatrix} f^1 \\ \dot{g}^1 \end{pmatrix}\ +\ \begin{pmatrix} f_{\dot{x}}^1 \\ \dot{g}_{\dot{x}}^1 \end{pmatrix}\ \delta\dot{x}\ \approx\ 0 \tag{15a}$$

$$\begin{pmatrix} \dot{f}^1 \\ \dot{g}^1 \end{pmatrix} + \begin{pmatrix} f^1_{\dot{x}} \\ \dot{g}^1_{\dot{x}} \end{pmatrix} \delta\dot{x} + \begin{pmatrix} f^1_{\dot{x}} \\ \dot{g}^1_{\dot{x}} \end{pmatrix} \delta\ddot{x} \approx 0 \tag{15b}$$

$$\begin{pmatrix} \ddot{f}^1 \\ \ddot{g}^1 \end{pmatrix} + \begin{pmatrix} \dot{f}^1_{\dot{x}} \\ \ddot{g}^1_{\dot{x}} \end{pmatrix} \delta\dot{x} + \begin{pmatrix} \ddot{f}^1_{\dot{x}} \\ \ddot{g}^1_{\dot{x}} \end{pmatrix} \delta\ddot{x} + \begin{pmatrix} f^1_{\dot{x}} \\ \ddot{g}^1_{\dot{x}} \end{pmatrix} \delta\dddot{x} \approx 0 \tag{15c}$$

and so on. It should be noted that, at each stage, the coefficient matrix of the highest derivative term is the same for each pair of equations.

Now we can apply Gaussian elimination to the coefficient matrix of $\delta\dot{x}$ in (15a) to yield:

$$L_1\, P_1 \begin{pmatrix} f^1_{\dot{x}} \\ \dot{g}^1_{\dot{x}} \end{pmatrix} Q_1 = \begin{pmatrix} f^2_{\dot{x}} \\ \cdot \end{pmatrix}, \qquad \text{defining} \quad \begin{pmatrix} f^2 \\ \dot{g}^2 \end{pmatrix} \equiv L_1\, P_1 \begin{pmatrix} f^1 \\ \dot{g}^1 \end{pmatrix} \qquad etc. \tag{16}$$

where the rank of $f^2_{\dot{x}}$ is $r_2 \geq r_1$, and again $\dot{g}^2_{\dot{x}} = 0$ (and, consequently, $\dot{g}^2$ is independent of $\dot{x}$) except possibly on a set of measure zero.

Again, if $f^2_{\dot{x}}$ is nonsingular, we can use (15a) to obtain $\dot{x}$, showing that the index is one. Otherwise, we form:

$$\begin{pmatrix} f^2 \\ \ddot{g}^2 \end{pmatrix} + \begin{pmatrix} f^2_{\dot{x}} \\ \ddot{g}^2_{\dot{x}} \end{pmatrix} \delta\dot{x} \approx 0 \tag{17a}$$

$$\begin{pmatrix} \dot{f}^2 \\ \dddot{g}^2 \end{pmatrix} + \begin{pmatrix} \dot{f}^2_{\dot{x}} \\ \dddot{g}^2_{\dot{x}} \end{pmatrix} \delta\dot{x} + \begin{pmatrix} f^2_{\dot{x}} \\ \dddot{g}^2_{\dot{x}} \end{pmatrix} \delta\ddot{x} \approx 0 \tag{17b}$$

etc.

The recursion can be continued in the same way until, after the Ith stage, we obtain a system defining $\dot{x}$:

$$f^I + f^I_{\dot{x}}\, \delta\dot{x} \approx 0 \tag{18}$$

Of course, the ranks r_1, r_2, etc. may change at certain points, even if the index does not, but no practical difficulty arises from terminating the factorization when there are no remaining pivots above a suitable threshold, since this merely provokes a further differentiation.

The successive differentiations of the equations (cf. (12b)) can easily be carried out using an automatic differentiation package. It is important to note that no further symbolic manipulation is required. We also note that it is not strictly necessary to differentiate all the equations in (12a). For the first phase, only those elements of the function vector that are involved in generating $\dot{g}^1_{\dot{x}}$ need be differentiated. Additional elements may be involved in subsequent phases to generate $\ddot{g}^2_{\dot{x}}$, $\dddot{g}^3_{\dot{x}}$, etc. The additional housekeeping involved in keeping track of which equations must be differentiated at each stage may well be justified in the case of large DAE systems.

Finally, we note that the system (2) may be such that $\dot{x}^j_i(t)$ does not appear for some individual variables $x^j_i(t)$. These "algebraic variables" can notionally be treated as derivatives of a new set of variables, i.e.,

$$\dot{\bar{x}}^j_i(t) \equiv x^j_i(t) \tag{19}$$

In the above procedure, this may save one differentiation. Of course, once the procedure determines the time derivatives $\dot{\bar{x}}_i^j(t)$, we do not really need to carry out the integration to obtain the $\bar{x}_i^j(t)$.

It is important to point out that, in forming (15a,b,c,...), we omitted the equation $g^1 = 0$, which is simply a relation between t, x, u, and v. Similarly, if a second differentiation is necessary, we also have $\dot{g}^2 = 0$, which is a similar relation — and so on for subsequent differentiations. We denote these equations collectively by:

$$G\,(t,\ x,\ u,\ v)\ =\ 0 \tag{20}$$

where $G : T \times X \times \bar{U} \times \bar{V} \to R^\mu$.

DAE System Integration

In this section, we consider the integration of the DAE system over each stage, assuming for the moment that consistent initial conditions are available.

Since we have already developed a procedure for determining $\dot{x}$ in terms of t, x, u, v, we could simply use this in conjunction with any standard ODE package, explicit or implicit, to obtain the required solution. Essentially the solution of the high index problem has been reduced to that of a set of ODEs. However, as we have already noted, this makes no use of the ancillary equations (20). Of course, (18) does include time derivatives of (20), and since the latter are explicitly satisfied by the initial conditions as described in the next section, in theory they should also be satisfied by the solution obtained by integrating (18). However, as discussed by Brenan *et al.* (1989), in practice the accumulation of errors during numerical integration may cause the constraints that are not explicitly taken into account to "drift", often rendering any solution thus obtained physically meaningless.

In fact, the issue of constraint stabilization is common to all index reduction algorithms, and several schemes for it have been proposed in the literature. In all cases, the equations (20) are incorporated explicitly in the system being integrated, together with equations (18) (or their equivalent, depending on the specific index reduction algorithm). This inevitably results in an overdetermined system of equations, and different algorithms differ in the way in which they handle this overdeterminacy. For instance, Gear (1988) introduces additional variables which he uses to adjoin the algebraic equations to the differential equations in the system, and demonstrates that the solution of the resulting index-2 system is such that the new variables take the value of zero.

The index reduction algorithm of Bachmann *et al.* (1990) deals with the overdeterminacy essentially by discarding some of the *differential* equations. In terms of the notation used in this paper, the first stage of their algorithm detects the purely "algebraic" equations $g^1(.) = 0$, differentiates them with respect to time, and then uses the "differential" equations $f^1(.)$ to eliminate any occurrences of the time derivatives $\dot{x}$ introduced by the differentiation. It then proceeds to discard some of the differential equations used in the elimination so as to preserve a well-determined system. This process is continued until an index one system is obtained.

Very recently, Mattsson and Söderlind (1993) have proposed a systematic procedure which is similar in spirit to that of Bachmann *et al.* (1990) in the sense that they retain all the information generated during the index reduction process, while effectively reducing the number of differential equations in the system. The main difference of their algorithm is that, instead of discarding differential equations, they retain them but treat one or more of the $\dot{x}_i$ as independent algebraic variables rather than as time derivatives of variables x_i already occurring in the system.

Mattsson and Söderlind rely on Pantelides' (1988) structural algorithm to identify the complete set of equations that have to be taken into account in the index reduction process. Once this is achieved, their algorithm determines which $\dot{x}_i$ should be treated as independent variables ("dummy derivatives" in their terminology) so as to remove the overdeterminacy and leave an index one system to be integrated.

The index reduction algorithm presented earlier in this paper is based on purely numerical information, and as such, may detect necessary equation differentiations that are missed by structural algorithms. We now proceed to consider the following overdetermined DAE system:

$$f^I \left(t,\ x,\ \dot{x},\ u,\ v\right)\ =\ 0 \tag{21a}$$

$$G\left(t,\ x,\ u,\ v\right)\ =\ 0 \tag{21b}$$

with the corresponding Jacobian matrix:

$$\left[\begin{array}{cc} f^I_{\dot{x}} & f^I_x \\ \cdot & G_x \end{array}\right] \tag{22}$$

Recalling that $f^I_{\dot{x}}$ is unit upper-triangular, we can continue the Gaussian elimination process on the rows of G_x to produce the matrix

$$\left[\begin{array}{ccc} f^I_{\dot{x}} & f^I_y & f^I_z \\ \cdot & \bar{G}_y & \bar{G}_z \\ \cdot & \cdot & \cdot \end{array}\right] \tag{22'}$$

where $[y,\ z]$ is a partition of x such that $y \in R^{\bar{\mu}}$, $z \in R^{n-\bar{\mu}}$, $\bar{\mu} \leq \mu$, and $\bar{G}_y$ is a $\bar{\mu} \times \bar{\mu}$ unit upper triangular matrix.

Thus, given values of z, (21a,b) can be solved uniquely for $\dot{y}$, $\dot{z}$, and y, which clearly defines an index-1 system with z as its differential variables, and y, $\dot{y}$ as its algebraic variables. We have therefore effectively replaced $\dot{y}$ by "dummy derivatives" similar to those introduced by Mattsson and Söderlind's (1993) algorithm.

The integration of the index 1 system using standard techniques is straightforward. Thus, the values of the variables at the kth integration step are determined by solving the equations:

$$f^I\left(t_k,\ x_k,\ \dot{x}_k,\ u(t_k),\ v\right)\ =\ 0 \tag{23a}$$

$$\bar{G}\ (t_k,\ x_k,\ u(t_k),\ v)\ =\ 0 \tag{23b}$$

$$z_k\ -\ \gamma\,\Delta t_k\,\dot{z}_k\ +\ \chi_k\ =\ 0 \tag{23c}$$

Here x_k, $\dot{x}_k$, z_k, $\dot{z}_k$ are approximations to $x(t_k)$, $\dot{x}(t_k)$, $z(t_k)$ and $\dot{z}(t_k)$ respectively. Equation (23c) represents a general linear multistep integration formula where γ is a scalar, and χ_k is an expression which depends only on past values of $z_{k'}$, $\dot{z}_{k'}$, $k' < k$.

DAE System Initialization

The initial conditions $\dot{x}^j(t_{j-1})$, $x^j(t_{j-1})$, for each stage $j = 1, ..., s$ must be determined uniquely by equations (4a) or (4b), together with (12a,b) for $f^j(t_{j-1})$, for given t_{j-1}, $u^j(t_{j-1})$, v. Equivalently, they can be determined from (4a), or (4b), together with (21a,b).

From $(22')$, we see that a unique solution requires the nonsingularity of the Jacobian matrix:

$$\begin{bmatrix} f_{\dot{x}}^{I,j} & f_{\underline{y}}^{I,j} & f_{\underline{z}}^{I,j} \\ \cdot & \bar{G}_y & \bar{G}_z \\ J_{\dot{x}^j}^{j-1} & J_{y^j}^{j-1} & J_{z^j}^{j-1} \end{bmatrix} \qquad j = 1, ..., s \tag{24}$$

showing that $(n^j - \bar{\mu}^j)$ independent junction conditions are required for $j = 0, ..., (s-1)$.

A Numerical Illustration of the Index Reduction Algorithm

We illustrate our index reduction and constraint stabilisation procedure by applying it to the DAE system:

$$-\dot{x}_1\ +\ x_1\ +\ x_2\ -\ x_3\ =\ 0 \tag{25a}$$

$$-\dot{x}_2\ +\ 2x_1\ -\ x_2\ +\ x_3\ =\ 0 \tag{25b}$$

$$x_1\ +\ x_2\ +\ u(t)\ =\ 0 \tag{25c}$$

where $u(t)$ is twice differentiable.

First we note that x_3 is an algebraic variable, and therefore in order to avoid an unnecessary differentiation, we notionally replace it by a time derivative $\dot{\bar{x}}_3$, leading to the modified system:

$$-\dot{x}_1\ +\ x_1\ +\ x_2\ -\ \dot{\bar{x}}_3\ =\ 0 \tag{25a'}$$

$$-\dot{x}_2\ +\ 2x_1\ -\ x_2\ +\ \dot{\bar{x}}_3\ =\ 0 \tag{25b'}$$

$$x_1\ +\ x_2\ +\ u(t)\ =\ 0 \tag{25c'}$$

The matrix $f_{\dot{x}}$ for the above system is:

$$\begin{bmatrix} -1 & 0 & -1 \\ 0 & -1 & 1 \\ 0 & 0 & 0 \end{bmatrix} \tag{26}$$

and the corresponding lower triangular matrix L_0 is simply

$$L_0 = \begin{bmatrix} -1 & & \\ 0 & -1 & \\ 0 & 0 & 1 \end{bmatrix}, \quad P_0 = I \tag{27}$$

which identifies $(25a', b')$ with f^1, and $(25c')$ with g^1.

Since $f_{\dot{x}}$ is singular, we differentiate $(25')$ and consider the matrix

$$\begin{pmatrix} f_{\dot{x}}^1 \\ \dot{g}_{\dot{x}}^1 \end{pmatrix} = \begin{bmatrix} -1 & 0 & -1 \\ 0 & -1 & 1 \\ 1 & 1 & 0 \end{bmatrix} \tag{28}$$

Once again, (28) is singular, as can be seen by premultiplying it by $L_1 P_1$ where

$$L_1 = \begin{bmatrix} -1 & & \\ 0 & -1 & \\ 1 & 1 & 1 \end{bmatrix} \quad P_1 = I \tag{29}$$

Premultiplying $\begin{pmatrix} f^1 \\ \dot{g}^1 \end{pmatrix}$ by $L_1 P_1$, we obtain $(25a', b')$ and

$$\dot{g}^2 \equiv 3x_1 + \dot{u}(t) = 0 \tag{30}$$

We therefore apply a second differentiation of the equations, and consider the matrix

$$\begin{pmatrix} f_{\dot{x}}^2 \\ \ddot{g}_{\dot{x}}^2 \end{pmatrix} = \begin{bmatrix} -1 & 0 & -1 \\ 0 & -1 & 1 \\ 3 & 0 & 0 \end{bmatrix} \tag{31}$$

This is nonsingular, and therefore the procedure terminates with the main equations:

$$-\dot{x}_1 + x_1 + x_2 - \dot{\bar{x}}_3 = 0 \tag{32a}$$

$$-\dot{x}_2 + 2x_1 - x_2 + \dot{\bar{x}}_3 = 0 \tag{32b}$$

$$3\dot{x}_1 + \ddot{u}(t) = 0 \tag{32c}$$

forming an ODE system in x_1, x_2, and $\bar{x}_3$, and the ancillary equations

$$x_1 + x_2 + u(t) = 0 \tag{32d}$$

$$3x_1 + \dot{u}(t) = 0 \tag{32e}$$

i.e., with (32a,b,c) identified as (21a), and (32d,e) as (21b) in the general statement of the algorithm. In particular, the matrix G_x is given by:

$$G_x = \begin{bmatrix} 1 & 1 & 0 \\ 3 & 0 & 0 \end{bmatrix} \tag{33}$$

and it is obvious that the partitioning of x must be $y \equiv (x_1, x_2)^T$ and $z \equiv \bar{x}_3$. Hence, we have an index 1 system formed from (32a-e) by treating $\dot{x}_1$ and $\dot{x}_2$ as independent algebraic variables rather than as time derivatives of x_1 and x_2. The only differential variable is $\bar{x}_3$.[1]

Overall, two differentiations were needed to convert $(25')$ to an ODE system in x_1, x_2, $\bar{x}_3$. A third one would be required to obtain an expression for $\dot{x}_3$. Hence the index of the original system (25) is three.

It is interesting to note that structural index reduction algorithms would detect the need for the first differentiation of equation (25c), but would miss the second one. This is due to the fact that (30), formed from the sum of the differentials of (25a) and (25b), is *not* a function of x_3 as would have been deduced erroneously from structural information only. Consequently, the index-3 system (25) would have been classified incorrectly as an index-2 one. This would have given the wrong impression that one could specify one arbitrary initial condition of the form (4a), whereas in reality no such freedom exists.

Gradient Evaluation

It remains only to establish a procedure for determining the derivatives $\bar{P}_\eta(\eta)$, $\bar{F}_\eta^j(\eta)$. Differentiating (9) with respect to η, we obtain:

$$\bar{P}_\eta(\eta) = P_{\dot{x}}(t_s)\dot{x}_\eta^s(t_s) + P_x(t_s)x_\eta^s(t_s) + P_u(t_s)u_\eta^s(t_s) + P_\eta(t_s) \tag{34}$$

Similarly we obtain a DAE system defining $\dot{x}_\eta$, x_η by differentiating (2) with respect to η, :

$$f_{\dot{x}}^j \dot{x}_\eta^j + f_x^j x_\eta^j + f_u^j u_\eta^j + f_\eta^j = 0, \qquad j = 1, ..., s \tag{35}$$

and initial and junction conditions by differentiating (4a) and (4b) respectively:

$$J_{\dot{x}^1}^0(t_0)\dot{x}_\eta^1(t_0) + J_{x^1}^0(t_0)x_\eta^1(t_0) + J_{u^1}^0(t_0)u_\eta^1(t_0) + J_\eta^0(t_0) = 0, \tag{36a}$$

$$J_{\dot{x}^{j+1}}^j(t_j)\, \dot{x}_\eta^{j+1}(t_j) + J_{x^{j+1}}^j(t_j)\, x_\eta^{j+1}(t_j) + J_{u^{j+1}}^j(t_j)\, u_\eta^{j+1}(t_j) +$$
$$J_{\dot{x}^j}^j(t_j)\, \dot{x}_\eta^j(t_j) + J_{x^j}^j(t_j)\, x_\eta^j(t_j) + J_{u^j}^j(t_j)\, u_\eta^j(t_j) +$$
$$J_\eta^j(t_j) = 0 \quad j = 1, ..., s-1 \tag{36b}$$

---

[1] Of course, in this case, we are not really interested in the value of $\bar{x}_3$, only in that of $\dot{\bar{x}}_3$ (*i.e.*, the original x_3). Therefore, it is not actually necessary to append the integration formula for $\bar{x}_3$ to (32a-e) — at each time step, one need only solve (32a-e) as an algebraic system for x_1, x_2, $\dot{x}_1$, $\dot{x}_2$, $\dot{\bar{x}}_3$.

Obviously these sensitivity equations have the same structure as the linearized equations for the corresponding functions of the original system, and can therefore be initialized and integrated in a manner analogous to that described earlier in the paper. This similarity can readily be exploited to avoid unnecessary duplication of most of the necessary linear algebra operations.

We can now give a summary of the algorithm to determine $\bar{P}(\eta)$, $\bar{P}_\eta(\eta)$, $\bar{F}^j(\eta)$, and $\bar{F}^j_\eta(\eta)$, $j = 0, ..., s$ for a given η:

Algorithm

1.a Apply the differentiation/factorization procedure to system (2) for $j = 1$ to obtain (18) and ancillary equations (20).

b Determine consistent initial conditions $\dot{x}^1(t_0)$, $x^1(t_0)$ from (4a), (21a,b) and $\dot{x}^1_\eta(t_0)$, $x^1_\eta(t_0)$ from the corresponding equations of the sensitivity system.

c Integrate (2) for $j = 1$ from t_0 to t_1 using (23) and its analogue for $\dot{x}^1_\eta$, x^1_η to determine $x^1(t_1)$, $x^1_\eta(t_1)$, and, if required, $\dot{x}^1(t_1)$, $\dot{x}^1_\eta(t_1)$.

2 For $j = 2, ..., s$ do:

   a Determine consistent conditions $\dot{x}^j(t_{j-1})$, $x^j(t_{j-1})$ at t_{j-1} from (4b), (21a,b), and $\dot{x}^j_\eta(t_{j-1})$, $x^j_\eta(t_{j-1})$ from the corresponding equations of the sensitivity system.

   b Integrate system (2) from t_{j-1} to t_j using (23) and its analogue for $\dot{x}^j_\eta$, x^j_η, thus determining $x^j(t_j)$, $x^j_\eta(t_j)$, and, if required, $\dot{x}^j(t_j)$, $\dot{x}^j_\eta(t_j)$.

3 Compute integrals as required from (5c), then $\bar{P}(\eta)$ from (9), $\bar{P}_\eta(\eta)$ from (34), and similarly $\bar{F}^j(\eta)$, and $\bar{F}^j_\eta(\eta)$, $j = 0, ..., s$.

This algorithm can be used as a subroutine for a nonlinear programming package for solving (11).

General Comments

The control parameterization approach for the solution of optimal control problems has a long pedigree (see for example, Horwitz and Sarachik, 1968, Pollard and Sargent, 1970, Sargent and Sullivan, 1978, Goh and Teo, 1988), and this is not the place for a comprehensive review. The main innovation in the present paper is a detailed analysis of what is required for a general-purpose package to deal with multistage, high-index DAE systems. In particular, a general method for the consistent initialization and stable integration of DAE systems of arbitrary index has been presented. Automatic differentiation is the only type of non-numerical operation required by the method.

There have been two earlier papers suggesting a purely numerical procedure for solving high-index DAE systems. Campbell (1988) assumes that an initial value

$x(t_0)$ is given, and that the index of the system is known and constant over the domain of interest. Thus he can generate system (12a,b) by successive differentiations. Then, for given x_k, he obtains the minimum-norm least-squares solution of (12a,b) for all other variables including $\dot{x}_k$. Therefore, this procedure can be used in conjunction with any ODE integration package. The use of a minimum-norm least-squares solution allows the user to specify any equations believed to be relevant to the problem, and still provides a unique "solution". However, there is no mechanism for determining whether the system is under-determined, or over-determined and inconsistent. Of course, in either of these cases, the user will not know exactly *what* problem has been solved!

The approach of Chung and Westerberg (1990) is similar to ours as far as the index reduction procedure is concerned. Their algorithm is stated informally, and requires the detection of the rank of subsets of equations. Our algorithm can be regarded as a systematization of their procedure, in which a threshold limit on pivots in the Gaussian elimination makes the required detection. At the same time, our systematic successive elimination avoids the need for arbitrary assignment of values for undetermined variables and solution for all the remaining variables. Like Campbell, they too advocate direct integration of the low index system generated by their procedure without explicit enforcement of the ancillary constraints (20). In contrast, constraint stabilization is a key feature of the approach described in this paper, and we show how this can be achieved with little additional cost.

There has been a long-standing controversy over the relative merits of generating derivatives using finite differences, sensitivity equations, or the adjoint system. Finite differences involve repeated integrations of the system equations for small perturbations of each parameter in turn, and this is clearly inefficient unless advantage is taken of the fact that the perturbations are small and hence the repeated integrations can solve the linearized system using the same sequence of step-lengths and Jacobian matrices as the base trajectory. But this system is identical to that used for the sensitivity equations, which are exact and avoid the need for estimating an appropriate perturbation size.

The choice between forward integration of sensitivity equations and backward integration of adjoints is partly a matter of coding convenience and partly a balance between the numbers of parameters and the number of end-stage constraints (see, for example, Rosen and Luus, 1991).

Gritsis *at al.* (1992), previously gave an analysis for semi-explicit index-two systems using adjoints. In this special case, the matrix reductions used for the system integration can also be used in the adjoint system. However, this is no longer possible in the general case, and a similar sequence of eliminations would have to be applied to the *transpose* of the system matrix. In contrast, the sensitivity equations use exactly the same matrices as the original system, and there is the added convenience of simultaneous integration of all equations in the forward direction. Hence, for general higher index systems, there is an obvious advantage for the sensitivity approach.

An apparently quite different approach to optimal control is the solution by collocation, proposed independently by Tsang *et al.* (1975), Biegler and his coworkers (Cuthrell and Biegler, 1987, Vasantharajan and Biegler, 1990), and Renfro *et al.* (1987). This involves the parameterization of both state and control trajectories using basis functions. However, if the block-diagonal structure of the equations is exploited, the two approaches are, in fact, quite similar, the collocation approach being essentially equivalent to the use of a fixed-step, fully implicit Runge-Kutta integration method. In the context of the integration of DAE systems, such methods have not so far proved competitive in comparison to the BDF method. In any case, it seems obviously preferable to use the flexibility of choosing the optimal order and step-length of integration at each step during the integration to maintain a specified accuracy, rather than making such adjustments between successive integrations. To our knowledge, the collocation approach has not yet been applied to higher index problems.

References

[1] Bachmann, R., L. Brüll, T. Mrziglod, and U. Pallaske, "On methods for reducing the index of differential-algebraic equations", *Comput. chem. Engng.*, **14**, pp. 1271–1273 (1990).

[2] Brenan, K.E., S.L. Campbell, and L.R. Petzold, "Numerical solution of initial-value problems in differential-algebraic equations", North Holland, New York (1989).

[3] Campbell, S.L., "A computational method for general higher index nonlinear singular systems of differential equations", *IMACS Transactions Sci. Computing*, pp. 178–180 (1988).

[4] Chung Y. and A.W. Westerberg, "A proposed numerical algorithm for solving nonlinear index problems", *Ind. Eng. Chem. Res.*, **29**, pp. 1234–1239 (1990).

[5] Cuthrell, J.E. and L.T. Biegler, "On the optimization of differential-algebraic process systems", *AIChE Journal*, **33**, pp. 1257–1270 (1987).

[6] Gear, C.W., "Differential-algebraic equation index transformations", *SIAM J. Sci. Stat. Comput.*, **9**, pp. 39–47 (1988).

[7] Gear, C.W. and L.R. Petzold, "ODE methods for the solution of differential/algebraic systems", *SIAM J. Numer. Anal.*, **21**, pp. 716–728 (1984).

[8] Goh, C.J. and K.L. Teo, "Control parameterization: a unified approach to optimal control problems with general constraints", *Automatica*, **24**, pp. 3–18 (1988).

[9] Gritsis, D.M., C.C. Pantelides, and R.W.H. Sargent, "Optimal control of systems described by index-two differential-algebraic equations", submitted to *SIAM J. Sci. Stat. Comput.* (1992).

[10] Horwitz, L.B., and P.E. Sarachik, "A computational technique for calculating the optimal control signal for a specific class of problems", Record of the *2nd Asilomar Conference on Circuits and Systems*, pp. 537–540 (1968).

[11] Mattsson, S.E. and G. Söderlind, "Index reduction in differential-algebraic equations using dummy derivatives", *SIAM J. Sci. Stat. Comput.*, bf 14 pp. 677–692 (1993).

[12] Morison, K.R., and R.W.H. Sargent, "Optimization of multistage processes described by differential-algebraic equations", *Lecture Notes in Mathematics*, **1230**, pp. 86–102, Springer-Verlag, Berlin (1986).

[13] Pantelides, C.C, "The consistent initialization of differential-algebraic equations", *SIAM J. Sci. Stat. Comput.*, **7**, pp. 720–733 (1988).

[14] Pollard, G.P. and R.W.H. Sargent, "Off-line computation of optimal controls for a plate distillation column", *Automatica*, **6**, pp. 59–76 (1970).

[15] Renfro, J.C., A.M. Morshedi, and O.A. Absjornsen, "Simultaneous optimization and solution of systems described by differential-algebraic equations", *Comput. chem. Engng.*, **11**, pp. 503–517 (1987).

[16] Rosen, O. and R. Luus, "Evaluation of gradients for piecewise constant optimal control" *Comput. chem. Engng.*, **15**, pp. 273–281 (1991).

[17] Sargent, R.W.H. and G.R. Sullivan, "The development of an efficient optimal control package", Proc. 8th IFIP Conference on Optimization Techniques, Würzburg, 1977. Springer-Verlag, Berlin (1978).

[18] Tsang, T.H., D.M. Himmelblau, and T.F. Edgar, "Optimal control via collocation and nonlinear programming", *Int. J. Control.*, **21**, pp. 763–768 (1975).

[19] Vasantharajan, S. and L.T. Biegler, "Simultaneous strategies for optimization of differential-algebraic systems with enforcement of error criteria", *Comput. chem. Engng.*, **14**, pp. 1083–1100 (1990).

International Series of Numerical Mathematics, Vol. 115, © 1994 Birkhäuser Verlag Basel

A New Class of a High Order Interior Point Method for the Solution of Convex Semiinfinite Optimization Problems

György Sonnevend[*]

Introduction

We shall deal with the numerical solution of optimization problems of the following type

$$\min\{f_0(y) \mid f_i(x_i, y) \geq 0, \quad x_i \in [a_i, b_i], \quad i = 1, \ldots, m, \quad y \in R^n\} \tag{1.1}$$

$$Ly = d, \quad L \in R^{k \times n} \tag{1.2}$$

where

1. L is a constant matrix of rank $k < n$ and
2. the semi-infinite constraint functions $f_1, \ldots, f_m$ are assumed to be *concave* in y (at least on the domain where they are positive, see the function $(\det(y))^{1/n}$ on $y \geq 0$, i.e. on the set of positive semidefinite matrices). Note that "finite" constraints $f_j(x) \geq 0$ are formally incorporated by just selecting $a_j = b_j$.
3. The functions f_i, $i = 0, \ldots, m$ are assumed to be algebraically simple and rather smooth, e.g. analytic on their domains of definition, both in x and y. This condition may seem rather stringent at first, it is however of great importance. Analyticity is, of course, a very broad property. We shall show that for important classes of constraints the arising analytic functions are of special type, e. g. like the Stieltjes functions, such that they can be approximated very effectively by suitable, low complexity algorithms.

The next examples show the simple but basic technique to transform many types of — at the first sight — "nonsmooth" constraints into a number of "elementary" analytic constraints.

Examples A nonsmooth constraint of the type, say

$$\max_i g_i(y) \leq 1, \quad g_i(y) := \sum_{1 \leq j \leq k} \mid f_{ij}(y) \mid, \quad i = 1, \ldots, n \tag{1.3}$$

will be replaced by a set of smooth constraints

$$\sum_j \eta_{ij} < 1, \quad -\eta_{ij} \leq f_{ij}(y) \leq \eta_j, \quad 1 \leq j \leq k, \quad i = 1, \ldots, n$$

[*]Eötvös University, Institute of Mathematics, Department of Numerical Analysis, Budapest, Hungary

i. e. we replace y by $(\eta_1, \ldots, \eta_k, y)$.

To give another characteristic example, note that when using smooth homotopies (to reach the optimum) a norm constraint $\|A(\lambda, x)\| \leq \beta$ for all values of the homotopy and semi-infinite parameter $\lambda_0 \leq \lambda$ and x, will be accounted by the single concave constraint

$$\det(\beta^2 I - A(\lambda, x)A^*(\lambda, x) > 0 \quad \text{for all} \quad \lambda \geq \lambda_0 \quad \text{and} \quad x, \tag{1.4}$$

whenever $\|A(\lambda_0, x)\| \leq \beta$ holds for all x.

In fact the most common convex optimization problems — arising e. g. in engineering control systems design and signal processing — can be reformulated in the above form, using for $f_i(x, y)$ just quadratic functions in y and (linear transforms of) the function $\det(y)$, $y \geq 0$ to incorporate norm and singular value constraints.

In passing we note that by solving (1.1) we also solve the "dual" problem, e.g. for constraint functions linear in y

$$f_0(x, y) = c^T y \quad f_i(x, y) = b_i(x) - a_i(x)y, \quad i = 1, \ldots, m,$$

the latter are the so called *moment problems* — also studied extensively in connection to control and design problems —

$$c = -\sum_i \int_{a_i}^{b_i} a_i(x)dx \tag{1.5}$$

$$\max_{\mu \geq 0}(-\sum_i \int b_i(x)\mu_i(x)dx). \tag{1.6}$$

Here $c_1, \ldots, c_n$ are interpreted as fixed "moments" and μ is the unknown mass distribution.

Also there is a large class of problems which, at the first sight, are not convex, but using special parametrizations, like the next example, turn out to be convex (or can be fairly well approximated by convex ones). Of course, before to do this, one has first to understand the structure of the problem and find the right parametrizations, see [11], [12].

In control system design the following model problem is of great importance: given interpolation data

$$H(z_i) = c_i, \quad i = 1, \ldots, k, \quad H(z) = \frac{p(z)}{q(z)}, \quad \deg p \leq \deg q = n \tag{1.7}$$

for a transfer function, find whether the otherwise free coefficients of p and q can be chosen such that

$$|H(z) - f(z)| \leq \varepsilon \tag{1.8}$$

for a given function f, and a fixed $\varepsilon > 0$, where I is a system of intervals or domains on the complex line (unit circle) or plane. Noting that (1.7–1.8) is, say in the case of a line segment, equivalent to

$$q(z)(f(z) - \varepsilon) \leq p(z) \leq (f(z) + \varepsilon)q(z) \text{ and } q(z) > 0 \ \forall z \in I,$$

which is a set of linear constraints, like (1.5) above. Even if the system of inequalities is not convex in $(\varepsilon, p, q) = (\varepsilon, y)$, the homotopy line $p(\varepsilon), q(\varepsilon)$ can be traced in the same way, as explained below; the general reason behind this last example is that if each level set of the homotopy (or cost) parameter ε =fixed is convex (a similar property is often called quasiconvexity), we can apply central paths.

The Main Features of the Present Method

To build a general algorithm for convex semi-infinite optimization is, of course a task with manifold difficulties, here we present a number of points which explain the basic new features of our approach, but we tried to make the paper accessible for readers having no previous experiences with interior point methods.

One of the main advantage offered by the methods proposed below is that they can exploit (i.e. make use of) the smoothness, (analyticity) of the elementary constraints. This is realized in our implementation in two ways. First, by the use of high order quadrature formulae which is important for effective solutions since the "traditional approaches" using a discretization of the semi-infinite domain G by a grid reducing the semi-infinite problem to a standard "finite" one (a survey of these methods is given in [5]) always regarded, when solving the latter, the grid values as "independent", i.e. arbitrary, and needed fine grids to get an accurate approximate solution, from which adaptive methods — to select the new discretization nodes looking to the local extrema — were started. These steps are in some sense analogous of the "pivot selection" procedures in linear programming, the potential index set being a continuum. This local nonconcave maximization, especially in two- or more-dimensional cases, is a rather nontrivial time consuming step better to be avoided, see below. The difficulties in finding these local extrema are "solved" only in the case of special, Tschebyshev-Markov type systems.

An equally important advantage of the methods proposed below (partly for the first time) is that they allow a drastic reduction of the dimension: every step of the algorithms is fulfilled in R^n, i.e. we need no increasing of the dimension for more accurate solution, but we have to compute the arising set of $O(n^2)$ one-dimensional integrals more and more accurately, where the integrands are algebraically simple functions of the data $f_0, \ldots, f_m$ and the current iterate $x = x_k$, when we wish to approximate the optimum more accurately, i.e. for $k \to \infty$.

The effectivity of interior point methods based on first order extrapolations of the central path in comparison with simplex type methods (or more generally with methods using active set strategies) for the solution of "finite", convex programs have already been enlightened in earlier works and partially demonstrated in successful implementations for large (i.e. high-dimensional) linear programs. A second basic new feature in our approach consist in exploiting the high degree of smoothness of the elementary constraints (data functions). More precisely, this possibility is offered in the methods presented in Section 3, which are based on following the central path using high-order extrapolations (up to 20) of this path via central paths associated to lower dimensional problems arising by restricting the original problem to a subspace. We describe an algorithmic realization and present experiences of implementation of this new perspective. Other important features of interior point methods are:

1. *invariance under scaling* of the constraints, this is just the consequence of using the logarithmic "penalty" function $\log(k \cdot f) = \text{const} + \log(f)$

2. *invariance under affine transformations*

$$y = T\hat{y} + w, \quad w \in R^n, \quad T \in R^{n \times n}, \quad \det(T) \neq 0.$$

In the semi-infinite case a set of constraint functions $\{f(x, \cdot), \ x \in [a, b]\}$

$$f(x, y) \geq 0, \quad x \in [a, b] \tag{2.1}$$

correspond to a set

$$P_f = \{y | f(x, y) \geq 0, \quad y \in [a, b]\} \tag{2.2}$$

It is desirable to ascertain further, that the main object of the interior point methods, the analytic center $c(P_f)$ defined as

$$\arg \max P(y) \quad \text{where} \quad P(y) = \exp\left(\int_a^b \log f(x, y) dx\right) \tag{2.3}$$

be an affine invariant point in P_f. This means that the measure dx on $[a, b]$ and the system of functions describing the set P_f should be selected carefully, on this point see [11], [9]. In control system applications, where [a,b] is a segment of the unit circle or imaginary axis (for transfer function constraints), or a time interval (for state space constraints) the choice of the usual Lebesgue measure is natural. In some uniform approximation problems we found that Tschebyshev weights $dx \to dx/((b - x)(x - a))^{1/2}$ often give significantly better results. The choice of optimal weights is studied in [15].

It should be kept in mind, that in interior point methods not the set P_f itself, but its analytic description (the "machine readable data ") is important. Thus the problem with affine invariance is that the method of describing $P = P_f$ by the given system of functions may, or may not be affine invariantly associated to P.

Note that $P(y)$ is a concave (barrier) function of y, if $\int_a^b dx = 1$, (this condition, a normalization will be used for all indices $i = 1, \ldots, m$, here and in the implementation) which is positive on (the interior of) P and vanishes on its boundary, see [11] for a proof. In fact our method of dealing with semi-infinite constraint is to "replace" the infinitely many constraints by one single constraint, since

$$P_f = \{y | P(y) \geq 0\} \tag{2.4}$$

Thus our method is "just" an interior point method for convex programming, which uses special quadrature (approximation) algorithms for evaluating the gradients and Hessian matrices of the "integrated" semi-infinite constraint functions, the integration being realized with the help of the logarithmic potentials, whose values are harmonic averages of the individual values for each

$$P = P_i, \quad i = 1, \ldots, m;$$

Besides the above invariance properties the use of centers (logarithmic penalty functions) is motivated by their appearance in the analysis of the classical polynomial (Nevanlinna Pick type) moment problems. These are easily computable solutions with a rich structure. The latter problem is at the heart of the theory of H^∞ feedback optimization, see [14], [15].

The *central d-path*, where d is a vector in R^n, of the problem (1.1) is defined as the solution (curve) of the problem

$$\min_y (f_0(x) - r \sum_i \int_{a_i}^{b_i} \log f_i(x, y) dx + d^T y), \quad r > 0 \tag{2.5}$$

It solves the implicit equation

$$E(r, x) = \frac{\partial f_0}{\partial x} - r \left(\sum_i \int \frac{\partial f_i(x, y)}{f_i(x, y)} dy + d \right) = 0 \tag{2.6}$$

its uniqueness (and analiticity) following from the concavity (and analiticity) of $f_i(\cdot)$. The function $y'(r)$ resp. $\bar{y}(t) = y(\frac{1}{t})$ satisfies the differential equation

$$y'(r) = H(y)\frac{c}{r^2}, \quad \bar{y}' = -H(\bar{y})c, \tag{2.7}$$

where

$$H(y)_{jk} = \sum_i \int_{a_i}^{b_i} \left(\frac{1}{f_i^2} \frac{\partial f_i(x, y)}{\partial y_j} \frac{\partial f_i(x, y)}{\partial y_k} - \frac{1}{f_i} \frac{\partial^2 f_i(x, y)}{\partial y_j \partial y_k} \right) dx. \tag{2.8}$$

We see that in the parameter t this is an autonomous differential equation. When $d = 0$ we obtain what is usually termed as the central path (path of "analytic" centers).

There are two main classes of methods for moving in the interior of the feasible set towards the optimal set by using the above equations. In the first we follow the central path rather closely by a sequence of *extrapolating* and *recentering* steps. A basic feature of the central path is that Newton's method used in the recentering steps (to reach $x(r_k)$ from x_k) has a relatively large domain of quadratic convergence (at least for the basic elementary constraints, like (1.3)–(1.4)). This domain can be constructively described and its relative size is in fact independent of their individual numerical constants, see below and [6], [12].

For a structural analysis it is convenient to rewrite the equation of the central path in a primal-dual form and to replace the (convex) cost function by a linear one at the cost of having to impose an additional (convex) constraint and a new variable $\lambda = y_{n+1}$, and $c = (0, \ldots, 0, 1)$

$$\lambda^* = \min\{ \ c^T(y, \lambda) \ | \ P_i(y) \geq 0, \quad i = 1, \ldots, m, \quad \lambda \geq f_0(y), \quad y \in R^n\} \quad (2.9)$$

this means that as an (internal) canonical form of the problem (1.1) we assume $f_0(y) = c^T y$, for an arbitrarily given vector c.

Note that one could allow a "semi-infinite" cost function e.g. a max-type one:

$$f_0(y) := \max_x f_0(x, y), \quad x \in [a_0, b_0]$$

by adding a further semi-infinite constraint

$$\lambda \geq f_0(x, y), \quad x \in [a_0, b_0] \tag{2.9'}$$

This is in fact the case in the uniform approximation problems considered here.

The dual variables are denoted by $\mu_i(x)$, $i = 1, \ldots, m$, $x \in [a_i, b_i]$ and are generated along the central path by $\mu_i(x) = r f_i^{-1}(x, y(r))$, so that they satisfy

$$\mu_i f_i(x, y) = r \quad \forall x \in [a_i, b_i], \quad i = 1, \ldots, m \tag{2.10}$$

$$c - \sum \int_{a_i}^{b_i} \mu_i \frac{\partial f_i(x, y)}{\partial y} dx = 0 \tag{2.11}$$

An immediate consequence of these equations is the following rather simple estimation of the distance to the optimum (in the cost function value): if $y = y(r_k)$ is a point on the central curve (generated in the algorithm at step k), then

$$c^T y(r_k) - \lambda^* = r_k \sum_i \int_{a_i}^{b_i} dx = r_k \sum_i (b_i - a_i). \tag{2.12}$$

This shows at least one reason to follow the central path. It turns out however, that following the central path rather closely is not necessary, at least when one uses a low order extrapolation method. Therefore we present another basic estimation method for the distance to the optimum for an arbitrary feasible point, y. For any fixed interior point y, a local "dual" linear programming problem arises

$$F(y) := \max(-\sum_i \int_{a_i}^{b_i} \mu_i(x) f_i(x, y) dx), \tag{2.13}$$

$$c - \sum_i \int_{a_i}^{b_i} \mu_i(x) \frac{\partial f_i(x, y)}{\partial y} dx = 0, \quad \mu_i(x) \geq 0, \quad i = 1, \ldots, m \tag{2.14}$$

We remind the saddle point property of the Lagrange function

$$L(y, x) := c^T y - \sum_i \int_{a_i}^{b_i} \mu_i f_i(x, y) dx, \qquad (2.15)$$

$$\lambda^* = \max_{\mu \geq 0} \min_{y \in P} L(y, \mu) = \min_{y \in P} \max_{\mu \geq 0} L(y, x), \qquad (2.16)$$

from which it follows, that for arbitrary fixed y and a just feasible, i.e. not necessarily optimal, solution μ of the above linear programming problem

$$c^T y \geq \lambda^* \geq c^T y - F(y) \geq c^T y - \sum \int_{a_i}^{b_i} \mu_i f_i(x, y) dx. \qquad (2.17)$$

Of course by computing the maximum in μ, $F(y)$ we get a much tighter bound.

Since the dual variables are functions (measures) in x, i.e. "high-dimensional", we do not want to generate them. However the adaptive integration rules provide us with nodes $[x_1, \ldots, x_N]$ and we can generate atomic approximate solutions concentrated at $(x_1, \ldots, x_N)$ of the local dual problem (at $y = y_k$: the current iterate).

In passing we note that the local dual problem arises by linearization of the constraints at $y = y_k$, and by the convexity of the constraints its primal feasible set is larger than the set P_f, (it might not be bounded even for bounded P). It is a further topic of research to find out whether it is advisable to use this local finite primal-dual problem for several iterations, i.e. to solve it by the same algorithm and using a local relaxation of constraints, similar to (4.2), instead of maintaining feasibility for the semi-infinite constraints (our experience about this is rather negative).

The second method of moving towards the optimum generates points $y = y_k$ which are in general not near to the central path, is provided by the following method (which is often named as "affine scaling") At an arbitrary point y we use the same direction $v(y)$ of the d-central curve going through that point. We do not need to identify d, since it does not influence the current direction at y. We determine on the half line issued from y (in direction $v(y)$) that point, $y_k + v(y_k) * s_k$, $s_k \in R_+$, at which this line hits the boundary of the feasible set. Note that while the computation of $s_k^i(x)$ for fixed x, from $0 = f_i(x, y_k + v(k)s)$ is in most cases easy, to find the global minimum of $s_k^i(x)$ would be more difficult. In our implementation we replace the global minimum with the minimum on the currently generated grid.

The next point is y_{k+1} determined as

$$y_{k+1} = y_k + \gamma v(y_k) \cdot s_k,$$

where γ is a positive constant. We recommend the choice $\gamma = 0.6$. It is known, see [16], that for arbitrary linear programs the above algorithm with $\gamma < 0.665$ converges *globally* (i.e. from an arbitrary feasible point), so that the *asymptotic* rate of convergence (for $k \to \infty$) in the value of the cost functions is $1 - \gamma$ and this holds irrespective of the dimension of y or the number of constraints.

The simple example $\min(c^T x \mid \|x\|^2 \leq 1$ $c = (0,1)$ shows, that starting the algorithm from $(0.99, 0)$ with $\gamma = 0.6$ we have no global, but a premature convergence to a non-optimal point, but already from $(0.7, 0)$ we have the usual fast convergence. Of course such premature convergence is not possible for the close following of the central path. We recommend therefore to use a phase 1 procedure, even for a feasible starting point to get a more centered starting point, see below.

Moreover, practical experience shows that this algorithm converges, in many cases if it is started from a well centered point, even for the value $\gamma = 0.9$ and is in general remarkably efficient (fast) also in the sense of global convergence for moderate requirements of accuracy.

Since many convex semi-infinite programs are just linear programs with infinitely many constraints, the convergence statements proved in [16] also hold. More precisely this class is given by the condition: each of the elementary constraint function $f(y) = f_i(x, y)$ (i.e. for i and x arbitrary), should have a representation

$$f(y) = \exp(\int \log \varphi(\alpha, y) p(\alpha) d\alpha)$$

for some measure $p(\alpha)d\alpha$. Note that for the example just above, with the simple quadratic constraint, the function $f(y)$, corresponding to the uniform distribution of α over $(0, 2\pi)$ is not quadratic, but only nearly quadratic. (Thus, formally it is not a counterexample to the above mentioned global convergence result of Tsuchia).

The partial effectivity of the affine scaling method is partly explained by the "stability" of the central path, the "parallel" d-trajectories converge to the same point as $t \to \infty$.

On the other hand the global convergence of the affine scaling method is enhanced, esp. when it is applied with $\gamma = 0.9$ instead of $\gamma = 2/3$ when we start from a point near to the center of the feasible set. Even though, after some time the iterates may get very close to the boundary. Obviously a curved boundary cannot be traced arbitrarily fast with a piecewise linear path. This and the complexity of computing integrals motivate the use of higher order extrapolation methods, see below.

It should be emphasized that the feasibility issue is a rather important and delicate one, if the generated grid is not fine or approximative enough, e.g. in the case of using a nonadaptive integration routine, even $\gamma = 0.5$ may lead to infeasibility and breakdown. In contrast to the central path, the trajectory of the affine scaling algorithm gets increasingly close to the boundary, which means that along it, say at the same distance from the optimum, the integrals (which have to be computed more and more accurately to get the right direction) are more singular than for points near the central path. Therefore we recommend the first method with continuous recentering to remain near to the central path. In fact one of the main advantage of using interior, esp. central path following method is their good *global* convergence property, and following the central path closely yields a globally uniform speed of convergence, and quadratic convergence even for a degenerate optimum, see [11].

The methods for following the central path closely are based on the notion of a "distance" $d(y)$ from that path, defined for an arbitrary point y, a most simple one being:

$$d(y) :=< H^{-1}(y)E(r,y), E(r,y) >, \tag{2.18}$$

where we already assumed, that in (2.6) $f_0(y) = c^T y$, and use the parametrization of that curve via

$$s = -\log(r). \tag{2.19}$$

Two numbers $0 < d_0 < d_1$ are fixed and the current (last) steplength $\triangle_{k-1}$ is enlarged by a factor $1 + \delta_+$ if for the current point $y = y_k$ $d(y) \leq d_0$; $\triangle_{k-1}$ is reduced by a constant factor $1 - \delta_-$ if $d_0 \leq d(y) \leq d_1$. If for a linearly extrapolated trial point $y = \tilde{y}_{k+1}$ $d(\tilde{y}_{k+1}) > d_1$, then $\tilde{y}_{k+1}$ is not accepted. The parameter $s = s_k$ is frozen and a number of recentering steps are made, i.e. Newton iterations to solve $E(s_k, y) = 0$. For the correct realization of this method, as specified in the main iteration file, we need to choose the above 4 constants $\delta_+, \delta_-, d_1, d_2$ and the whole algorithm rather carefully. In fact the steplength $\triangle_k$ has to be reduced also if the extrapolated point is infeasible, more precisely, if the ratio of the consecutive slacks for constraints which corresponds to the selected nodes becomes too small.

More theoretical details can be found in [5], [11]. There are several reasons to use this parametrization, see [11], one of the simplest is just that it leads — for the correct algorithmic choices — to roughly constant stepsizes, which do not depend on the problem parameters, but rather weakly on its size see [11].

There is a further important advantage of the close path following method: one can select the stepsizes without checking feasibility. In the semi-infinite case the latter task i.e. checking feasibility (along a line) is not easy at all. It requires a difficult line search (along the extrapolation line), more precisely the computation of the zero of the concave function (to be restricted for $s \geq 0$)

$$f(s) = \exp(\int \log f(y + sv, \alpha)d\alpha)$$

which, besides of being not quite cheap, is — even for one-dimensional domains of α — connected with instabilities (i.e. when v is parallel to the boundary near y). Note that other, traditional methods would require a nonconcave maximization procedure to check feasibility (compute the moments of hits with the boundary).

In [16] a rule for selecting exactly and easily computable stepsizes is described and analyzed for a primal-dual central path following method applied to linear programs. It is based on the primal-dual approach only formally, it can be applied without computing and updating dual variables, since these variables are easily generated (approximately) on (resp. near to) the central path.

This rule is — using $r = -\log(t)$ in the parametrization —

$$f(r_k)(r_k - r_{k+1})\sqrt{\frac{r_k}{r_{k+1}}} = \sqrt{\alpha_0}, \qquad (2.20)$$

where

$$f(r) = \frac{1}{r} \parallel \varphi(r) \parallel^{1/2}, \quad \varphi(r) := \sigma(r) \circ \tau(r) = \sigma^2(r) - \sigma(r)$$

$$\sigma(r) = r\frac{d}{dr}\log s(r), \quad \tau(r) := r\frac{d}{dr}\log \mu(r) \qquad (2.21)$$

$$\sigma(r) = M(r)e, \quad \tau(r) = (I - M(r))e = e - \sigma(r)$$

$$M(r): = \tilde{A}^T(\tilde{A}\tilde{A}^T)^{-1}\tilde{A}, \quad \tilde{A} := AS^{-1}(r)$$

$S(r)$ is the diagonal matrix associated to $s(r) := s(x(r)) = b - A^T x(r)$. The value $M(0)$ is a basic, affine invariant of the LP problem, See [11]. Moreover, in terms of the vectors and matrices

$$z(r) := (s(r), \mu(r)), \quad H(r) := H(x(r), \eta(r)), \quad \xi(r) := (x(r), \eta(r))$$

the function $f(r)$ is also given by

$$f(r) = \frac{1}{\sqrt{2}}\|\ddot{\xi}(r)\|_{H(r)}^{1/2} = \frac{1}{\sqrt{2}}\|z^{-1}(r) \circ \ddot{z}(r)\|_2^{1/2}. \qquad (2.22)$$

The quantity $f(r)$ is a "weighted" curvature of the primal-dual path at $x(r), \eta(r)$, where $\eta(r)$ is the variable (analog to y) of the dual program

$$f(r) = \frac{1}{\sqrt{2}}\|\xi(r)\|_{H(r)}^{1/2}, \quad H(r) = H_{xx} \oplus H_{\eta\eta}.$$

While this rule is easily generalized to (and seems to work quite well even on some classes of the modified) central paths of linear semi-infinite programs, we do not have yet a complexity analysis for its (easy) generalization to convex problems.

High Order Extrapolation of the Central Path

We now present a new class of high order extrapolation algorithms for following the central path. These algorithms can be easily generalized to the case of convex analytic programs. Here we assume, just for simplicity, that the constraints are linear, more precisely, that we have a linear programming problem of the following standard form

$$\min\{c^T y | A^T y \leq b\}.$$

For the parametrization of the central path we use $t = \exp(1/r)$, where r is the usual penalty parameter, thus

$$\text{res}(t, y) := mc\exp(t) + \sum_{i=1}^{m}\frac{A_i}{b_i - a_i y} = 0. \qquad (3.1)$$

Suppose we already computed $y(t_i)$, $i = 0, 1, \ldots, k$, which are *very accurate* approximations of the central path. In order to select t_{k+1} and compute y_{k+1}, we construct a $d(k)$-dimensional approximation $y_k(t)$ of $y(t)$, where $d(k) \leq k - 1$, may depend on k, so that $d(k) \to \infty$ when $k \to \infty$. The new parameter value t_{k+1} is determined algorithmically as the smallest (first) value t, for which

$$\mathcal{D}(y_k(t), t) \geq \alpha.$$

Here α is a constant and $\mathcal{D}(u, v)$ is a constructively defined "distance" function, defined for the points in $\mathcal{P}$, to measure their distance from the point of the central path corresponding to the parameter value t

$$\mathcal{D}(y, t) := \text{res}^T(t, y) H^{-1}(y) \text{res}(t, y),$$

where

$$H(y) = \sum_{i=1}^{m} \frac{a_i a_i^*}{(b_i - a_i y)^2}. \tag{3.2}$$

The constant α is selected in such a way that Newton's method to solve (3.1) converges quadratically in the domain

$$\mathcal{N}\mathcal{Q}_\alpha(t) = \{y | \mathcal{D}(y, t) \leq \alpha\}$$

Below we shall give alternative, more constructive i. e. easier to compute ways of determining t_{k+1} with the latter property.

Newton's method is applied with a frozen value of the inverse:

$$H^{-1}(y) \cong H^{-1}(y_k(t_{k+1})),$$

a few times to get the value $y(t_{k+1})$ with great accuracy. The latter is needed because we use $[y(t_{k+1}), \ldots, y(t_{k+1-d(k+1)})]$ — quite contrary to (earlier implementations of) first order extrapolation methods — to define the next high order extrapolation, which should use accurate node values.

Now the extrapolating curve $y_k(\cdot)$ will be defined as the central path corresponding to a linear program of smaller size, arising by *restriction* of the original linear program to the $d(k)$ dimensional subspace L_k spanned by $[y(t_{k-d(k)}), \ldots, y(t_k)]$. Another possibility would be to use the derivatives up to order $d(k)$ at some point $t = t_k$ to generate the subspace L_k with eventual rank, singular value analysis.

The defining data $[c_k, A_k, b_k]$ of this linear program are easily obtained from $[c, A, b]$. The easiest way is, say for $d(k) = k$, to introduce in L_k the local coordinates $\alpha_1, \ldots, \alpha_k$ via

$$y(\alpha) = y_k + \sum_{j=0}^{k-1} \alpha_{j+1}(y_k - y_j).$$

The complexity of the computation of the path segment $y_k(t)$ $t \in [t_k, t_{k+1}]$ is much less than that of $y(t)$ because one needs only the inversion factorization of

$k \times k$ matrices (instead of $n \times n$ ones) (resp. computation of $O(k^2)$ integrals), and $k \ll n$.

Note however that the computation ("tracking") of the distance, as it is stated in (3.2), requires the value of H^{-1}. There are two remedies. The first (cheaper one) is to use a fixed sequence $t_1, t_2, \ldots, t_k, \ldots$. We found for small (below 200) values of n and m the rather preliminary, heuristic or experimental law

$$t_{k+1} = t_k + \exp(0.1 * t_k) \max\{0.5\sqrt{(k)}, \frac{0.2k}{\sqrt{m}\sqrt{mn}}\}$$

quite satisfactory in the sense that the objective to need about 3–6 frozen Newton steps to reach $y(t_{k+1})$ from $y_k(t_{k+1})$. This could be achieved on randomly generated as well as on specially chosen difficult test problems. The method was able to reduce r from 1 to 10^{-10}, in 10-15, i. e. about *2-3 times* less iterations, i. e. new factorizations of H, than for first order extrapolation methods. In these tests we had several rather badly conditioned problems where the condition number at the optimum was over 10^7.

Another, more satisfactory solution — i. e. to get a less complex distance function — is to follow the primal and dual path simultaneously, for additional reasons to do this see [11]. Note that the dual problem

$$\max\{-b^T\mu | A\mu + c = 0, \mu \geq 0\}$$

can be written in "primal" form (by representing $\mu = -L^T\xi + \beta$, where $AL^T \equiv 0$, $\text{rank}L = m - n$)

$$\min\{\gamma^T\xi | L^T\xi \leq \beta\}, \quad \gamma = -Lb.$$

It is known that the following primal-dual function

$$d(y, \xi) := \|(\beta - L^T\xi) \circ (b - A^Ty) - \exp(-t)e\| \exp(t)$$

is a suitable distance function, see [11]. Of course, here y and ξ together constitute the state, "y" variable, the primal-dual problem is a direct sum problem, and the matrices to be inverted are of size n resp $m - n$. Without going into more or less known details of implementation we concentrate on describing only some of the new features. We do not recommend to solve the local, restricted linear program "till the end" i. e. similarly as recommended for $d(k) = 2$ in [3], because this introduces instability.

It is very instructive to consider first how this scheme works, and what are the theoretical reasons for its success on a simpler class of problems. The analogy (connection) with the method of conjugate gradients and rational Pade type approximation will be apparent if we consider the following simple example (which could be interpreted as a generalization: linear→quadratic or as a special case when $m \to \infty$ and A, b, c are selected approximately):

$$\min_{\frac{1}{2}\|x\|^2 \leq 1} (\frac{1}{2}x^T Ax + b^T x)$$

The central path here can be parametrized by the Langrange type parameter s

$$s = s(r) = \frac{r}{1 - 0.5\|x(r)\|^2}$$

such that $x(s) = -(A+sI)^{-1}b$, the final value $r = 0$ corresponding to the Lagrange multiplier $s^* = s(0)$ which solves

$$R(s) = 0.5\|(A + sI)^{-1}b\|^2 - 1 = 0.$$

Let

$$L(s) = b^T(A + sI)^{-1}b, \quad L_k(s) = b^T(A_k + sI)^{-1}b.$$

The method described above for linear programs corresponds now to the sequential $(k = 1, 2, \ldots)$ restriction, A_k of the operator A to the subspaces L_k and building $y_k(s) = (sI + A_k)^{-1}b$. If we let r go to zero in the subproblems then s_{k+1} is determined by

$$R_k(s_{k+1}) = 0.5\|(s_{k+1}I + A_k)^{-1}B\|^2 - 1 = 0 \tag{3.3}$$

The type of approximation provided by $y_k(\cdot)$ is further characterized by the next

Proposition: The functions $L_1, L_2, \ldots, L_k, \ldots$, for $k = l, 2 \ldots$, are multipoint, diagonal Pade interpolants for $L(s)$ sequentially built on the interpolation nodes $s_1, \ldots, s_k, \ldots$ in the sense, that

$$L_k(s_j) = L(s_j), \quad L'_k(s_j) = L'(s_j), \quad j = 1, \ldots, k. \tag{3.4}$$

Proof: By differentiation we have

$$R(s) = 0.5 < (A + sI)^{-2}b, b > -1 = 0.5L'(s) - 1, \quad R_k(s) = 0.5L'_k(s) - 1,$$

from which the statement follows by construction.

Note that our optimization problem is equivalent — except the trivial case $\|A^{-1}b\| < 1$ — to find s^*, such that

$$L'(s^*) = 2. \tag{3.5}$$

The effectivity of the algorithms depends in this case on the ideal approximability of Stieltjes functions, like $L(\cdot)$, by x multipoint Pade methods, a well known fact in the theory of rational approximation. Selecting r to be different from zero e. g. like in the general method proposed above does not change the type (speed) of convergence significantly. On the other hand it is not good to select $r = 0$ (and eventually using a factor γ like in the affine scaling method, arising formally for $d(k) = 1$) there, because in that general case the corresponding value $x_k(r_{k+1})$ would not be an analytic (stable enough) function of the local data.

The connections with Pade approximation of Stieltjes functions has already been observed in [13], where the first variant of the above high order algorithm was proposed for the special case considered now. Later a 2- and 3-dimensional subspace algorithm similar but essentially different from the ones presented here has been proposed and studied in [3].

We have written Matlab program not only to realize the above method i. e. to compute s^*, as a solution of the equation (3.5) using the sequence $s_1^*, \ldots s_k^*, \ldots$ and $L_1, \ldots, L_k$, such that

$$L_k'(s_{k+1}^*) = 2, \quad L_j^{(i)}(s_j^*) = L^{(i)}(s_j^*), \quad i = 0, 1, \quad k = 1, 2, \ldots$$

the values s_k^* being computed via Newton's method (started from s_k), but also to use interpolation points with multiplicity 1

$$R_k(s_i) = R(s_i), \quad i = 1, \ldots, N, \quad R_k(s_{k+1}^*) = 0, \quad k = 1, 2, \ldots$$

The results of comparing the cpu times arising when a standard quasi-Newton (SQP type) method (Matlab's constr.m supplying explicit gradients) is used, showed a remarkable speedup between 10-20 times (!) for problems $n \leq 200$. Of course, in our programs we needed only one eigenvalue decomposition to get the constants of the functions L and R. The rest, i. e. building up the interpolants and performing Newton steps is "one-dimensional".

Description of the Implementation

We shall concentrate on the specific points of our implementation, each variant of which uses the following type of files:

1. Startfile; here the basic initial constants, parameters, global variables, e.t.c. are specified, the result of the computation, outputs of the main file are gathered here. It invokes the

2. Main iteration file, which in turn invokes the extrapolation or (and) direction finding files, implements the stepsize rules and the stopping rule, it also gathers the variables, e.g. outputs for display of results to be returned to the start file. Note that in the extrapolation file the program in fact could invoke itself (for the generated lower dimensional data).

3. Direction finding file. Here the direction of the next move is computed, more precisely the right hand side of the basic ODE (with possible stabilizing, modifying terms) which requires the invocation of the routines computing the function values, gradients and Hessians of the constraints. In the semi-infinite case these correspond to "potentials" formed from the integrated constraints.

4. The files providing the values, gradients and Hessians of the constraint functions.

5. Integration (quadrature) routines. In the case of semi-infinite constraints the files of type 4 invoke integration routines, since the function whose values, gradients and Hessians are needed, are integrals. It is required that the user provides such a program for calculation of these components which allows to compute the values on an arbitrary k-tuple of arguments in $x_i \in [a, b]$ for all $i = 1, \ldots, m$

For efficient implementation one should try to "parallelize" the computation of the components of the gradients and Hessians as well as those functions, which are of the same type (using vector operations). Note that in most applications there are usual only a few groups formed by functions of the same algebraic type (differing only in parameters) which specify the constraints.

There are two main methods (options) for computing the integrals:

1. In the first method (option, which we recommend only for moderate requirements of final accuracy) we use an increasingly high order (composite) Gauss quadrature formula, where the order (and the number of composition subintervals) are raised linearly with the iteration number.

2. In the second more accurate and generally applicable method an adaptive selection of the nodes is realized using a fixed (order) Gaussian formula and diadic subdivision algorithm is constructed according to the principle of "equal local errors".

We recommend to use high order Gaussian rules. But the following should be kept in mind if a constraint function (or the cost function) is only piecewise analytic, like

$$f_0(x, y) = 0.5(\text{sign}(x - \alpha) + \text{sign}(\beta - x))$$

the characteristic function of a subinterval $[\alpha, \beta]$ of $[a_0, b_0]$, one should use three (or in a better method two) separate quadrature formulae for the arising subintervals. This is absolutely necessary when using nonadaptive quadratures, but makes a lot of difference for adaptive quadratures also.

Instead of the relative error rule for subintervals $[\alpha, \beta]$ of $[a, b]$ used for adaptive quadrature (e. g. in Matlab)

$$|Q - (Q_1 + Q_2)| \leq \text{tol} * Q$$

we prescribed, for any subinterval $[\alpha, \beta]$ of $[a, b]$,

$$|Q - (Q_1 + Q_2)| \leq \text{tol} * I_{old}, \tag{4.1}$$

where I_{old} is an estimate of the integral on $[a, b]$, which is kept from the previous iteration. The point is that for the intervals, where the integrand is small (relatively) but varying (e.g. for the function $\exp(x)$, the "standard" method generates an obviously "noneffective" uniform grid). In our problems the integrands, more precisely the dual variables (Lagrange multipliers)

$$\frac{r dx}{f_k(x, y(r))} \longrightarrow \mu_k(x) dx$$

tend to a measure on $[a, b]$. Therefore nonadaptive methods cannot be very effective.

The second method thus generates in an adaptive (sequential) way nearly optimal approximations of the semi-infinite problem by finite problems, more precisely the dual problem is implicitly approximated by a sequence of finite (even "bounded") dimensional problems. This approximation problem is rather complicated. We can give here only a few hints of the arising difficulties and possibilities. In the case of a one-dimensional interval $[a, b]$ the limiting measure is concentrated — in general — in n points $x_1^{i_1}, \ldots, x_n^{i_n}$, counted with multiplicities $(n = \dim(y))$, such that the problem

$$\min\{f_0(y)|f_{i_j}(x_j^{i_j}, y) \geq 0, \quad i_j \in [1, \ldots, m], \quad j = 1, \ldots, n\}$$

has the same optimal solution(s) as the original one. In the case of a multidimensional domain of x'-s such finite "sufficient" set of points — while existing by the well-known theorem of Caratheodory — is, in general not determined uniquely and just is the main problem for the "classical" approaches to the solution of semi-infinite programs. On the other hand the unique smooth measures corresponding to point $y(r)$ on the central line converge for $r \searrow 0$ to a unique measure. This is very important in favor of interior point central path following methods. We do not know whether this is — in the two-dimensional case (at least in a generic sense) — atomic or not, it is not difficult to construct simple examples of two-dimensional uniform approximation problems where the optimum (i. e. maximal distance) is reached on a one-dimensional set, say a circle, and then the limit measure is concentrated e. g. uniformly on this set (curve).

An important issue — by the choice of the integration routine — is detecting infeasibility. The generated system of nodes $X_{(k)}$ is used not only for estimating the integrals but also for checking whether the candidate for the next iteration point z is feasible. This is because the quantity

$$r(z) = \min_{x \in [a,b]} f(x, z) \quad \text{is replaced by} \quad \min_{x \in X_{(k)}} f(x, z).$$

For fixed z the function $f(x, z)$ has in general no special structure (e.g. it is neither convex or concave), so that $r(z)$ cannot be approximated easily. In fact one has to compute for given iterate z_k and direction $v = v_k$ the stepsize $s = s_k$ as to satisfy

$$\min_{x \in [a,b]} f(x, z_k + sv) \geq 0.$$

Since the equation, in s, $f(x, z_k + sv) = 0$, for given z_k, x and v has by the concavity of f in y (in case when f has bounded level sets) a unique solution $s = s(x) > 0$, which can be computed — in most cases — easily, we can select for arbitrary $0 \leq \alpha < 1$

$$s_k = \alpha \cdot \min\{ s(x) \mid x \in [a, b]\} \cong \alpha \min\{ s(x_j) \mid x_j \in X_{(k)}\}$$

There is another way of computing the moment s_k of hit with the boundary. We use the overall barrier function

$$F(y) = \exp(\sum_{i=1}^{m} \sum_{j=1}^{k_i} \log f(x_j^i, y))$$

and compute the "positive" root s_k of the function

$$\varphi(s) = F(z_k + sv_k),$$

i.e. that value $s = s_k$, for which

$$\varphi(s_k) = 0, \quad s_k > 0.$$

Here we have to suppose that the set $\{y|F(y) > 0\}$ is bounded. One can show that this boundedness condition makes in fact no problem, since the search direction v is always a descent direction for $c^T y$ and the level sets (2.9) must be bounded if the existence and boundedness of the optimal set (λ^*, Y^*) is assumed. Note that $\varphi(s)$ is analytic and concave in s. Thus special methods can be constructed to find a good approximation of s_k. The most simple is of course a combination of bisection and linear extrapolation. We assume that an arbitrary point $y0$ is fixed, which should (ideally) be, of course, an initial approximation of the optimum, however nothing will be assumed about it.

Let replace the constant 0 in the constraint $f_i(y) \geq 0$ by an exponentially vanishing term

$$f_i(y, t) = f(y) - \exp(-t)(Snegpart(f_i(y0)) - d_i) \geq 0 \qquad (4.2)$$

where $d_i > 0$ and $S_i > 1$, $i = 1, \ldots, m$ are arbitrary. This means that the equation for the modified central line becomes

$$c(t) - \sum_{i=1}^{m} \frac{\partial f_i(y, t)}{f_i(y, t)} = 0.$$

We can make the point $y0$ to be the startpoint of this homotopy path — corresponding to the value $t = 0$ — by selecting

$$c(t) = \exp(-kt)c0 + (\exp(t) - 1)c,$$

by choosing

$$c0 = \sum_{i=1}^{m} \frac{\partial_y f_i(y0, 0)}{f_i(y0, 0)}$$

where $k > 0$ may be arbitrary, (by default $k = 1$). It is to be noted that up to now — as far as we know — there are no complexity results available about following the above "modified" phase 1 - phase 2 central line. It may turn out that following two or three central paths is more effective than following one modified path. It is interesting to note that for $k = 1$ and a single quadratic constraint $\|y\|^2 \leq 1$, the arising curves for $t \in (-\infty, \infty)$ are just the extremals i. e. "lines" of the hyperbolic geometry (Poincare's model).

In the presence of linear equality constraints the same routines are used as described up to this point with the following simple modifications. Suppose we have for y the equality type constraint

$$Ky = v, \quad K : R^n \to R^{n_0}.$$

We can represent the vector y in the form

$$y = y_+ + Lw,$$

where $L = \text{null}(K)$ and y_+ is the minimum norm solution

$$y_+ = (K^*K)^{-1}K^* = \text{pinv}(K)v,$$

and w is arbitrary. Of course one should first make a singular value analysis of K. In the main iteration file only the value of $w = w(y)$ will be updated (correspondingly in the startfile $y0$ is replaced by $(K^*K)^{-1}K^*(y0 - y_+)$), in the direction finding file $y' = H(y)$ is replaced by

$$v' = (K * K)^{-1}H(y_+ + Lw).$$

We need not to replace (or modify) the function, gradient and Hessian files, but only to use their values "modified" with the L matrix in the direction finding file as follows

$$f(w) = f(y_+ + Lw), \quad g(w) = \sum_i \frac{\partial f(y_+ + Lw)}{\partial y_i} L_j^i, \quad h_{ij}(w) = \sum_{k,l} L_i^k \frac{\partial f(y_+ + Lw)}{\partial u_k \partial y_l} L_j^l$$

References

[1] N.S. Bahvalow, Methodes Numeriques, ed. Nauka (Mir) Moscow, 1980.

[2] P.J. Davis, P. Rabinovitz, Methods of Numerical Integration, Academic Press, New York, 1975.

[3] P.D. Domich et al., Optimal 3-dimensional methods for linear programming, National Institute of Standards and Technology, Gaithersburg, NISTIR 89-4225.

[4] M.C. Ferris, A.B. Philpott, An interior point method for semi-infinite programming, Mathematical Programming (43) 1989 pp. 257–276.

[5] R. Hettich, ed., Semi-Infinite Programming, Lect. Notes in Control and Inf Sciences, Vol. 15, Springer Verlag 1979.

[6] F. Jarre, Interior Point Methods for convex programming, to appear in Applied Math. and Optimization, 1993.

[7] J.C. Lagarias, M.J. Todd, eds., Mathematical Developments Arising from Linear Programming, Contemporary Mathematics, Vol. 114, Amer. Math. Soc., Providence, 1989.

[8] G. Lopes, Conditions for convergence of multipoint Padé approximations for functions of Stieltjes type, Math. USSR Sbornik, Vol. 35 (1979), No. 3.

[9] I.J. Lustig, R.E. Marsten, D.F.Shanno, Computational Experience with a Primal-Dual Interior Point Method for Linear Programming Techniques, Industrial and Syst. Engineering Rep. Ser., Rep. J-89-11, Inst. of Technology, Atlanta, Georgia.

[10] U. Schättler, An interior point method for semi-infinite programming problems, Doct. Dissertation, Univ. Würzburg, Inst. f. Angew. Math. 1992.

[11] G. Sonnevend, J. Stoer, G. Zhao, On the complexity of following the central path by linear extrapolation in linear programs, Mathematical Programming, 1991, pp. 527–553.

[12] G. Sonnevend, Constructing feedback control in differential games by the use of central trajectories, DFG Report, Nr. 385/1992, Inst. für Angewandte Mathematik, Univ. Würzburg (July 1992), 37 p., to appear in ZAMM.

[13] G. Sonnevend, J. Stoer, Global ellipsoidal approximations and homotopy methods for smooth, convex, analytic programs, Applied Math. and Appl., 21 (1980) pp. 139–165.

[14] G. Sonnevend, Applications of Analytic Centers, NATO ASI Ser. F, Vol. 70, "Numerical Linear Algebra and Digital Signal Processing" (P. van Dooren, and G.Golub eds.), Reidel 1988.

[15] G. Sonnevend, Application of analytic centers for the numerical solution of semi-infinite, convex programs arising in control theory, 15 p., DFG Report , *Anwendungsbezogene Optimierung und Steuerung*, Nr. 170/1989, Inst. für Angewandte Mathematik, Univ,. Würzburg.

[16] T. Tsuchiya, M. Murarmatsu, Global convergence of a long-step affine scaling algorithm for degenerate linear programming problems, Res. Memo. Nr. 423, Inst. of Statistical Math., Tokyo 1992.

International Series of Numerical Mathematics, Vol. 115, © 1994 Birkhäuser Verlag Basel

A Structured Interior Point SQP Method for Nonlinear Optimal Control Problems

Marc C. Steinbach*

Abstract

Direct boundary value problem methods in connection with SQP iteration have proven very successful in solving nonlinear optimal control problems. Such methods use parameterized control functions, discretize the state
differential equations by, e.g., multiple shooting or collocation, and treat the
discretized BVP as an equality-constraint in a large nonlinear constrained
optimization problem. In realistic applications several thousands of variables can appear in the NLP. Solution by standard techniques is therefore
impractical. A careful choice of the discretization leads to QP subproblems
possessing a very special m-stage block-sparse structure, where m is the grid
size. The paper presents a recursive solution algorithm that fully exploits
this QP sparseness to generate a factorization of the inverse of the KKT
matrix in $\mathcal{O}(m)$ operations. A structure-preserving primal-dual barrier method is proposed for treating the generally large number of state and control
inequality constraints.

Introduction

Consider the Mayer control problem with multi-point boundary conditions and
path constraints (usually including state and control box constraints):

$$\phi(x(T)) = \min, \tag{1}$$
$$\dot{x}(t) - f(x(t), u(t)) = 0, \tag{2}$$
$$r(x(t_1), \ldots, x(t_k)) = 0, \tag{3}$$
$$g(x(t), u(t)) \geq 0. \tag{4}$$

As described below, we will discretize the problem by direct multiple shooting
or collocation schemes and solve the resulting nonlinear optimization problem by
an SQP iteration. This numerical approach has proven very effective during the
last decade, and several methods based on it have been developed [2, 8, 1, 16].

Applying the Karush-Kuhn-Tucker necessary conditions to the QP subproblems yields very large, sparse KKT systems with many inequality constraints.

---

*Interdisciplinary Center for Scientific Computation, University of Heidelberg, Im Neuenhei-
mer Feld 368, D-69120 Heidelberg, Germany.

More specific, the numbers of variables and of nonzero matrix elements both grow linearly with the number m of discretization stages. Therefore QP solution dominates the computations by far when standard $\mathcal{O}(m^3)$ algorithms are applied, and exploitation of the sparseness is crucial for an efficient SQP method.

The first structured SQP method for optimal control problems was developed in [14, 2] and implemented in the direct multiple shooting code MUSCOD. Here a generalized condensing algorithm eliminates all but the initial state variables from the QP in a recursive projection. An active set QP solver then operates on the condensed system. MUSCOD has been successfully applied, e.g., to solve moderate size robot trajectory optimization problems [10, 11, 15]. A different sparse SQP approach is reported in [1] for direct collocation. The authors apply a multifrontal algorithm to obtain a sparse KKT matrix factorization, which is then re-used in a Schur complement method to account for active set changes.

Here we consider a specific m-stage block-sparse structure that can be fully exploited by the SQP method. This rather natural structure is obtained under mild restrictions on the discretization whenever the boundary conditions satisfy a second order *separability condition*. In this paper we restrict ourselves to the case of first order decoupling, $r(x_1, \ldots, x_k) = (r_1(x_1), \ldots, r_k(x_k))$, which covers an important class of application problems. The more general case of second order decoupling, $r(x_1, \ldots, x_k) = r_1(x_1) + \cdots + r_k(x_k)$, includes almost all practical nonlinear optimal control problems (in particular, periodic ones), and is already implemented in our recursive solver. However, this case exhibits basic differences in theory and practice, and will be treated in a subsequent paper. (See also the remarks after Theorem 1 and after the description of the solver.)

The direct BVP approach

The infinite dimensional optimal control problem is converted to a finite dimensional nonlinear optimization problem (NLP) by parameterizing the controls and discretizing the states.

1) Choose a grid $0 = \tau_1 < \cdots < \tau_m = T$ of $m \geq k$ points which include all instances where boundary conditions are evaluated: $t_i = \tau_{j(i)}$ for some $j(i)$.

2) Choose a finite-dimensional space of admissible controls, defined by fixed *local* base functions v_j and free control parameters u_j:

$$u(t) = v_j(t; u_j) \quad \text{on } I_j = [\tau_j, \tau_{j+1}). \tag{5}$$

3) Define a discrete trajectory $x_j = x(\tau_j)$ and *local* approximations y_j to the ODE solution. Here $y_j(\tau_j; x_j, z_j) = x_j$, and the additional *collocation variables* z_j satisfy *collocation conditions* $c_j(x_j, z_j, u_j) = 0$. Couple the stages by matching conditions

$$h_j(x_j, z_j, x_{j+1}) = y_j(\tau_{j+1}; x_j, z_j) - x_{j+1} = 0. \tag{6}$$

In the multiple shooting case y_j solves the IVP $y_j(\tau_j; x_j, u_j) = x_j$, $\dot{y}_j = f(y_j, v_j)$ on I_j, and u_j replaces z_j in y_j, h_j.

4) Discretize the remaining problem functions by substituting u_j, v_j, x_j, z_j into the definitions, and possibly include collocation conditions: [1]

$$\min \phi(x_m), \quad e_j(x_j, z_j, u_j) = \begin{pmatrix} r_j(x_j) \\ c_j(x_j, z_j, u_j) \end{pmatrix}, \quad g_j(x_j, u_j) = g(x_j, v_j(\tau_j; u_j)). \quad (7)$$

Replacing continuous path constraints by pointwise constraints may lead to small violations between grid points. This can affect the SQP convergence order depending on the problem, parameterization, and grid choice. Note that the discretization scheme must not include any stage-coupling terms, except for x_{j+1} in h_j satisfying $\partial^2 h_j / \partial(x_j, u_j) \partial x_{j+1} = 0$. Precisely this restriction yields our special structure; it is usually not satisfied for collocation schemes applied by other authors. Details of our collocation approach will be given in a forthcoming paper.

SQP method for the structured NLP

Define $F_1 = \phi$, $F_2 = (e_1, h_1, \ldots, e_{m-1}, h_{m-1}, e_m)$, $F_3 = (g_1, \ldots, g_m)$. A numerical solution $y = (x_1, z_1, u_1, \ldots, x_m, z_m, u_m)$ of the constrained NLP

$$\min F_1(y) \qquad \text{subject to} \qquad F_2(y) = 0, \quad F_3(y) \geq 0, \qquad (8)$$

is computed by applying a damped SQP iteration, $y^{k+1} = y^k + t^k \Delta y^k$, $t^k \in I\!R$, where Δy^k solves an approximating quadratic subproblem with linear constraints:

$$\min_{\Delta y} \frac{1}{2} \Delta y^T H^k \Delta y + J_1^k \Delta y \qquad \text{s.t.} \qquad \begin{cases} J_2^k \Delta y + F_2^k = 0, \\ J_3^k \Delta y + F_3^k \geq 0. \end{cases} \qquad (9)$$

Here $F_i^k = F_i(y^k)$, $J_i^k = F_i'(y^k)$, and $H^k \approx L_{yy}(y^k, w^k)$ approximates the Hessian of the Lagrangian. The following theorem establishes our special QP structure.

Theorem 1 *The Jacobians J_2^k, J_3^k and Hessian L_{yy}^k are block-diagonal, with the only exception of stage-coupling superdiagonal blocks $\partial h_j / \partial x_{j+1} = -I$ in J_2^k.*

Proof. This follows from the definitions of functions F_i and constituents ϕ, h_j, g_j, e_j. Except for h_j they involve only local variables x_j, z_j, u_j. The Hessian L_{yy}^k is a linear combination of Hessians of ϕ, h_j, g_j, e_j. ∎

Remark: In the case of second order decoupled boundary conditions we redefine $F_2 = (c_1, h_1 \ldots, c_{m-1}, h_{m-1}, r)$ and get a full row of blocks in J_2^k representing globally coupled constraints. All other structures are preserved.

From now on we assume that collocation variables and conditions have been eliminated from the discrete QP. After appropriate partitioning of the Jacobian

---

[1]For notational simplicity assume $\tau_j = t_j$ and introduce zero-dimensional variables z_m, u_m.

and Hessian blocks the QP then fits into the general m-stage block-sparse structure $({}^tJ = J^T = \text{transpose}(J))$:

$$\sum_{j=1}^{m} \left[\frac{1}{2} \begin{pmatrix} \Delta x_j \\ \Delta u_j \end{pmatrix}^T \begin{pmatrix} H_j & {}^tJ_j \\ J_j & K_j \end{pmatrix} \begin{pmatrix} \Delta x_j \\ \Delta u_j \end{pmatrix} + \begin{pmatrix} f_j \\ d_j \end{pmatrix}^T \begin{pmatrix} \Delta x_j \\ \Delta u_j \end{pmatrix} \right] = \min_{\Delta x, \Delta u}, \quad (10)$$

$$\begin{pmatrix} G_j & E_j \end{pmatrix} \begin{pmatrix} \Delta x_j \\ \Delta u_j \end{pmatrix} + h_j - \Delta x_{j+1} = 0, \quad (11)$$

$$F_j^x \Delta x_j + e_j^x = 0 \text{ or } \geq 0, \quad (12)$$

$$D_j^u \Delta u_j + e_j^u = 0 \text{ or } \geq 0, \quad (13)$$

$$\begin{pmatrix} F_j^c & D_j^c \end{pmatrix} \begin{pmatrix} \Delta x_j \\ \Delta u_j \end{pmatrix} + e_j^c = 0 \text{ or } \geq 0. \quad (14)$$

In addition to the minimization and matching conditions we distinguish three categories of (local) constraints: separated state constraints (12), separated control constraints (13), coupled constraints on state and control (14). This is necessary for full efficiency of the QP solver; see next section.

Exploitation of the structure is also possible in the nonlinear algorithmic parts. Generation and storage of function values, gradient and Hessian matrices is *completely parallel* for the SQP and interior point iterations, since all computations are local to the individual discretization stages. When the exact Hessian is not available or indefiniteness must be avoided, block-wise rank-2 BFGS updates yield a rank-$2m$ Hessian update. This leads to superlinear SQP convergence with an essentially mesh-independent convergence order. To increase efficiency, the accuracy of interior point QP solution should be adapted to the SQP needs, if possible also the accuracy of function and gradient evaluation. For details see [14].

Recursive $\mathcal{O}(m)$ solution of block-sparse QP

In this section we treat the purely equality-constrained QP. Assume for simplicity that a unique minimum exists, i.e., J_2^k has full rank and H^k is positive definite on $\ker(J_2^k)$. Standard solution techniques for the associated symmetric, indefinite KKT system

$$\Omega z + a := \begin{pmatrix} H^k & {}^tJ_2^k \\ J_2^k & \end{pmatrix} \begin{pmatrix} \Delta y \\ -w_2 \end{pmatrix} + \begin{pmatrix} {}^tJ_1^k \\ F_2^k \end{pmatrix} = 0 \quad (15)$$

include the null space or *Schur complement* (SC) method and range space or *projected Hessian* (PH) method. A structured SC method would have to factorize H^k and resulting block-diagonal and block-tridiagonal matrices. Partial projection (on the boundary conditions) would generate a similar but smaller block-tridiagonal matrix. Both techniques require positive-definiteness of H^k on a larger space than $\ker(J_2^k)$, and, more severely, both destroy the block-sparseness when boundary conditions are only second-order decoupled.

Our solution algorithm combines PH and SC methods in an alternating fashion to obtain a solution of the m-stage KKT system (15). In a backward recursion, $k = m(-1)1$, a sequence of similar k-stage KKT systems

$$\Omega_k z_k + a_k = 0 \qquad (16)$$

is generated.[2] A recursion step "cuts off the last stage" by applying a local projection and computing the Schur complement of the projected Hessian. This generates a block-sparse factorization

$$\Omega^{-1} = \Lambda^T \Pi \Lambda, \qquad \Lambda = \Lambda_1 \ldots \Lambda_m, \quad \Pi = \Pi_1 \ldots \Pi_m. \qquad (17)$$

(without producing any fill-in) in the factorization phase, and a transformed right hand side $a_1 = \Lambda a$ in an independent transformation phase. Finally, a forward recursion generates $z = \Lambda^T z_1$ from $z_1 = \Pi a_1$ in the solution phase. Each phase requires $\mathcal{O}(m)$ operations.

The whole procedure is based on the optimal control nature of the KKT system. The initial state and all controls are "free" variables, which determine all following states via the matching conditions. Hence, satisfaction of any constraint must be "controlled" either by local free variables or by free variables in *previous* stages (via the matching conditions). Higher order state constraints, for instance, always produce such a backward effect. When free variables in the *last* stage are not fixed by local constraints, they can be determined by *local* minimization according to the basic concept of optimality of sub-arcs. Our algorithm takes these facts into account when eliminating each variable as early as possible in the backward recursion.

Each recursion step involves six transformation steps. Each transformation reduces the local system size by direct or formal elimination of some variables. Transformations 1 and 2 are independent w.r.t. the stages. They eliminate the local separated constraints and may be performed as parallel pre- (post-)processing before backward (after forward) recursion. Transformation steps 3–6 constitute the genuinely recursive part of the algorithm, where steps 3 and 4 eliminate local coupled constraints, and steps 5 and 6 eliminate matching and control minimization conditions respectively. Steps 3 and 4 of stage $j - 1$ can run parallel to steps 5 and 6 of stage j.

All projection steps except 5 involve the following operations in the backward recursion. A constraint matrix is factorized, the system is accordingly transformed and partitioned. Some variables are either immediately determined or formally eliminated using the constraint equation. Corresponding multipliers are always formally eliminated. In the forward recursion formally eliminated variables are actually computed and partitions are re-transformed to obtain the original variables. Orthogonal (LQ type) and Cholesky (LL^T) factorizations are used in projection and Schur complement steps respectively.

---

[2]Strictly speaking, all systems are m-stage systems, with the last $m - k$ stages remaining unchanged in steps $j \leq k$.

The recursion—step by step

Let us introduce the formal definitions $\lambda_0 := 0$, $(G_m\ E_m) := 0$, $\lambda_m := 0$, $h_m := 0$, and $\Delta x_{m+1} := 0$, with appropriate dimensions. When stage coupling terms are included in the right hand side, the local original KKT system at each stage j reads

$$
\begin{bmatrix} H_j & {}^tJ_j & {}^tF_j^x & & {}^tF_j^c & {}^tG_j \\ J_j & K_j & & {}^tD_j^u & {}^tD_j^c & {}^tE_j \\ F_j^x & & & & & \\ & D_j^u & & & & \\ F_j^c & D_j^c & & & & \\ G_j & E_j & & & & \end{bmatrix} \begin{bmatrix} \Delta x_j \\ \Delta u_j \\ \mu_j^x \\ \mu_j^u \\ \mu_j^c \\ \lambda_j \end{bmatrix} + \begin{bmatrix} f_j - \lambda_{j-1} \\ d_j \\ e_j^x \\ e_j^u \\ e_j^c \\ h_j - \Delta x_{j+1} \end{bmatrix} = 0. \tag{18}
$$

Steps 1 and 2. *Project onto* $\ker(D_j^u)$ *and* $\ker(F_j^x)$ *of local separated control and state constraints.*

Factorize D_j^u, F_j^x, transform and partition $\Delta u_j \to (\Delta u_j^1, \Delta u_j^*)$, $\Delta x_j \to (\Delta x_j^1, \Delta x_j^*)$. Determine $\Delta u_j^1, \Delta x_j^1$ immediately, eliminate μ_j^u, μ_j^x formally. Eliminating Δx_{j+1}^1 results in *splitting* of the matching conditions h_j: rows corresponding to Δx_{j+1}^1 are converted to local coupled constraints, and $\lambda_j \to (\lambda_j^1, \lambda_j^*)$ is partitioned accordingly. These operations and recursion in higher stages produce the local KKT system

$$
\begin{bmatrix} H_j^{**} & {}^tJ_j^{**} & {}^tC_j & {}^tG_j^3 \\ J_j^{**} & K_j^{**} & {}^tA_j & {}^tE_j^3 \\ C_j & A_j & & \\ G_j^3 & E_j^3 & & \end{bmatrix} \begin{bmatrix} \Delta x_j^* \\ \Delta u_j^* \\ \nu_j \\ \lambda_j^3 \end{bmatrix} + \begin{bmatrix} f_j^{**} - \lambda_{j-1}^* \\ d_j^{**} \\ c_j \\ h_j^3 - \Delta x_{j+1}^3 \end{bmatrix} = 0, \tag{19}
$$

$$
S_{j+1}\Delta x_{j+1}^3 + [\alpha_{j+1} - \lambda_j^3] = 0. \tag{20}
$$

Here the third row combines original local coupled constraints with such resulting from splitting in steps 2 and 4 of stage $j+1$, and the additional equation remains from stage $j+1$ after the final step 6. The whole recursion is started in stage m with a zero-dimensional augmentation $S_{m+1} := 0$, $\alpha_{m+1} := 0$.

Steps 3 and 4. *Project onto* $\ker(A_j)$ *of control part and onto* $\ker(\bar{C}_j^x)$ *of resulting state part of projected local coupled constraints.*

Some care must be taken in this part of the algorithm. We wish to eliminate as many control variables as possible. However, there may be more local constraints than controls, or the matrix A_j may be ill-conditioned. Hence, the factorization of A_j should include pivoting and a rank decision to determine which rows can be used safely to eliminate control variables. The operations are as follows:

Factorize A_j, transform and partition $\Delta u_j^* \to (\Delta u_j^2, \Delta u_j^3)$, $\nu_j \to (\nu_j^u, \nu_j^x)$. Eliminate Δu_j^2 and ν_j^u formally. Factorize the resulting state constraint matrix $\bar{C}_j^x$, transform and partition $\Delta x_j^* \to (\Delta x_j^2, \Delta x_j^3)$. Determine Δx_j^2 immediately, eliminate ν_j^x formally. As in step 2, eliminating Δx_j^2 splits matching conditions and yields partitioning $\lambda_{j-1}^* \rightarrow (\lambda_{j-1}^2, \lambda_{j-1}^3)$. The reduced local system can be written

$$\begin{bmatrix} H_j^{33} & {}^tJ_j^{33} & {}^tG_j^{33} \\ J_j^{33} & K_j^{33} & {}^tE_j^{33} \\ G_j^{33} & E_j^{33} & \end{bmatrix} \begin{bmatrix} \Delta x_j^3 \\ \Delta u_j^3 \\ \lambda_j^3 \end{bmatrix} + \begin{bmatrix} \tilde{f}_j^3 - \lambda_{j-1}^3 \\ \tilde{d}_j^3 \\ \tilde{h}_j^3 - \Delta x_{j+1}^3 \end{bmatrix} = 0, \tag{21}$$

$$S_{j+1}\Delta x_{j+1}^3 + [\alpha_{j+1} - \lambda_j^3] = 0. \tag{22}$$

Steps 5 and 6. *Project onto* $\ker\left(G_j^{33}\ E_j^{33}\ -I\right)$ *of projected matching conditions and compute Schur complement of projected control minimization conditions.* Eliminate $\Delta x_{j+1}^3, \lambda_j^3$ formally by substituting the third equation into the fourth one and the resulting expression into the first two equations. Eliminate Δu_j^3 formally. This completes recursion step j, leaving the projected state minimization condition which couples to stage $j-1$:

$$S_j\Delta x_j^3 + [\alpha_j - \lambda_{j-1}^3] = 0. \tag{23}$$

After the final recursion step 1 this "1-stage KKT system" is solved directly.

Remarks. In the case of second order decoupled boundary conditions the QP solver must deal with globally coupled constraints. Optimality of sub-arcs does not apply, and local minimization requires positive-definiteness of the Hessian on a larger space than $\ker(J_2^k)$. Projection steps 1 and 2 are not completely parallel in the right hand side transformation and solution phases. Fill-in is produced as Schur complement of the matrix of global constraints, and the recursion should include an additional SC step to exploit its sub-structure.

So far we assumed that the QP has a unique minimum. However, conflicting or rank-deficient constraints and semidefinite Hessians can be treated using more sophisticated factorizations like singular value and spectral decompositions.

Interior point QP solution

Any phase inequality constraint appears in each stage of the discrete problem. Typical applications may include box constraints on all state and control variables. Efficient handling of that many inequalities remains a central difficulty.

In general we suggest a primal-dual barrier method as described below. However, an active set strategy with warm start may be competitive, especially when a good estimate of the correct active set is known and only a few exchange steps are expected. Note that large numbers of binding box constraints reduce the effort of recursive KKT solution, since such constraints are pre-eliminated in recursion steps 1 and 2, and that active set changes can be handled by a Schur complement method [5]. Improving a given solution by mesh refinement may therefore be very efficient with an active set strategy, and a good code should offer this alternative.

The concept of interior point or barrier methods for constrained optimization was introduced in [4] and theoretically developed in [3]. After the appearance of Karmarkar's paper [9] barrier methods for linear programs were developed by many authors, and extended to convex quadratic programs shortly thereafter. See, e.g., [13, 17, 7, 6]. A course classification of barrier methods both for LP and QP cases includes the primal, primal-dual, and predictor-corrector methods.

We adopt a primal-dual version of the QP barrier method described in [6], with the recursive KKT solver substituted. In the following the essentials of the approach to our problem formulation are sketched. Consider the general primal QP (9) with Lagrangian

$$L(y, w_2, w_3) = \frac{1}{2} y^T H y + J_1 y - (J_2 y + F_2)^T w_2 - (J_3 y + F_3)^T w_3 \qquad (24)$$

(we drop superscript k and write y instead of Δy) and its dual problem

$$\max_{z, w_2, w_3} -\frac{1}{2} z^T H z - F_2^T w_2 - F_3^T w_3 \quad \text{s.t.} \quad \begin{cases} H z + J_1^T - J_2^T w_2 - J_3^T w_3 = 0, \\ w_3 \geq 0. \end{cases} \qquad (25)$$

Introducing slack variables s we convert QP (9) to an equality-constrained barrier function form with potential parameter μ:

$$\min_y \frac{1}{2} y^T H y + J_1 y - \mu \sum_i \ln s_i \quad \text{s.t.} \quad \begin{cases} J_2 y + F_2 = 0, \\ -s + J_3 y + F_3 = 0. \end{cases} \qquad (26)$$

With $S = \mathrm{diag}(\{s_i\})$, $W_3 = \mathrm{diag}(\{w_{3i}\})$, $e = (1, \ldots, 1)$, the first order necessary conditions can be written

$$Hy + J_1^T - J_2^T w_2 - J_3^T w_3 = 0, \qquad (27)$$
$$-S^{-1}\mu e + w_3 = 0 \quad \text{or} \quad \mu e = S W_3 e, \qquad (28)$$
$$J_2 y + F_2 = 0, \qquad (29)$$
$$-s + J_3 y + F_3 = 0. \qquad (30)$$

A Newton step $(\delta y, \delta s, \delta w_2, \delta w_3)$ is obtained by solving the KKT system

$$\begin{pmatrix} H + J_3^T \Theta J_3 & J_2^T \\ J_2 & 0 \end{pmatrix} \begin{pmatrix} \delta y \\ -\delta w_2 \end{pmatrix} + \begin{pmatrix} (Hy + J_1^T - J_2^T w_2) - J_3^T \theta \\ J_2 y + F_2 \end{pmatrix} = 0 \qquad (31)$$

and equations

$$\delta s = J_3 \delta y, \quad \delta w_3 = \theta - \Theta \delta s - w_3, \qquad (32)$$

where $\Theta = S^{-1} W_3$, $\theta = S^{-1} \mu e$. Note that all iterates must satisfy equation (30) exactly, but need not satisfy (27,29). When (y, s, w_2, w_3) satisfy the necessary conditions for any μ, then y is feasible for the primal QP, (y, w_2, w_3) is feasible for the dual QP, and the duality gap is given by $s^T w_3 \geq 0$.

The primal equations are only slightly different. No information on the dual variable w_3 is maintained, and the definition of Θ must be changed to $\Theta = \mu S^{-2}$.

Mehrotra's predictor-corrector approach [12] defines $\hat{\theta} = 0$ first to compute the affine predictor step $(\delta\hat{y}, \delta\hat{s}, \delta\hat{w}_2, \delta\hat{w}_3)$; the corrector step as final search direction is then obtained for $\theta = S^{-1}(\mu e - \delta\hat{S}\delta\hat{W}_3 e)$.

We suggest the primal-dual approach because LP case experience of other authors shows that it is often more robust and takes fewer iterations than the primal approach, whereas the computational effort is only a little higher. On the other hand, a predictor-corrector method is considered superior only when matrix factorization clearly dominates the computations. This is not the case here, since any of the interior point variants respects our m-stage block-sparse QP structure:

Theorem 2 *The modified Hessian $H + J_3^T \Theta J_3$ is block-diagonal.*

Proof. This follows since Θ is diagonal (for both definitions), and H, J_3 are block-diagonal according to Theorem 1. ∎

Theoretical results indicate that a QP solution accuracy of d digits can be achieved after $\mathcal{O}(\sqrt{N}d) = \mathcal{O}(\sqrt{m}d)$ iterations. Hence, the total effort would be $\mathcal{O}(m^{1.5}d)$ operations. It should be pointed out, however, that the $\sqrt{N}$ iteration count applies to arbitrary QP's of dimension N. But under certain regularity assumptions the switching structure of the solution of the nonlinear optimal control problem (i.e., the sequence of constraint arcs and free arcs) will be reproduced on any sufficiently fine grid. In that case we expect a better convergence behavior.

Conjecture 1 *For a "regular" optimal control problem the QP subproblems are solved in $\mathcal{O}(d)$ interior point iterations and $\mathcal{O}(md)$ operations.*

Conclusions

We have outlined a structured SQP method for discretized nonlinear optimal control problems with path constraints. It was shown that a suitable choice of the discretization leads to QP subproblems resembling the very nature of the nonlinear optimal control problem in form of a specific m-stage block-sparse structure which is preserved when inequality constraints are treated by a primal-dual barrier method. A new recursive solver for the purely equality-constrained QP was presented which makes full use of the block-sparseness to produce a solution in $\mathcal{O}(m)$ operations.

Acknowledgement. The research was supported in part by the Deutsche Forschungsgemeinschaft (DFG).

References

[1] J. T. Betts, W. P. Huffman: *Path Constrained Trajectory Optimization Using Sparse Sequential Quadratic Programming*, Boeing Computer Services, 1991.

[2] H. G. Bock, K.-J. Plitt: *A Multiple Shooting Algorithm for Direct Solution of Optimal Control Processes*, Proc. 9th IFAC World Congress, Budapest, 1984.

[3] A. V. Fiacco, G. P McCormick.: *Nonlinear Programming: Sequential Uncons-trained Minimization Techniques,* John Wiley and Sons, New York, 1968.

[4] K. R. Frisch: *The Logarithmic Potential Method of Convex Programming,* Unpublished manuscript, University Institue of Economics, Oslo, 1955.

[5] P. E. Gill, W. Murray, M. A. Saunders, M. H. Wright: *A Schur-Complement Method for Sparse Quadratic Programming,* Report SOL 87-12, Department of Operations Research, Stanford University.

[6] P. E. Gill, W. Murray, D. B. Ponceleón, M. H. Wright: *Solving Reduced KKT Systems in Barrier Methods for Linear and Quadratic Programming,* Report SOL 91-7, Department of Operations Research, Stanford University.

[7] C. Gonzaga: *An Interior Trust Region Method for Linearly Constrained Op-timization,* MPS Newsletter 19 (1991) 55–65.

[8] C. R. Hargraves, R. W. Paris: *Direct Trajectory Optimization Using Nonlinear Programming and Collocation,* AIAA J. Guidance 10 (1987) 338–342.

[9] N. Karmarkar: *A New Polynomial Time Algorithm for Linear Programming,* Combinatorica 4 (1984) 373–395.

[10] J. Konzelmann, H. G. Bock, R. W. Longman: *Time Optimal Trajectories of Polar Robot Manipulators by Direct Methods,* Modeling and Simulation, 20/5 (1989) 1933–1939, Instrument Society of America.

[11] J. Konzelmann, H. G. Bock, R. W. Longman: *Time Optimal Trajectories of Elbow Robots by Direct Methods,* Proc. AIAA Guidance, Navigation and Control Conference, Boston (1989) AIAA Paper 89-3530-CP.

[12] S. Mehrotra: *On the Implementation of a (Primal-Dual) Interior Point Me-thod,* Technical Report 90-03, Department of Industrial Engineering and Ma-nagement Sciences, Northwestern University, Evanston, IL, 1990.

[13] R. D. C. Monteiro, I. Adler: *Interior Path Following Primal-Dual Algorithms. Part II: Convex Quadratic Programming,* Mathematical Programming 44 (1989) 43–66.

[14] K.-J. Plitt: *Ein superlinear konvergentes Mehrzielverfahren zur direkten Be-rechnung beschränkter optimaler Steuerungen,* Diploma Thesis (in German), Department of Applied Mathematics, University of Bonn, 1981.

[15] M. C. Steinbach, H. G. Bock, R. W. Longman: *Time Optimal Control of SCARA Robots,* Proc. AIAA Guidance, Navigation and Control Conference, Portland (1990) AIAA Paper 90-3394-CP.

[16] O. von Stryk: *Numerical Solution of Optimal Control Problems by Direct Collocation,* Report No. 322, Department of Mathematics, Munich University of Technology, 1991.

[17] Y. Ye, E. Tse: *An Extension of Karmarkar's Projective Algorithm for Convex Quadratic Programming,* Mathematical Programming 44 (1989) 157–179.

4 Software for Optimal Control Calculations

International Series of Numerical Mathematics, Vol. 115, © 1994 Birkhäuser Verlag Basel

Automated Approach
for Optimizing Dynamic Systems

Dieter Bestle*   and   Peter Eberhard*

Abstract

The optimal design of nonlinear dynamic systems can be formulated as a multicriteria optimization problem. On the basis of a multibody system model integral type objective functions are defined evaluating the dynamic behavior of the system under consideration. Multicriteria optimization methods reduce the problem to nonlinear programming problems which can be solved with standard algorithms like the SQP method. The gradients required for such an efficient optimization procedure are computed by solving additional differential equations resulting from an adjoint variable approach. The whole design process can be highly automated by using computer algebra packages.

Introduction

Due to the complexity of technical systems and the wide variety of conflicting specifications for their dynamical behavior, dynamic systems have been designed by engineers with help of experience and intuition for a long time. The design process has been based on experimental studies of prototypes resulting in rather long development cycles due to their time-consuming and costly construction.

Only recently, production companies have started to switch to a computer-aided design process to shorten development cycles and improve their products. In most cases, however, computers are used for parameter studies only, whereas the design itself is still found by intuitive changes of the design variables. On the other hand, optimization algorithms for solving standard nonlinear programming problems are highly developed.

It is the aim of this paper, therefore, to describe an integrated modeling and design approach consisting of four phases [1]: (i) formulation of a mathematical model, (ii) choice of design variables, (iii) definition of criteria, and (iv) optimization. A multibody system approach will be used for generating models for complex dynamic systems. Parameters of the model will serve as design variables and two types of criteria will be defined. Finally, a multicriteria approach will be applied to account for the presence of conflicting performance criteria in applications to

---

*University of Stuttgart, Institute B of Mechanics, D-70550 Stuttgart, Germany

real systems. In an interactive design process, the design engineer can provide information on the importance of each criterion, and he can choose between several multicriteria methods for reducing the problem to one or a recursive sequence of nonlinear programming problems which are solved by SQP methods.

Formulation of the design problem

Computer-aided design of dynamic systems has to be based on mathematical models. If we can neglect small deformations of the individual parts, the multibody system approach has shown to be a good representation of the system. A multibody system model consists of rigid bodies connected by ideal links and coupled by ideal force elements like springs, dampers or actively controlled elements, Fig. 1. Multibody system models have been used with success in vehicle dynamics, robotics, satellite dynamics and biomechanics.

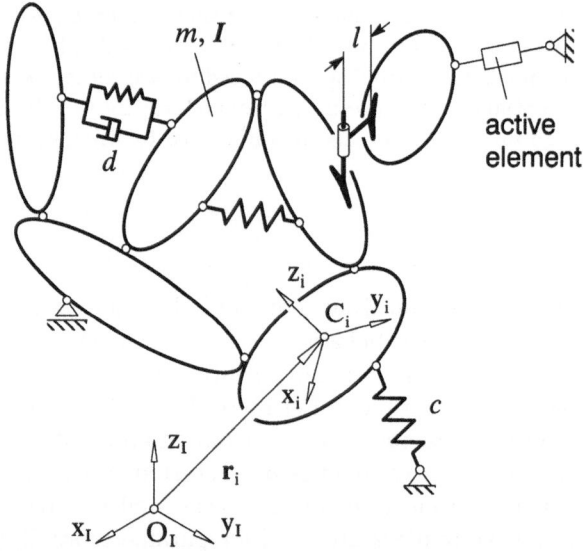

Figure 1: Multibody system model

Modeling technical systems as multibody systems involves an implicit parameterization. The dynamic behavior of the model is completely determined by parameters like the mass and moments of inertia of each body, geometrical dimensions, and damping and stiffness coefficients of coupling force elements. The parameters which can be changed within given ranges for optimizing the dynamical behavior are considered as design variables and summarized in a vector

$$\boldsymbol{p} \in I\!\!R^h\,, \quad p_k^l \leq p_k \leq p_k^u\,, \quad k = 1(1)h\,, \tag{1}$$

where p_k^l and p_k^u are lower and upper bounds, respectively, due to technical restrictions or physical meaning.

The dynamic behavior of a multibody system is described by differential equations of motion:

$$\begin{aligned}
\dot{\boldsymbol{y}} &= \boldsymbol{v}(t,\boldsymbol{y},\boldsymbol{z},\boldsymbol{p}) \\
\boldsymbol{M}(t,\boldsymbol{y},\boldsymbol{p})\,\dot{\boldsymbol{z}} + \boldsymbol{k}(t,\boldsymbol{y},\boldsymbol{z},\boldsymbol{p}) &= \boldsymbol{q}(t,\boldsymbol{y},\boldsymbol{z},\boldsymbol{p})
\end{aligned} \tag{2}$$

where $\boldsymbol{y}(t) \in I\!\!R^f$ and $\boldsymbol{z}(t) \in I\!\!R^g$ are vectors of generalized coordinates and velocities, respectively, and f and g are the numbers of degrees of freedom for position and velocity, respectively. The equations of motion result from Newton's and Euler's laws and d'Alembert's or Jourdain's principle for eliminating reaction forces and moments, e.g. [9]. The mass matrix $\boldsymbol{M}(t) \in I\!\!R^{g \times g}$ summarizes mass properties of the individual bodies, vector $\boldsymbol{k}(t) \in I\!\!R^g$ centrifugal and Coriolis forces, and $\boldsymbol{q}(t) \in I\!\!R^g$ applied forces resulting from gravity and coupling elements. For holonomic multibody systems we have $g = f$ and we can use $\boldsymbol{z} = \dot{\boldsymbol{y}}$.

The equations of motion for models of technical systems are already too complex for generating them by hand. Therefore, computer codes have been developed for a computer-aided modeling and generation of equations of motion in symbolical or numerical form [10]. Although such codes exist now for several decades, they are still subject of intensive research [11].

For a complete description of the motion, initial conditions for $\boldsymbol{y}$ and $\boldsymbol{z}$ have to be provided. This can be done by implicit conditions

$$\boldsymbol{y}^0: \quad \boldsymbol{\phi}^0(t^0, \boldsymbol{y}^0, \boldsymbol{p}) = 0, \qquad \boldsymbol{z}^0: \quad \dot{\boldsymbol{\phi}}^0(t^0, \boldsymbol{y}^0, \boldsymbol{z}^0, \boldsymbol{p}) = 0 \tag{3}$$

for some fixed starting time t^0.

In [2] the design problem has been stated for a single criterion. But generally, dynamic systems have to be optimal with respect to several specifications. Often such problems are simplified to nonlinear programming problems by choosing one criterion as objective function and the others as constraints. It is more natural, however, not to distinguish between objective functions and constraints in such an early design phase, and consider the decision on the importance and type of each criterion as part of a multicriteria optimization process.

Mainly, two types of criteria are used: we will call a criterion to be explicit if it is an algebraic function of the design variables:

$$\psi_i^E = \psi_i^E(\boldsymbol{p}), \quad i = 1(1)n_E. \tag{4}$$

A second type of performance criterion evaluating the dynamic behavior of multibody systems can be formulated as an integral type performance function

$$\psi_i^I = G_i^1(t^1, \boldsymbol{y}^1, \boldsymbol{z}^1, \boldsymbol{p}) + \int_{t^0}^{t^1} F_i(t, \boldsymbol{y}, \boldsymbol{z}, \dot{\boldsymbol{z}}, \boldsymbol{p})\, dt, \quad i = 1(1)n_I, \tag{5}$$

which is also known from optimal control problems. The first term accounts for cases where special values for the final state $\boldsymbol{y}^1$, $\boldsymbol{z}^1$ or a minimum time t^1 must be achieved, the second term evaluates the dynamic behavior within an interesting time interval $[t^0, t^1]$. The final time t^1 may be fixed or given implicitly by the final state:

$$t^1 : \quad H^1(t^1, \boldsymbol{y}^1, \boldsymbol{z}^1, \boldsymbol{p}) = 0. \tag{6}$$

Although the functions G_i^1 and F_i depend on state variables, the functions ψ_i^I are determined entirely by the values of the design variables $\boldsymbol{p}$ due to Eqs. (2) and (3).

Multicriteria optimization

The problem of optimizing dynamic systems with respect to several conflicting criteria does not have a single optimal solution. The theory of multicriteria optimization has shown that the optimum depends on additional decisions of the designer.

At the beginning of the optimization phase, the designer has to classify all the criteria (4) and (5) as objective functions or constraints. Objective functions are criteria which should be minimized with respect to the design variables. Constraints are criteria which should have a special value or be less than an upper bound. Summarizing all objective functions in a vector function $\boldsymbol{f}(\boldsymbol{p}) \in I\!\!R^n$, all equality constraints in $\boldsymbol{g}(\boldsymbol{p}) = 0$, and all inequality constraints in $\boldsymbol{h}(\boldsymbol{p}) \leq 0$, we end up with the optimization problem

$$\underset{\boldsymbol{p} \in \mathcal{P}}{\text{minimize}} \quad \boldsymbol{f}(\boldsymbol{p}) \quad \text{where} \quad \mathcal{P} := \left\{ \boldsymbol{p} \in I\!\!R^h \mid \boldsymbol{g}(\boldsymbol{p}) = 0,\ \boldsymbol{h}(\boldsymbol{p}) \leq 0 \right\}. \tag{7}$$

If only a single criterion is left, $n = 1$, problem (7) is called a nonlinear programming problem [4]. More realistic, however, is that more than one criterion has to be minimized simultaneously, $n > 1$. Then it is called a multicriteria or vector optimization problem, e.q. [8].

In the latter case, we cannot expect a feasible design point $\boldsymbol{p} \in \mathcal{P}$ where all objectives become minimal. Therefore, a design point $\boldsymbol{p}^P \in \mathcal{P}$ is defined to be Pareto-optimal if there is no other feasible point $\boldsymbol{p}$ with $f_i(\boldsymbol{p}) \leq f_i(\boldsymbol{p}^P)\ \forall i$ and $f_j(\boldsymbol{p}) < f_j(\boldsymbol{p}^P)$ for at least one j [13]. In general, Pareto-optimal solutions are not unique, and the designer has to choose a special Pareto-optimal point as desired solution due to additional information on the design problem. For finding such points the multicriteria optimization problem has to be reduced to a scalar one for which efficient routines exist, Fig. 2. This reduction is based on two principles: scalarization and hierarchization.

In case of scalarization, Fig. 3a, the objective functions are combined to a new utility function $u(\boldsymbol{p}) \in I\!\!R$ which will be optimized instead of the vector criterion. A well known approach using scalarization is the weighting objectives method:

$$u(\boldsymbol{p}) := \sum_{i=1}^{n} w_i \frac{f_i(\boldsymbol{p})}{\hat{f}_i}, \qquad \sum_{i=1}^{n} w_i = 1 \tag{8}$$

a) b)

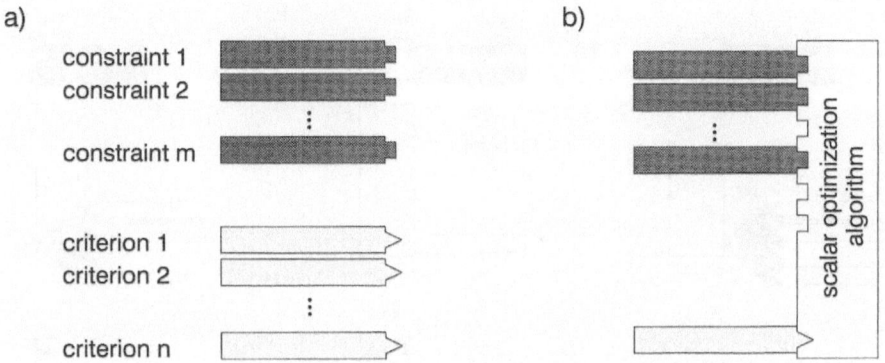

constraint 1
constraint 2
⋮
constraint m

criterion 1
criterion 2
⋮
criterion n

scalar optimization algorithm

Figure 2: Difference between vector optimization problems (a)
and scalar optimization problems (b)

where $w_i \in [0,1]$ are weighting coefficients and $\hat{f}_i$ are scaling factors. Other possibilities are absolute and relative distance functions with respect to a predefined design point [5].

For hierarchical methods the designer has to assign a level of importance l_i to each objective function $f_i(\boldsymbol{p})$ where one is the level of most important criteria. Especially, level zero is assigned to the equality and inequality constraints in Eq. (7), i.e.,

$$\mathcal{P}_0 := \mathcal{P} \quad \longleftrightarrow \quad l_0 = 0. \tag{9}$$

If there is only a single criterion on each level, we define in the first step a scalar optimization problem by neglecting the objectives on lower levels, and taking into consideration only the objective function on level one and the constraints, Fig. 3b:

$$f_i^* = \min_{\boldsymbol{p} \in \mathcal{P}_0} f_i(\boldsymbol{p}) \quad \text{where} \quad i: l_i = 1. \tag{10}$$

For the next step we can use the information on the optimal value f_i^* of the most important objective to define a constraint on $f_i(\boldsymbol{p})$:

$$\mathcal{P}_i := \left\{ \boldsymbol{p} \in \mathbb{R}^h \mid f_i(\boldsymbol{p}) \leq (1+\varepsilon_i) f_i^* \right\} \tag{11}$$

where $\varepsilon_i > 0$ is a user defined tolerance for function increase. Then we formulate a new scalar optimization problem for the objective function on level two similar to Eq. (10). The whole procedure is a recursive sequence of n scalar optimization problems:

$$f_i^* = \min_{\substack{\boldsymbol{p} \in \bigcap_{k: \, l_k < j} \mathcal{P}_k}} f_i(\boldsymbol{p}) \quad \text{where} \quad i: l_i = j, \quad j = 1, 2, \dots \; . \tag{12}$$

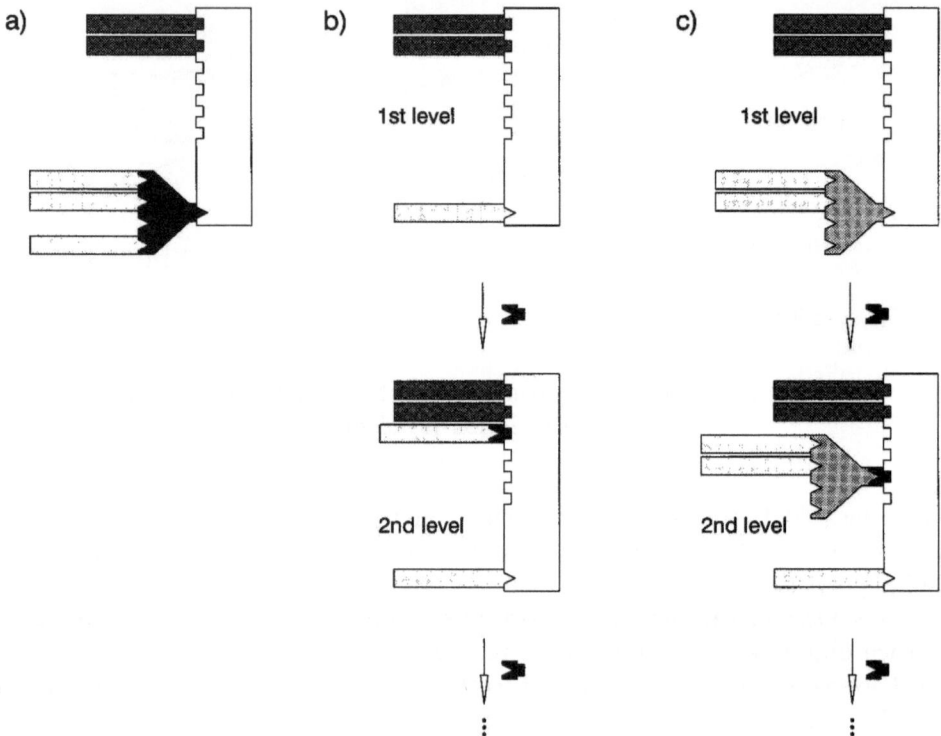

Figure 3: Multicriteria optimization principles: (a) scalarization,
(b) hierarchization, and (c) combination

The scalarization and hierarchization principles can be combined if several objectives are on the same level, Fig. 3c. For example, goal programming becomes a flexible tool and gives good insight into the problem if both principles are used together. The user has to define goals $\hat{f}_i$ to be reached and assign levels l_i to each objective $f_i(\boldsymbol{p})$. Then, instead of the objectives the deviations from the goals are minimized. Objectives on the same level j can be combined to utility functions similar to Eq. (8):

$$u_j(\boldsymbol{p}) := \sum_{i:\ l_i=j} w_i \max\{0, f_i(\boldsymbol{p}) - \hat{f}_i\}, \qquad \sum_{i:\ l_i=j} w_i = 1. \qquad (13)$$

The utility functions can then be handled according to Eqs. (10) to (12).

Solution of the scalar optimization problem

The scalar optimization problems resulting from a multicriteria approach have themselves to be solved in an iterative procedure. Due to the high computational effort for evaluating integral type performance functions by numerical integration, optimization algorithms like the SQP methods with nice convergence properties should be used. The drawback of such methods, however, is the use of gradients.

A simple way of computing gradients is the use of finite differences. Applied to integral type criteria these approximations cause several problems. Numerical experience has shown that due to the limited accuracy of the function values finite differences are not very reliable near the optimum. On the other hand, we need one additional function evaluation for each perturbed design variable which is a time consuming numerical simulation of the dynamic behavior of the multibody system.

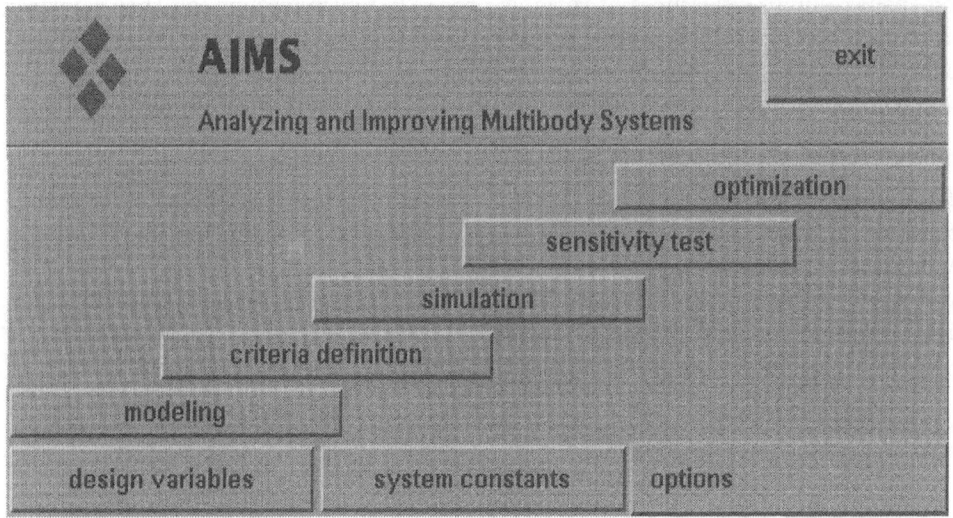

Figure 4: Graphical user interface AIMS

Therefore, a semi-analytical approach is used for computing gradients of this type of criteria which is called adjoint variable method [2]. This results in additional differential equations for the gradient where the computational effort is almost independent of the number of design variables. Numerical studies have shown high reliability and about the same accuracy for the gradients as for the function values.

Automated optimization approach

Due to the complexity of models for technical systems the design process has to be supported by approved computer programs. These programs can be integrated in a graphical user interface which also helps to organize the whole design process and especially the interactive and iterative optimization phase, Fig. 4. The user interface AIMS (Analyzing and Improving Multibody Systems) integrates several numerical and computer algebra programs for modeling, simulation and optimization.

Modeling is supported by the computer program NEWEUL [7] which generates symbolical equations of motion for multibody systems. Criteria of type (4) or (5) have to be defined by the user in a MAPLE-compatible form. The computer algebra package MAPLE [3] will then generate problem-specific INCLUDE-files which can be linked together with problem-invariant FORTRAN-code for simulation and optimization. Simulation is performed with a multistep integration algorithm [12]. As already mentioned, the gradients are computed from additional differential equations which can also be generated by MAPLE. These gradients should always be checked on consistency with the problem definition by comparing them to finite differences of variable order [2]. Optimization can then be performed interactively using multicriteria optimization methods where the resulting nonlinear programming problems are solved by SQP methods [6].

Application to Vehicle Control

The application to a plane vehicle model shows some principal effects of different optimization approaches. The model has 6 degrees of freedom described by the generalized coordinates $y = [y, z, \alpha, \phi, w, z_D]^T$, and consists of four bodies: the car body, the driver, and the two wheel sets, Fig. 5. The vehicle has to be optimized with respect to comfort and riding safety.

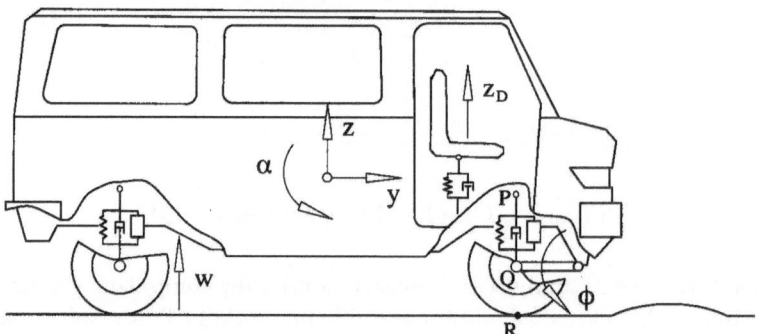

Figure 5: Plane vehicle model

A frequently used measure for comfort is the vertical acceleration $\ddot{z}_D$ of the driver. Since driving over a bump is considered as a test, accelerations are penalized by time to avoid long term vibrations:

$$f_1 := \int_{t^0}^{t^1} (t\,\ddot{z}_D)^2 \, dt. \tag{14}$$

Optimal comfort is then expressed by a minimal value of f_1. Riding safety is related to the dynamic variation of the load between the wheels and the road. If the tire is considered as a linear spring, the load is proportional to the relative displacement between the wheel and the road surface:

$$f_2 := \int_{t^0}^{t^1} (z_Q - z_R)^2 \, dt. \tag{15}$$

A constraint on the design of suspension systems is the limited space for relative displacement between wheel and car-body. A criterion like

$$f_3 := \int_{t^0}^{t^1} \left(\frac{z_P - z_Q}{s_0}\right)^6 \, dt \tag{16}$$

may be used where s_0 is a predefined amplitude which should not be exceeded to much. For improving the dynamic behavior of the vehicle the stiffness and damping parameters of the front and rear suspension are used as design variables.

Fig. 6 shows some results for the weighting objectives method for different weighting coefficients. If only riding comfort is considered as criterion it can be improved drastically compared to the initial design. But this improvement is achieved at the expense of riding safety and it requires a large suspension displacement. For a more realistic design optimization all three criteria have to be taken into consideration. As Fig. 6 shows, riding comfort still can be improved but the improvement depends highly nonlinear on the weighting coefficients.

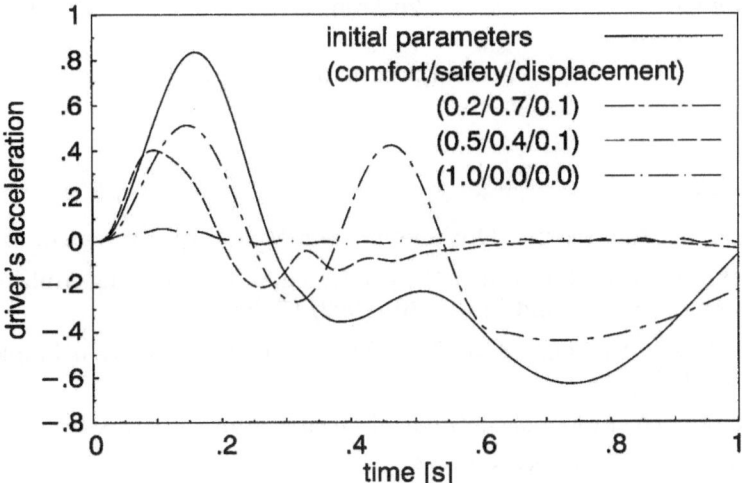

Figure 6: Weighting objectives method

Goal programming can give better insight into the problem. As a measure of importance, level one is assigned to riding comfort, level two to riding safety, and level three to relative displacement. Fig. 7 shows the results of three different runs where the values of the criteria are normalized with respect to the initial design. In the first run, the goals are set very cautiously and can be achieved. In the second run, the goal for riding safety is decreased and cannot be achieved anymore. Therefore, in the third run the goal for riding safety is increased and the relative displacement can be decreased at the expense of comfort. In a practical design process, the designer has to explore the design space with further runs and find a suitable balance for the conflicting criteria.

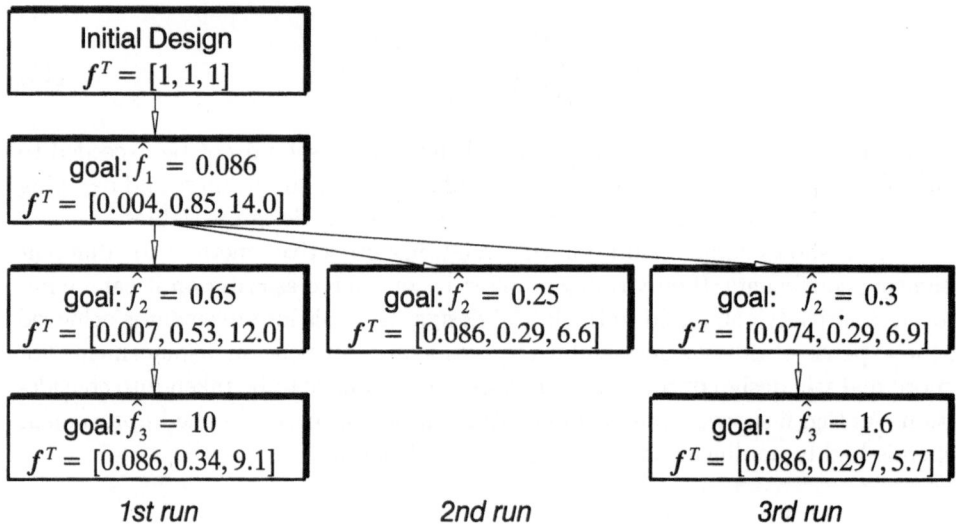

Figure 7: Goal programming method

References

[1] BESTLE, D., Analyse und Optimierung von Mehrkörpersystemen, to appear.

[2] BESTLE, D., AND EBERHARD P., Analyzing and Optimizing Multibody Systems, Mech. Struct. and Mach. **20** (1992) 67–92.

[3] CHAR, B.W., ET. AL., MAPLE-Reference Manual, Waterloo Maple Publ., Waterloo 1990.

[4] FLETCHER, R., Practical Methods of Optimization, Wiley, Chichester 1987.

[5] HWANG, C.-L., AND MASUD, A.S., Multiple Objective Decision Making: Methods and Applications, Springer, Berlin 1979.

[6] IMSL MATH LIBRARY, User's Manual, IMSL Inc., Houston 1989.

[7] KREUZER, E., AND LEISTER, G., Programmsystem NEWEUL'90, Anleitung AN-24, Institute B of Mechanics, University of Stuttgart, Stuttgart 1991.

[8] OSYCZKA, A., Multicriterion Optimization in Engineering, Ellis Horwood, Chichester 1984.

[9] SCHIEHLEN, W., Technische Dynamik, Teubner, Stuttgart 1986.

[10] SCHIEHLEN, W. (ed.), Multibody Systems Handbook, Springer, Berlin 1990.

[11] SCHIEHLEN, W. (ed.), Advanced Multibody System Dynamics, Kluwer, Dordrecht 1993.

[12] SHAMPINE, L.F., AND GORDON, M.K., Computer Solution of Ordinary Differential Equations: The Initial Value Problem, Freeman, San Francisco 1975.

[13] STADLER, W., Multicriteria Optimization in Mechanics: A Survey, Appl. Mech. Rev. 37 (1984) 277–286.

International Series of Numerical Mathematics, Vol. 115, © 1994 Birkhäuser Verlag Basel

ANDECS: A Computation Environment for Control Applications of Optimization*

G. Grübel[†]   R. Finsterwalder[†]   G. Gramlich[†],

H.-D. Joos[†]   and   S. Lewald[†]

Abstract

An engineering control design environment is reported on which integrates optimal control synthesis within a vector-optimization based engineering control design frame. Dealt with in particular are: a conceptual frame of a design-process feedback loop, a modular software realization thereof, and features of a human-interaction facility for computational design experimenting.

Keywords. Optimal control synthesis, vector-optimization control design, software environment for design experimentation.

Introduction

'Control Applications of Optimization' can be seen under the perspective of mathematical control synthesis and under the perspective of engineering control design. Although worked on in different disciplines, mathematical control synthesis i.e. the development of existence conditions, algorithms, and their numerically efficient and reliable software implementations, and engineering control design i.e. the iterative task of finding a best-possible compromise solution among conflicting goals in a specific application, must be seen as complementary activities. Even more so, we need a coherent environment where results of both activities play together in a synergetic way. ANDECS[1] of DLR was developed in this spirit to provide a coherent environment where results of mathematical control synthesis can be modularly integrated and investigated within a goal-attainment framework for computational control-engineering design.

In the following sections we report on optimal control synthesis as part of the conceptual frame for control design, the software frame to produce application results, and the human-interaction frame for the set-up, execution, and monitoring of computation experiments for both synthesis and design. Throughout the paper

---

*Invited Paper

[†]Control Design Engineering, Institute for Robotics and System Dynamics, DLR – German Aerospace Research Establishment, D-82234 Oberpfaffenhofen

[1]ANDECS®(**AN**alysis & **DE**sign of Controlled **S**ystems) is a registered trademark of DLR

we see the particular role of mathematical optimization techniques from a user's point of view.

Control Synthesis

A control synthesis algorithm either generates a control function $u(t)$, or a control law $u = f(\dot{u}, y, t)$ where $y(t)$ are the available feedback variables of the plant which is to be controlled.

The best known optimization based control law is $u(t) = K \cdot x(t)$, i.e. linear feedback of the plant state $x(t)$ solving the so-called optimal LQ problem [1]:

$$\min_{u} \frac{1}{2} \int_{0}^{\infty} [x^T(t)Qx(t) + u^T(t)Ru(t)] \, dt$$

$$\text{s.t. } \dot{x} = Ax + Bu, \quad x(0) = x_0.$$

The feedback gain K can be computed in closed form using the Riccati equation:

$$\begin{aligned} K &= -R^{-1}B^T S \\ 0 &= SA + A^T S + SBR^{-1}B^T S - Q. \end{aligned}$$

The feedback gain K depends on the 'design parameters' Q, R, which determine the dynamic behaviour of the closed-loop feedback system, but is independent of the system initial conditions x_0. A solution exists if $\{A, B\}$ is completely controllable (stabilizable) and $R > 0, Q \geq 0$.

If we restrict the control law to output feedback $u = F \cdot y(t)$, with $y = Cx$, we may solve the LQ parameter optimization problem

$$\min_{F} \int_{0}^{\infty} x_0^T e^{[A+BFC]^T t}[Q + C^T F^T RFC]e^{[A+BFC]}x_0 dt$$

$$\text{s.t. } Re \, \lambda(A + BFC) < 0.$$

This is called the Parametric Linear Quadratic control problem (PLQ). For this optimization problem there is no general closed form solution and the optimal feedback matrix F depends on the system initial conditions x_0. Again, Q, R, are the free 'design parameters' to tune the feedback system dynamics.

Various methods are available to iteratively solve the PLQ problem, e.g. [18], starting from a stabilizing F_0. A recent study [20] comparing the efficiency of advanced implementations of the Anderson-Moore descent algorithm [19], a problem-specific Newton algorithm [3], a Quasi-Newton algorithm with various update formulae [3], and a Least-Change-Quasi-Newton algorithm [3], led to a combined Anderson-Moore/Quasi-Newton algorithm as being most efficient. This algorithm makes a few Anderson-Moore iterations before starting the Powell-symmetric-Broyden update of the Quasi-Newton algorithm.

Both problems, the LQ and the PLQ problem, can be generalized to a dynamic feedback control law $u = f(\dot{u}, y)$. The generalization of the LQ problem is the well known optimal Linear Quadratic Gaussian (LQG) problem [1]. Here two Riccati equations have to be solved for control synthesis: the dual 'filter' Riccati equation to compute a Kalman-filter gain, in addition to the 'control' Riccati equation for computing a state-feedback gain as above. The free 'design parameters' are the fictitious covariance matrices V, W, for process and measurement noise, and Q, R, as above.

The PLQ problem is the core computation problem for the synthesis of structured, e.g. low-order, LQ dynamic compensators. This is explored in detail in [11], where also synthesis formulae are given to deal with specific control structures and various control engineering demands such as parameter sensitivity minimization.

Linear feedback control laws are used for coping with the disturbance regulator problem, i.e. disturbance attenuation around a nominal system behaviour where a linear error model can be assumed. The nominal, nonlinear behaviour itself is to be controlled by a ('feedforward') control function $u(t)$.

For the numerical synthesis of optimal control functions under control- and state constraints, we use essentially the TOMP method, i.e. Trajectory Optimization by Mathematical Programming [15]. This method nicely fits into the modular (vector-) optimization-based control design frame as dealt with in the next section. In that context, the control synthesis algorithm is just the approximation of a control function $u(t) = A(p_i, t_i)$ from vertices $[p_i, t_i]$ by a suitable approximation scheme A.

Control Design Frame

Engineering control design in itself is an iterative feedback process which requires interrelations [5] with the customer's goals and needs for which a control system is to be designed, with the established expertise embodied in the various control analysis and synthesis methods, and with the real world of physics and control realization constraints. Correspondingly, this requires computational experimentation on three strata [7]:

- Parametric experimentation for trade-off exploration and goal attainment to best satisfy the customer's needs.
- Method experimentation for demand modelling and performance evaluation to make best use of scientific expertise.
- Model experimentation for system modelling and result validation to best cope with physical reality in a computationally tractable manner.

In this section we deal with a conceptual model [8] of how design proceeds. In particular we deal with the task of parametric experimentation for trade-off exploration and goal attainment. The approach which we call 'demand-driven experimental search' is characterized by explicit decision making in the demand space and automatic search-iterations in the design space using vector-optimization. The structure of this approach is depicted later in figure 2.

In parametric control-dynamics design the system dynamics to be designed, e.g. a control function or a control law, can be parameterized either by an analytic synthesis algorithm $P = f(T)$ with free tuners T or directly by the dynamics model parameters P. We call $\{T\}$ the design space and $\{P\}$ the parameter space. In this view a control function $u(t) = A(p_i, t_i, \tilde{p}_i, \tilde{t}_i)$ with interpolation parameters $[p_i, t_i], [\tilde{p}_i, \tilde{t}_i]$ and an approximation scheme A belongs to $\{P\}$, whereas the free 'tunable' interpolation parameters $[\tilde{p}_i, \tilde{t}_i]$ belong to $\{T\}$.

The mathematical model M of the overall dynamical system, e.g. {plant $\oplus$ control}, belongs to a class of 'Control-Dynamics Objects' (CDO) as defined and dealt with in [13]. Possible CDO-classes for dynamic systems representations are 'general nonlinear system', 'generic state-space linear system', 'transfer matrix', etc..

Then as depicted in figure 1, we can structure a computational chain:

A *dynamics synthesis* algorithm $P = f(T)$ maps an analytically defined system property (e.g. the dynamics laws of mechanics, the generic properties of linear optimal controllers, or hyperstability) on the design object, i.e. the *dynamics model M*.

Tuners T may be the mechanical system parameters of bodies, joints, or force elements if the dynamics synthesis algorithm is a multibody algorithm; in optimal LQ control law design tuners are the weighting coefficients of a Riccati control law synthesis; in designing positive-real (hyperstable) control laws the tuners are the free elements in the equations of the Kalman-Yacubovich lemma [2]; in optimal control via Trajectory Optimization by Mathematical Programming [15], tuners are the interpolation parameters of the input functions to the dynamics model.

Indicators I such as eigenvalues, time responses, frequency responses, characterize the behaviour of a design object in the state space, time domain, or frequency domain, respectively. The indicators show attributes of design objects: Whether a linear time-invariant system is stable or unstable is seen by the real part of the eigenvalues; whether such a system is hyperstable can be seen from its frequency response behaviour.

Dynamics analysis and simulation modules are the means to compute indicators.

In this framework, design is the task to choose a 'well-balanced' specimen among all instances in the chosen class of design-object models $\{M(P)\}$. This requires to determine an appropriate candidate in the design space $\{T\}$, while judging the object's behaviour in the indicator space $\{I(M)\}$. Since in the beginning there is uncertainty of what constitutes a suitable 'well-balanced' trade-off, this is a task of re-iterating search.

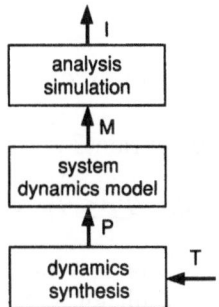

Figure 1: The computation chain from tuners T to indicators I.

A systematic approach of search we call experimenting. In this sense, an experimental approach most naturally fits the designer's role in dealing with multi-objective design tasks. This is true in particular if the experimentation work frame for the designer is confined to be just a decision process in demand space whose outcome directs an automatically running parametric search loop in the design space. Such a work frame combines the human strength, i.e. the ability to abstract, simplify and conceptualize in setting and refining goals, with the strength of a computer, i.e. super-fast data- and information processing. For that the design logic has to be suitably augmented, cf. figure 2:

The designer has to define measures for attributes $C = \{c_i\}$ which serve as *criteria* to quantitatively judge whether a design is well balanced. Design decisions w.r.t. searched-for performance levels for $C = \{c_i\}$ can now be made by corresponding *design directors* $D = \{d_i\}$. For that, a *design-comparator* α is required to relate C and D. The comparator outcome α must steer a *design synthesis* module which generates tuner candidates in the design space $\{T\}$ and thereby closes this feedback loop via $\{P(T)\}$, $\{M(P)\}$, $\{I(M)\}$.

To formalize further, we use criteria c_i which are judged the better the smaller their value is. Based on this assumption, as a *design comparator* we use the max-function $\alpha = \max\{c_i/d_i\}$, d_i being the particular design director corresponding to c_i. As a consequence, if $\alpha < 1$ we have $c_i \leq \alpha d_i < d_i$, or in vector notation $C \leq \alpha D < D$:

$$\begin{bmatrix} c_1 \\ \vdots \\ c_i \\ \vdots \end{bmatrix} \leq \alpha \begin{bmatrix} d_1 \\ \vdots \\ d_i \\ \vdots \end{bmatrix} < \begin{bmatrix} d_1 \\ \vdots \\ d_i \\ \vdots \end{bmatrix}.$$

A design is the better, the smaller the criteria vector C is compared to the design-director vector D. Hence $\alpha = \max\{c_i(T)/d_i\}$ should be minimized as a function of the tuners T. This is the purpose of *design synthesis*:

$$\min \ \alpha(T) \to T$$

subject to performance and tuning restraints r

$$r: \ g_j(T) \geq 0; \quad \underline{T}_k \leq T_k \leq \overline{T}_k.$$

This min-max formulation with α is a zero-value goal-attainment approach.

By this approach, each design result for a chosen set of design directors $\{d_i\}$ is a *pareto-optimal* design, which means that no criterion value can be decreased without increasing at least one other criterion value in the set $\{c_i\}$. Hence by varying the design directors d_i, we are examining the set of "best-possible compromise" solutions. For the ν-th design step we solve

$$\min_{T^{(\nu)}} \ \max_i \ \{c_i^{(\nu)}(T^{(\nu)}) \, / \, d_i^{(\nu)}\} \, ,$$

where the designer chooses each $d_i^{(\nu)}$ depending on the criterion value $c_i^{(\nu-1)}$ of the previous design step. Choosing a posteriori $d_i^{(\nu)} > c_i^{(\nu-1)}$, yields the design systematic of Kreisselmeier [16] for feasible design directions.

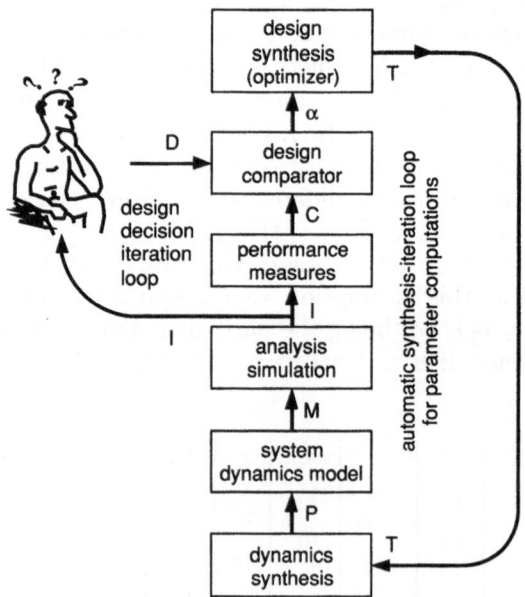

Figure 2: The human design-decision loop and the "background" vector-optimization loop.

Vector-Optimization plays a central role in this design frame. The minmax scalarized problem is solved using various mathematical programming methods and algorithms. The mathematical programming problems to be solved in general are not convex, are not Linear Programs nor Quadratic Programs, they are not sparse and not very large. That is, contrary to the control synthesis part, we do not have any structure. In general we have non-convex, nonlinear, and often non-smooth problems. This requires to have a suitably broad palette of optimization methods and algorithms at one's disposal. The methods and algorithms palette available in ANDECS is shown in table 1.

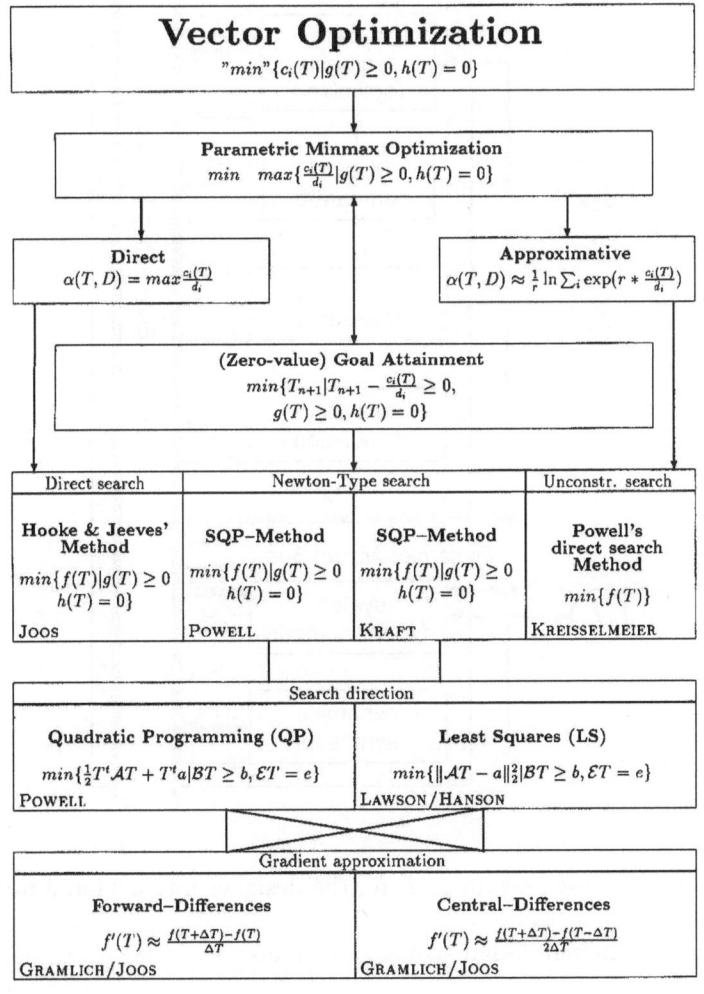

Table 1: The numerical methods for vector optimization used in ANDECS.

Modular Software Frame

The conceptual control design frame as described in the preceding section is implemented in a modular, application-flexible software frame called ANDECS-MOPS (Multi-Objectives Programming Systems) [14] [7]. Figure 3 shows the software modularity corresponding to the design loop of figure 2. Also indicated are the three design strata I, J, K an engineering control design has to be iterated on.

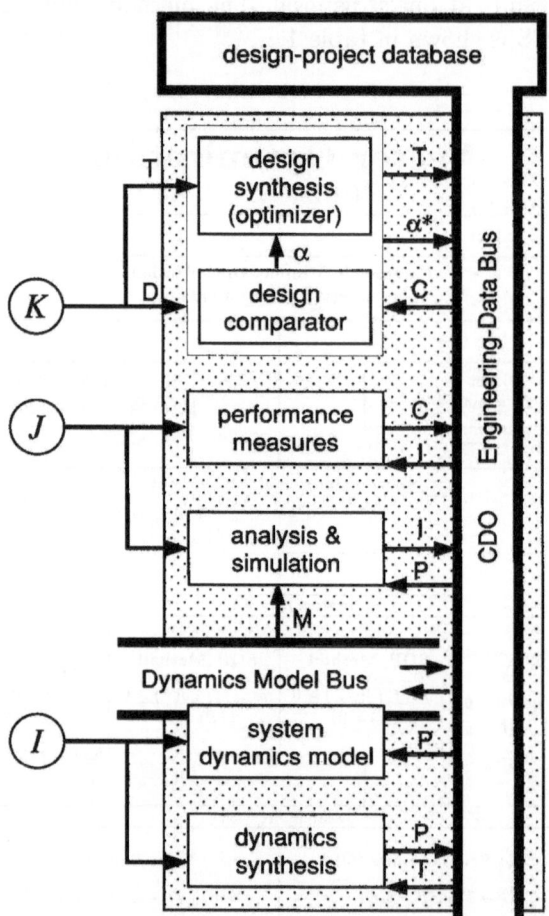

Figure 3: The 3 design strata I, J, K, the designer interacts and iterates on.

The three principal design strata I, J, K are:

(I) The designer creates or modifies the model M or its parameterization P. This task is supported by available system modelling and synthesis tools.

(*J*) The designer defines and modifies the performance measures.
The design specifications have to be expressed by performance criteria C based on the indicators I, which are computed by analysis or simulation of the dynamics model.

(*K*) The designer weights the various performance criteria by the design directors D to achieve the trade-off that satisfies the chosen demand level best.

The 3 decision strata I, J, K form a hierarchy. Hence a hierarchical database is best suited to record and reflect the design history, figure 4:

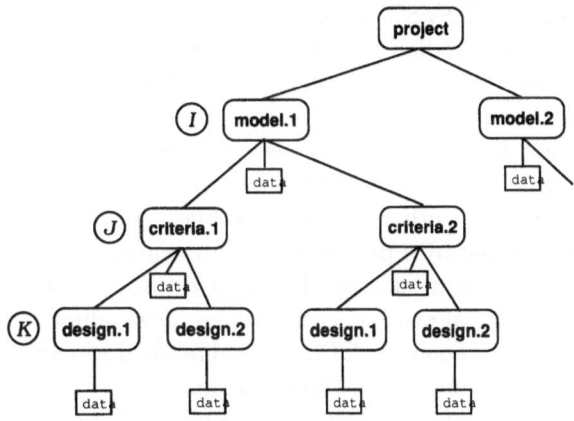

Figure 4: Structure of a design-history database.

The variability of possible design decisions and the amount of computational data necessary to get the proper indicators for these decisions, make it necessary to support the designer in

- automatically recording the various design steps and the pertinent data,

- allowing to compare any design outcomes
  by making the necessary indicator data easily recoverable,

- allowing backtracing and branching of the design process.

Based on the formal structuring I, J, K, of the design process the data structure of figure 4 yields the required formal design support in that automatically all design steps are recorded for backtracing and comparing different design steps. Branching of the design process is possible by choosing an already existing design $(I, J, K)_a$ as the first branch of a new design-history data structure.

Variability and flexibility during design requires application-adaptable software. Figure 3 shows the data flow in a MOPS design iteration. The corresponding software architecture is based on tool abstraction: For each of the function blocks, such as system dynamics model evaluation; analysis and simulation; performance criteria computation; design synthesis (i.e. vector optimization); dynamics synthesis (e.g. PLQ optimization); a bundle of function modules (toolies) is available. These communicate among each other via the 'engineering-data bus'. The communication data are Abstract Data Types (1D-, 2D-, 3D-arrays, parameter sets of numeric and symbolic values, text, pixel-pictures), as well as Control Data Objects (CDO) which are higher-level data structures (e.g. 'signal', 'transfer matrix', 'generic linear state space model'), and which are formally defined via the class scheme of the underlying RSYST database [9].

From the available bundle of toolies, computational sequences can be composed by a dialog and macro facility. In essence, MOPS is a monitoring module which drives user-defined application macros.

Tool abstraction has also consequences for the *mathematical programming* algorithms. It requires, that these algorithms are also implemented as modules which are *independent* from other function modules, such as modules for criteria evaluation. This is achieved by a program structure called *reverse communication*, where the control of the computational flow is returned from the optimizer to the driver module MOPS on every occasion where new values of the performance measures or gradients are required. All the mathematical programming algorithms listed in Table 1 therefore have been (re-)implemented in reverse communication. They are available in the control numeric subroutines library RASP [9].

Human Interaction Frame

Human feedback has to close the design loop between the design indicators and the design directors, i.e. the performance demands D. In order to explore properties that may be difficult to understand or even remain unnoticed otherwise, all design relevant data are simultaneously displayed in multiple graphical views while the design process is in action. Since previous results remain on the screen, the designer gets an immediate visual feedback about the effect of changes in the problem data. This closes the loop between computation/visualization and the designer.

The design environment has been developed in such a way [6], that at any time the computation can be interrupted or stopped. Then the designer can examine actual and previous results by the following facilities:

- Selecting a particular design candidate by picking any point/curve on any design related display invokes that all curves belonging to this design are highlighted.

- Scanning the design history stepwise along any one of the performance measures c_i yields all data of an actualized candidate being visualized. The actual candidate is highlighted, while the highlighting of the previous chosen candidate is then reset.

- Tradeoff-analysis via the conflict-editor gives a design-history-based information of what criteria are competing how strongly.

- Information Zooming: Validation of results requires to look at data from different abstraction levels. In general, this implies that more detailed computations for the corresponding visualizations are needed. For example, in axis 3 of the criteria-conflict editor the maximum amplitude of a time simulation is displayed as criterion c_3. By clicking this performance item, a new simulation run is started and the complete trajectory together with previously shown trajectories is visualized in a new window.

- Interactive Steering: By acting on the design directors or tuners, a new computation is started and new results are immediately shown by the displays. If the computations are fast enough, this allows the designer to do on-line (multi-) parameter sensitivity explorations via 'analog' input-devices such as the DLR steering ball or mouse-driven sliders.

Two *optimization-related human interaction* aspects deserve a special mention. These are the need for a decision support facility - called criteria conflict editor - for specifying a locally demand-consistent vector-optimization problem set up in each design iteration, and an on-line monitoring of the optimizer runtime behaviour via a 'status face'. These two aspects are briefly dealt with in the following.

By the design directors D the designer opts for a specific 'best-compromise' solution in a pareto-optimal solution set. This means, no criterion in this set can be decreased any further without increasing the value of at least one conflicting criterion. Hence the designer should know, at least locally, which criteria are coherent and which criteria are in conflict, in order to make a consistent choice of demands in updating the design directors D for the next design step.

To visualize criteria conflicts is the purpose of the ANDECS *conflict editor*. Since in common engineering applications we usually have many more than 3 criteria, a criteria space visualization in cartesian coordinates is not feasible. Therefore 'parallel coordinates' [12] are used. Figure 5 shows the example of 3 points displayed in a cartesian criteria space and the corresponding 3 polygonal lines representing these 3 points in parallel coordinates. The editor function allows "klicking through" the already found pareto-optimal solutions in the different design steps, thereby showing the coherent or contrary behaviour of the various criteria, figure 6. The editor also allows to re-order coordinates for e.g. sorting the criteria in two sets of coherent criteria which are in conflict with each other.

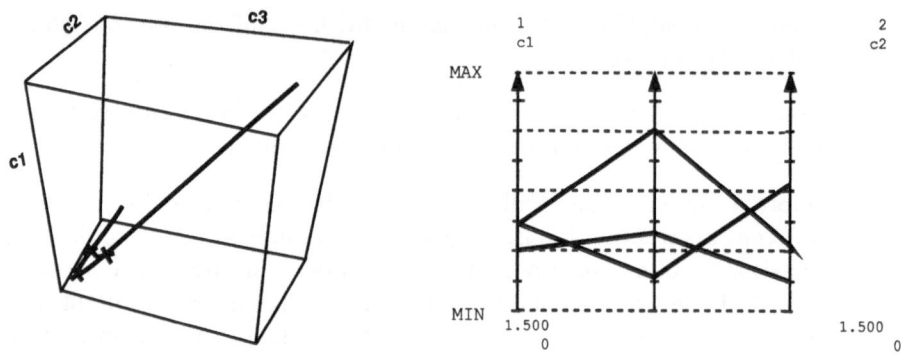

Figure 5: On the left: 3 points in cartesian coordinates,
on the right: display of these 3 points in parallel coordinates.

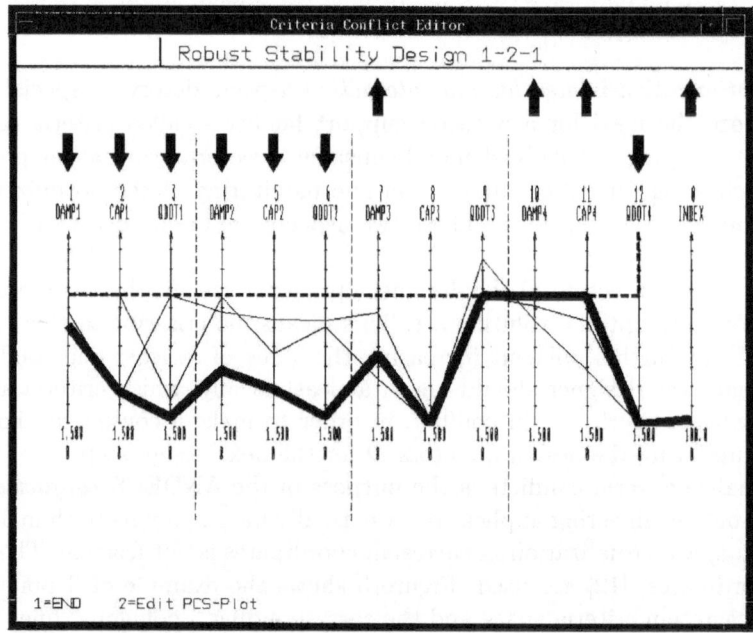

Figure 6: The arrows indicate coherent or contrary criteria behaviour in "klicking
through" 3 different design results of a flight-control problem. The problem is
characterized by 12 performance criteria, which result from considering 4 flight
conditions simultaneously.

The parallel coordinates display can also be used for monitoring the run-time behaviour of the vector-optimization process, seen by the step-by-step evolution of the individual criteria, figure 7.

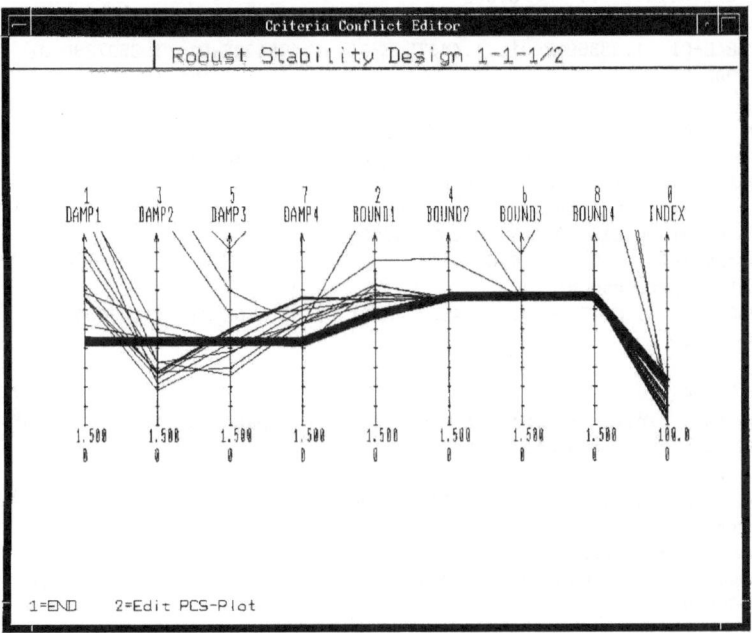

Figure 7: Monitoring an optimization run by the step-by-step changes of 8 criteria.

In engineering control design, numerical optimization is a tool to be run in the "background". This means the designer should be able to concentrate more on the application-specific design task rather than on details of the optimization algorithm's run-time behaviour. But since numerical optimization is not a fool-proof process, the designer must have the possibility of monitoring the optimizer in action. Our solution in ANDECS is the *status face*, which gives an easily intelligible qualitative information on how the optimizer proceeds: The smiling expression of the status face indicates that the optimization process is converging. A yellow color of the face indicates that no constraints are violated, otherwise the face's colour would be red and more sorrowful. In addition, an alpha-numeric status window is provided to give quantitative informations, if necessary; figure 8.

```
Iteration:    2
  Criteria (m - min./ c - constr./ e - equal./ p - passive)
   * = near maximum / violated constraint:
    1 m*    2 m    3 m    4 m    5 c*   6 c
    0.999  0.728  0.727  0.786  1.000  0.993
  maximum criteria (  1) =    0.999
  optimization parameters:
    1.597584E-01  1.633591E-01  1.440730E-01  1.495422E-01  1.350738E-01
    1.337799E-01

Iteration:    3
  Criteria (m - min./ c - constr./ e - equal./ p - passive)
   * = near maximum / violated constraint:
    1 m     2 m*   3 m    4 m    5 c    6 c
    0.471  0.681  0.615  0.415  0.965  0.980
  maximum criteria (  2) =    0.681
  optimization parameters:
    1.742007E-01  1.654777E-01  1.457792E-01  1.387536E-01  1.000000E-01
    1.089647E-01

Iteration:    4
  Criteria (m - min./ c - constr./ e - equal./ p - passive)
   * = near maximum / violated constraint:
    1 m     2 m*   3 m    4 m    5 c    6 c
    0.176  0.395  0.340  0.161  0.908  0.908
  maximum criteria (  2) =    0.395
  optimization parameters:
    2.200000E-01  2.200000E-01  2.100000E-01  2.100000E-01  1.000000E-01
    1.000000E-01
```

Figure 8: 'Status face' for monitoring the step-by-step vector-optimization process.

Application Customization

ANDECS is a general-purpose tool which can be easily customized for specific application areas. This is due to both its strict modularity, which makes the same analysis- and synthesis methods tool set directly reusable in different applications, and its openness to import in a neutral model description format computational models, which are set up in different external modelling environments like Dymola [4] or ACSL [10].

So far a range of different multi-criteria engineering control problems has been solved by using the vector-optimization facility of ANDECS:

- robust aircraft handling qualities control,

- active structural damping of a spacecraft antenna mast,

- satellite attitude pulse-width/pulse-frequency control

- pneumatic car suspension optimization for a heavy-duty truck,

- suspension-control optimization for a maglev passenger vehicle,

- industrial- and space robots trajectory optimization and simulation.

An application area we are currently working on, is robot trajectory optimization by optimal feedforward control in concurrence with multi-criteria parameter optimization of the individual robot-joint drive feedback control laws. This is a re-iterative two-stage process: (i) determine an optimal control for the basic multibody nonlinear dynamics, (ii) find best feedback-control parameter settings to make the actual robot dynamics, — including the motor-drive elasticities —, to behave as close as possible to the 'optimal' behaviour specified in stage (i). Re-iterate the weightings of a mixed time/energy optimization functional in stage (i) to design a nominal trajectory which takes account of the dynamic restraints found in stage (ii), until a well-balanced feedforward/feedback control is achieved. ANDECS is particularly well suited to tackle such a complex design problem because it provides a common frame for control function optimization and for feedback control-law optimization, and because a dynamic model behaviour of various complexity is easily handled and compared. The customized ANDECS design-experimentation environment for robotic control investigations is depicted in figure 9.

A graphical signal editor serves to initially specify, to modify, and to display, the vertices and the approximation scheme $P(p_i, t_i, A)$ for the control functions $u(t)$, and to decide which vertices $T(\tilde{p}_i, \tilde{t}_i)$ should be used as design variables ('tuners') to be optimized. A mouse-driven slider can be used to re-iterate the demand levels (weightings) in a multi-criteria (time/energy/...) performance functional as well as state- and control constraints. Various kind of information is graphically displayed in different windows to support a design decisions: Windows show a robot dynamics animation with force/torque arrows attached to a moving robot

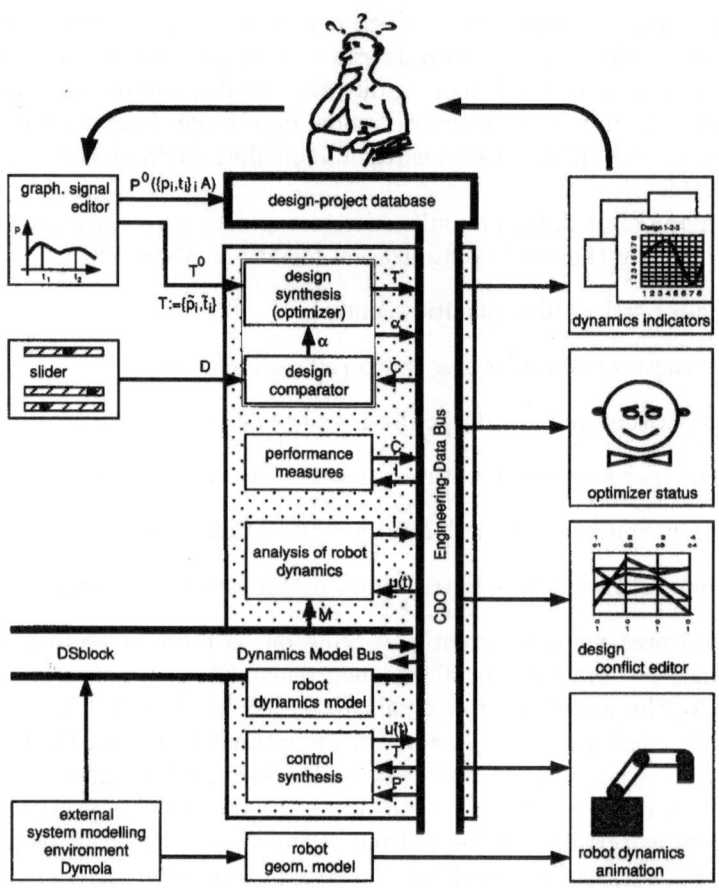

Figure 9: The ANDECS robotics-control experimentation environment.

geometric model, windows show chosen time responses and other dynamic indicators, and the conflict editor shows a quantitative view of what performance criteria are in conflict how much. Also, the status of the (vector-) optimization process can be monitored by the status face.

An important aspect is dynamics model experimentation, i.e. the task of deciding what dynamic effects have to be considered in a 'design-model' under a trade-off with computational load, and the task of design result validation using a more detailed evaluation model. For this, the object-oriented modelling environment Dymola is particularly well suited. It allows robot dynamics multibody/mechatronics modelling [17] via reusable model-component libraries and the simulation code generated by Dymola is coherently interfaced to ANDECS via the neutral 'model bus' DSblock.

Summary

A computation environment for integrating optimal control synthesis within an engineering-control design-automation system using vector optimization is reported on. Besides proper use of mathematical optimization methods and mathematical programming algorithms, other facilities are of prime importance for an efficient application of optimization techniques to solve complex engineering problems. These are design-decision support tools, such as an automatically evolving design history database and a 'conflict editor', as well as visual-display interaction tools for the set-up, execution, and monitoring in the tasks of application-specific dynamics modelling and computational experimention.

References

[1] B.D.O. Anderson and J.B. Moore. *Linear Optimal Control*. Prentice Hall, Englewood Cliffs, 1979.

[2] J. Bals. Parameter optimized hyperstable controllers. In *Proc. 30th CDC, Brighton, UK*, 1990.

[3] J.E. Dennis and R.B. Schnabel. *Numerical Methods for Unconstrained Optimization and Nonlinear Equations*. Prentice-Hall, Englewood Cliffs, 1983.

[4] Elmqvist, H.: Object-Oriented Modeling and Automatic Formula Manipulation in Dymola. SIMS'93 Scandinavian Simulation Society, Kongsberg, Norway, June 9–11, 1993.

[5] A.G.J. MacFarlane, G. Grübel, and J. Ackermann. Future design environments for control engineering. *Automatica*, vol. 25, pp. 165–176, 1989.

[6] R. Finsterwalder. Interaktiv-graphische Rechnerunterstützung beim parametrischen Entwurfs-Experimentieren, *Dissertation,* Ruhr-Universität Bochum, VDI-Fortschrittberichte, Reihe 20, Nr. 102, 1993.

[7] G. Grübel, H.-D. Joos, M. Otter, and R. Finsterwalder. The ANDECS design environment for control engineering. In *Proc. 12th IFAC 1993 World Congress, Sydney,* July 18–23, vol. 6, pp. 447–454, 1993.

[8] G. Grübel, R. Finsterwalder, H.-D. Joos, and J. Bals. Parametric-Design Modes of the ANDECS Control-Dynamics Design Machine. *Proc. Second European Control Conference ECC'93, Groningen,* pp. 1587–1592, 1993.

[9] G. Grübel and H.-D. Joos. RASP and RSYST - Two Complementary Program Libraries for Concurrent Control Engineering. *Proc. 5th IFAC/IMACS Symposium on Computer Aided Design in Control Systems,* University of Wales, Swansea, UK, July 15–17, 1991, Pergamon Press, Oxford, pp. 101–106, 1991

[10] G. Grübel, M. Otter, F. Breitenecker, A. Prinz, G. Schuster, I. Bausch-Gall, and H. Fischer. ACSL^{TM}–Model Translator to the Neutral FORTRAN DSblock-Model Format. Submitted to Joint CACSD'94/CADCS'94 Symposium Tucson, March 7–9, 1994.

[11] A.B. Höfler. Gradientenkettenoperatoren und ihre Anwendung bei der Reglerparameteroptimierung. *Dissertation,* Ruhr-Universität Bochum, 1980.

[12] A. Inselberg and B. Dimsdale. Parallel coordinates for visualizing multidimensional geometry. In: Kunii T.L., editor, *Computer Graphics 1987 - Proceedings of CG International '87,* pp. 69–91, 1987.

[13] H.-D. Joos and M. Otter. "Control engineering data structures for concurrent engineering". *In Proc. 5th IFAC/IMACS Symposium on Computer Aided Design in Control Systems,* Swansea, UK, pp. 107–112, 1991.

[14] H.-D. Joos. Informationstechnische Behandlung des mehrzieligen optimierungsgestützten regelungstechnischen Entwurfs. *Dissertation,* Universität Stuttgart, VDI Fortschrittberichte, Reihe 20, Nr. 90, 1992.

[15] D. Kraft. On converting optimal control problems into nonlinear programming problems, in: K. Schittkowski (ed.) *Computational Mathematical Programming,* Springer, Berlin, 1985.

[16] G. Kreisselmeier and R. Steinhauser. Application of vector performance optimization to a robust control loop design for a fighter aircraft. *Int. J. Control,* vol. 37, No. 2, pp. 251–284, 1983.

[17] M. Otter and G. Grübel. Direct Physical Modeling and Automatic Code Generation for Mechatronics Simulation. *Second Conference on Mechatronics and Robotics, Duisburg,* September 27–29, 1993

[18] H.T. Toivonen and P.M. Mäkilä. Newton's method for solving parametric linear quadratic control problems. *Int. J. Control,* pp. 897–911, 1987.

[19] H.T. Toivonen and P.M. Mäkilä. A Descent Anderson-Moore Algorithm for Optimal Decentralized Control. *Automatica* 21, p. 743 ff., 1985.

[20] T. Rautert, G. Gramlich and J. Bals. Optimierungsverfahren für parametrische LQ-Probleme zur Synthese linearer Regelungen mit Ausgangsgrößenrückkopplung, *Technical Report TR R120-93,* DLR, Institute for Robotics and System Dynamics, D-82234 Wessling, 1993.

International Series of Numerical Mathematics, Vol. 115, © 1994 Birkhäuser Verlag Basel

Application of Automatic Differentiation to Optimal Control Problems

Rainer Mehlhorn*   Michael Dinkelmann*   and   Gottfried Sachs*

Abstract

Automatic differentiation is applied for improving productivity when solving optimal control problems. A gradient operator is evaluated with an algorithm which uses automatic differentiation in reverse mode. This technique is combined with a dedicated compiler with the use of which a symbolic rather than a programming language can be applied for expanding the formal description of optimal control problems. Numerical results concerning the optimal ascent of an orbital stage are presented.

Nomenclature

C_D	drag coefficient	g	acceleration due to gravity
C_L	lift coefficient	$\mathcal{H}$	Hamiltonian
D	drag	h	altitude
$\mathbf{f}$	right hand side of ODE	I_{sp}	specific impulse
$\mathcal{J}$	performance criterion	α	intermediate variables
L	lift	α_T	thrust vector angle
m	mass	γ	gradient vector, flight path angle
$\mathbf{p}$	parameter vector	δ_T	throttle setting
$\mathbf{S}$	interior point and boundary conditions	η	latitude
		$\lambda_{\mathbf{x}}$	adjoint state vector
T	thrust	$\lambda_{\mathbf{p}}$	adjoint parameter vector
t	time	ξ	arguments of basic operations
$\mathbf{u}$	control vector	ρ	Boolean vector
V	velocity	ϱ	atmospheric density
$\mathbf{x}$	state vector	ς	switching variables
$\mathbf{y}$	vector of state and control variables and parameters	τ	normalized time
		φ	basic operations, bank angle
$\mathbf{z}$	vector of all variables	χ	heading angle

*Institute of Flight Mechanics and Flight Control, Technische Universität München, D-80290 München, Germany

Introduction

Solving optimal control problems with the use of indirect methods requires to derive the adjoint system. This derivation basically involves a differentiation process which is a time consuming task and a source of errors. It may require a significant effort when a complex mathematical model needs to be considered in order to realistically describe the properties of a system. Time and effort are also necessary when considering different models of a system or when evaluating different types of subsystems. It is the purpose of this paper to show that effort and time necessary for generating the adjoint system can be substantially reduced when applying an automatic differentiation technique. In addition, an automatic differentiation process is an efficient tool for avoiding errors.

There are different approaches to automatic differentiation which is subject of recent research [1-6]. Automatic differentiation provides exact numerical values of the derivatives of a function at given arguments. An important consideration for the use of an automatic differentiation technique in optimal control problems concerns a high computational efficiency. This is because a high number of gradient (differentiation) evaluations is required. The present paper addresses the point of high computational efficiency by presenting an adequate automatic differentiation technique.

Efficiency is further increased by the use of a dedicated compiler which is combined with the automatic differentiation technique. This compiler which transforms an input file expandable by TEX to a set of FORTRAN77 subroutines and functions recognizes and evaluates also gradient and partial derivative operators. Thus, the adjoint system can be automatically produced.

Automatic Differentiation Applied to Optimization

Optimization Problem Formulation

The type of optimal control problems considered may be formulated as

$$\min_{\mathbf{u} \in \Omega, t_1, \ldots, t_m} \mathcal{J} := \Phi\left(\mathbf{x}(t_f), \mathbf{p}\right) + \int_0^{t_f} \mathcal{L}\left(\mathbf{x}, \mathbf{u}, \mathbf{p}\right) \, dt \tag{1}$$

subject to

$$\frac{d\mathbf{x}}{dt} = \mathbf{f}_j(\mathbf{x}, \mathbf{u}, \mathbf{p}), \quad j = 1, \ldots, m \tag{2}$$

$$0 = \mathbf{S}_j(t_j, \mathbf{x}(t_j^-), \mathbf{u}(t_j^-), \mathbf{x}(t_j^+), \mathbf{u}(t_j^+), \mathbf{p}), \quad j = 1, \ldots, m-1 \tag{3}$$

$$0 = \mathbf{S}_m(t_m, \mathbf{x}(0), \mathbf{u}(0), \mathbf{x}(t_m), \mathbf{u}(t_m), \mathbf{p}) \tag{4}$$

where $\mathbf{x}$ denotes the state, $\mathbf{u} \in \Omega$ the control, and $\mathbf{p}$ the parameter vector. The optimization problem is to find the time history of the control variables $\mathbf{u}(t)$ and the breakpoints t_j, $j = 1, \ldots, m$, with $0 < t_1 < \ldots < t_{m-1} < t_m := t_f$, such that the performance criterion $\mathcal{J}$ is minimized subject to the differential equations $\mathbf{f}_j$, $j = 1, \ldots, m$, the boundary conditions $\mathbf{S}_m$ and interior point conditions $\mathbf{S}_j$, $j = 1, \ldots, m - 1$.

Solving this kind of problems with the use of the Minimum Principle, the Hamiltonian and adjoint variables $\lambda_\mathbf{x}$, $\lambda_\mathbf{p}$ have to be introduced. For optimal control problems of the type considered (Bolza type), the Hamiltonian denoted by $\mathcal{H}$ may be written as

$$\mathcal{H} := \mathcal{L}(\mathbf{x}, \mathbf{u}, \mathbf{p}) + \lambda_\mathbf{x}^\mathrm{T} \frac{d\mathbf{x}}{dt} \tag{5}$$

For the adjoint variables, the following relations hold

$$\frac{d\lambda_\mathbf{x}^\mathrm{T}}{dt} := -\frac{\partial \mathcal{H}}{\partial \mathbf{x}}, \qquad \frac{d\lambda_\mathbf{p}^\mathrm{T}}{dt} := -\frac{\partial \mathcal{H}}{\partial \mathbf{p}} \tag{6}$$

A first order necessary condition for a control $\mathbf{u}$ to be optimal can then be obtained by

$$\mathcal{H}(\mathbf{x}, \mathbf{u}, \mathbf{p}, \lambda_\mathbf{x}, \lambda_\mathbf{p}) \leq \mathcal{H}(\mathbf{x}, \mathbf{v}, \mathbf{p}, \lambda_\mathbf{x}, \lambda_\mathbf{p}) \quad \forall \mathbf{v} \in \Omega \tag{7}$$

Evaluation of these conditions leads to a differentiation operation for regular controls yielding an algebraic condition for the control variables.

$$\mathcal{H}_\mathbf{u} := \frac{\partial \mathcal{H}}{\partial \mathbf{u}} = 0 \tag{8}$$

Since in general the condition Eq. (8) can not be solved explicitly an index reduction of the algebraic condition to a differential condition is performed yielding differential equations for the control variables and further boundary conditions.

$$\frac{d\mathbf{u}}{dt} := -\mathcal{H}_{\mathbf{uu}}^{-1} \left(\mathcal{H}_{\mathbf{ux}} \frac{d\mathbf{x}}{dt} + \mathcal{H}_{\mathbf{u}\lambda_\mathbf{x}} \frac{d\lambda_\mathbf{x}}{dt} \right), \qquad \mathcal{H}_\mathbf{u}|_0 = 0 \tag{9}$$

Considering regular controls, this is always possible since the matrix $\mathcal{H}_{\mathbf{uu}}$ must be positive definite.

In other cases, the evaluation of the optimal control law may lead to values on the boundary of the control domain Ω or to the evaluation of singular controls.

With the use of Eqs. (2-4,6,9), the original optimization problem is reduced to a multiple-point boundary-value problem with additional boundary and interior point conditions concerning the Hamiltonian and the adjoint variables. A more detailed treatment including a description of an efficient technique for solving multiple-point boundary-value problems (multiple-shooting method) is provided by [7-11].

Automatic Differentiation Technique

Formulation of the optimization problem in the previous section shows that a significant effort may be concerned with differentiation operations, Eqs. (6,9). Such operations can be a time consuming task and also a major source of errors. This may particularly hold for realistic engineering problems which yield a complex mathematical modelling. Time and effort are further increased when the characteristics of the system are changed or when different types of subsystems are considered in the course of an evaluation process.

For reducing the effort described and for improving productivity, automatic differentiation is a promising technique. Automatic differentiation provides the exact value of the gradient of a scalar function for given arguments. Application of this method requires to split up the computation of the scalar function into basic steps which are either elementary binary (e.g., $+, -, *, /$) or unary (e.g., sin, cos) operations. It may be noted, that the selection of elementary operations is not restricted to the intrinsics of a certain programming language, in fact the computational efficiency can be increased when allowing for super elementary functions as for example polynomials.

There are several possibilities of designing the algorithm of automatic differentiation. Two basic approaches are called forward and reverse mode. Using the forward mode both function and gradient are computed at the same time. Each assignment to an intermediate quantity is followed by the corresponding calculation of the gradient. The complexity of the forward mode is therefore $O(m)$-times the complexity of the underlying scalar function, where m denotes the number of independent variables.

In case of using the reverse mode the scalar function is computed in the forward sweep and the gradient is computed afterwards in the reverse sweep. It has been shown that the complexity of the reverse mode is less than 5 times the complexity of the underlying scalar function, taking into account not only the floating point operations but also memory access [4].

For computing derivatives of higher order, several hybrid approaches are possible providing higher efficiency than pure forward or reverse mode [6].

In order to achieve a high level of efficiency for computing gradients, the technique of automatic differentiation in reverse mode is used [4]. Experience has shown that it is possible to calculate the gradient of a scalar function at about two times the cost of evaluating the scalar function. It may be pointed out that this ratio is not dependent on the number of independent variables.

However, the above mentioned task of splitting up the computation of the scalar function into basic steps is performed by the TEX to FORTRAN77 compiler. As a result, a list of intermediate values α_i; $i = 1, \ldots, n$ is used, where n is the number of elementary steps required to calculate the scalar function and α_n is its value. The value of each α_i; $i = 1, \ldots, n$ can accordingly be calculated in one of those two ways

$$\alpha_i := \begin{cases} \varphi_i(\xi_{i,1}) & \text{if } \varphi_i \text{ is a unary basic operation} \\ \varphi_i(\xi_{i,1}, \xi_{i,2}) & \text{if } \varphi_i \text{ is a binary basic operation} \end{cases}$$

where $\xi_{i,1}, \xi_{i,2}$ denote the arguments of step i (i.e., they are either arguments of the scalar function or intermediate values already calculated). If the scalar function

---

function reverse_mode $([\alpha_{i; \; i=1,\ldots,n}], \; [z_{i; \; i=1,\ldots,m}]) \; \mapsto \; [\gamma_{i; \; i=1,\ldots,m}]$

$$\gamma_{i; \; i=1,\ldots,m} \quad := \quad 0$$
$$\bar{\gamma}_{i; \; i=1,\ldots,n-1} \quad := \quad 0 \; ; \qquad \bar{\gamma}_n \quad := \quad 1$$
$$\rho_{i; \; i=1,\ldots,n-1} \quad := \quad \text{unused} \; ; \quad \rho_n \quad := \quad \text{used}$$

    for $i \; := \; n$ **down to** 1
      if $\rho_i \; \equiv \;$ used **then**
        if $\alpha_i \; = \; \varphi_i(\xi_{i,1})$ **then** diff $(\alpha_i, \; \xi_{i,1})$
        if $\alpha_i \; = \; \varphi_i(\xi_{i,1}, \xi_{i,2})$ **then** diff $(\alpha_i, \; \xi_{i,1})$;   diff $(\alpha_i, \; \xi_{i,2})$

    procedure diff $(\alpha_i, \; \xi_{i,j})$
      if $\exists_1 \; k \in \{1,\ldots,m\}$ $: \xi_{i,j} \equiv z_k$ **then** $\gamma_k := \gamma_k + \bar{\gamma}_i \; \frac{\partial \alpha_i}{\partial \xi_{i,j}}$
      if $\exists_1 \; k \in \{1,\ldots,i-1\} : \xi_{i,j} \equiv \alpha_k$ **then** $\bar{\gamma}_k := \bar{\gamma}_k + \bar{\gamma}_i \; \frac{\partial \alpha_i}{\partial \xi_{i,j}}$; $\rho_k \; := $ used

---

Table 1: Algorithm of automatic differentiation

is considered to be a function of m variables z_i; $i = 1, \ldots, m$ the components of the gradient can be referenced as γ_i; $i = 1, \ldots, m$. The basic algorithm for the calculation of the gradient can be improved by the use of two further variable lists:

1. A list for intermediate values used in the calculation of the gradient. They are denoted by $\bar{\gamma}_i$; $i = 1, \ldots, n$

2. A Boolean variable list ρ_i; $i = 1, \ldots, n$. It provides information about the usage of intermediate values

$$\rho_i := \begin{cases} \textbf{used} & \text{if the intermediate value } \alpha_i \text{ is} \\ & \text{necessary for the calculation of } \alpha_n \\ \textbf{not used} & \text{if the intermediate value } \alpha_i \text{ is} \\ & \text{not necessary for the calculation of } \alpha_n \end{cases}$$

Using this information the nonused intermediate values may be skipped on the reverse sweep.

With these declarations, the algorithm applied can be stated as shown in Table 1.

Computer Program Structure

The following section shows how the technique described may be used. In order to increase the readability of the formal description of the optimal control problem, the algorithm used for automatic differentiation is embedded in a compiler which accepts a subset of TEX and produces a FORTRAN77 code with a predefined

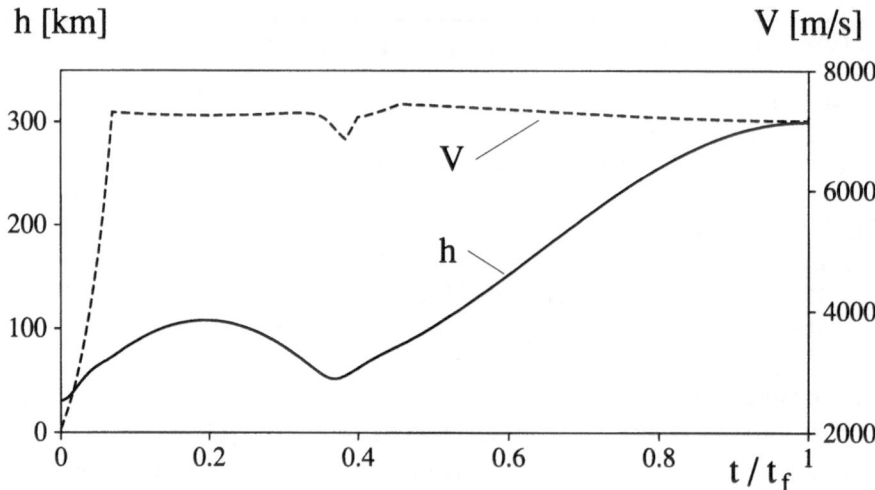

Figure 1: Nonplanar ascent, altitude and velocity, $t_f = 3802.6$ s

interface to a solver based on the multiple shooting method [7]. As a consequence, the same text file can be used for documentation and computation purposes.

An example is used to demonstrate the application of the proposed method. It concerns the nonplanar ascent of a lifting upper stage of a two-stage hypersonic flight system to an orbit. The rocket propelled upper stage is released from the airbreathing first stage at a hypersonic flight condition.

Modelling of System Dynamics

The dynamics of the systems can be described as a set of piecewise defined first order differential equations of the form of Eq. (2). The development of these equations represents the modelling process which is an important part when dealing with realistic engineering problems and provides fundamental knowledge for the optimization task. The TeX to FORTRAN77 programming tool is aimed at supporting the transfer of the results of the modelling process to a formal description. The basic idea is a transparent and modular design concept. The modules are used to break down the computation of the differential equations into steps which are directly correlated to the modelling process:

1. Declaration of variables to deal with state, adjoint state, control and parameter vector

2. Declaration of symbolic constants

3. Modelling functions

4. Differential equations for describing system dynamics

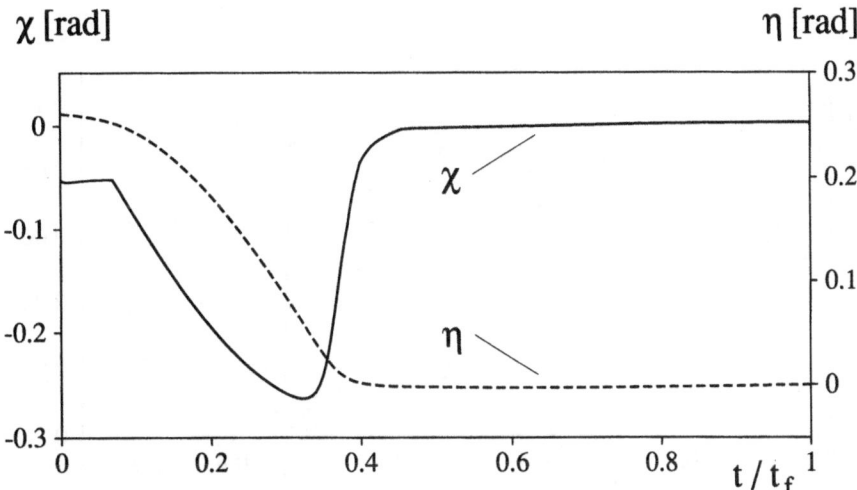

Figure 2: Nonplanar ascent, heading angle and latitude, $t_f = 3802.6$ s

These steps are described in more detail in the Appendix A.1-A.4 for the ascent optimization problem addressed above.

Modelling of Optimality Conditions (Hamiltonian, Adjoint Variables, Optimal Control Law)

For solving the optimization problem under consideration, the Hamiltonian $\mathcal{H}$ and the differential equations for the adjoint variables are formulated. For optimal control problems of Bolza type, reference must be made to the performance criterion (Appendix A.5). The Hamiltonian and the differential equations for the adjoint variables are depending on the individual characteristics of a system in a similar way as the modelling equations. Accordingly, they must be generated each time a new or modified system is dealt with. This may be a time consuming task and a source of errors, particularly when realistic engineering problems yielding a complex mathematical modelling are considered.

The proposed method greatly reduces the effort for the task addressed. A simple input provided by the software system is activated. This input is represented by a single statement as shown in Appendix A.6 (third expression). The equations presented include a transformation of the Hamiltonian and a vector formulation combining the state and adjoint variables.

The optimal control law is described in Appendix A.7. It implies a bang bang type behavior for one control which is handled by introducing interior point conditions using a switching function S_δ.

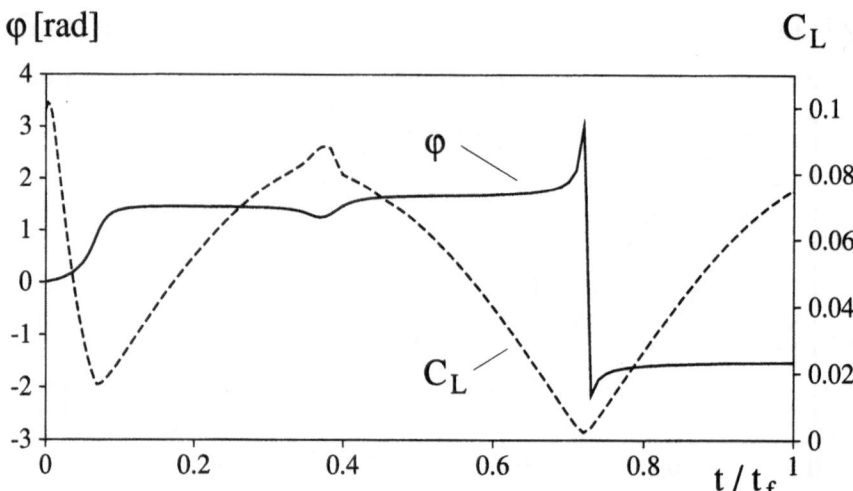

Figure 3: Nonplanar ascent, bank angle and lift coefficient, $t_f = 3802.6$ s

Boundary and Interior Point Conditions

Having specified the dynamics of the system as well as the differential equations for the adjoint system, the next task is to formulate and/or derive the boundary conditions.

Some boundary conditions may be determined from prescribed values or from properties of the trajectory (e.g. periodicity), others may be related to the performance criterion (Appendix A.5).

At the switching points t_j, $j = 1, \ldots, m - 1$ the trajectory may be discontinuous, therefore continuity conditions must be stated as interior point conditions. The multiple-shooting code available [11] is able to handle implicitly defined switching functions which can be understood as further interior point conditions. The definition of switching points leads to a partitioning of a period into phases. Correlated with this partitions, the switching variables may be defined as

$$\varsigma_i \quad := \quad \begin{cases} 1 & \text{inside phase } i \\ 0 & \text{outside phase } i \end{cases}$$

$$\overline{\varsigma_i} \quad := \quad 1 - \varsigma_i$$

$$i \quad = \quad 1, \ldots, 1 + \text{number of phases}$$

Thus, the switching variables are implicitly defined and can be used like modelling functions. They may be used for constructing modelling functions which are defined as functions which are piecewise different.

For the nonplanar ascent problem considered, interior point and boundary conditions are described in Appendix A.8 and A.9.

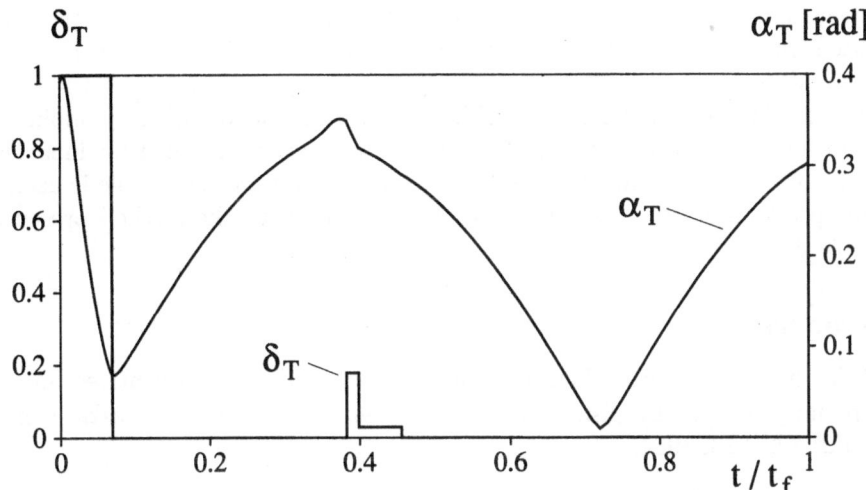

Figure 4: Nonplanar ascent, throttle setting and thrust vector angle, $t_f = 3802.6$ s

	Optimal control problem			
Method applied	Planar ascent	Dynamic soaring	Nonplanar ascent	Singular control
Programming FORTRAN77 Differentiation by hand	6.74 sec	13.80 sec	1044.2 sec	58.2 sec
Programming TEX Automatic differentiation in reverse mode	6.48 sec	21.94 sec	1683.4 sec	63.0 sec

Table 2: Comparison of computational costs

Numerical Results

Numerical results are presented in Figs. 1-4 for the ascent problem described in a previous section. The ascent of the rocket propelled upper stage begins at a hypersonic flight condition (Mach number: 6.8, altitude: 31 km) at a rather small flight path angle. The vehicle rapidly accelerates to its orbital speed level while altitude is built up more slowly. Thrust is primarily applied in the initial phase of the trajectory. Aerodynamic lift is also utilized for controlling the flight path.

Conclusions

A method is presented for improving productivity when solving optimal control problems. This method utilizes an efficient automatic differentiation technique for evaluating optimality conditions. The differential equations for the adjoint variables are generated automatically with a gradient operator. The gradient operator is evaluated with an algorithm which uses automatic differentiation in reverse mode.

In addition, a dedicated compiler is applied with the use of which the formal description of the optimal control problem can be expanded with a symbolic language rather than a programming language. This feature is aimed for increasing efficiency in problem formulation and debugging. A numerical example is used for showing productivity increase which can be achieved with the method presented, c.f. Table 2.

References

[1] K.V. Kim et.al. An efficient algorithm for computing derivatives and extremal problems, English translation. *Ekonomika i matematicheskie metody*, 20(2):309–318, 1984.

[2] M. Iri and K. Kubola. Methods of Fast Automatic Differentiation and Applications. Research memorandum RMI 87-02, Department of Mathematical Engineering and Instrumentation Physics, Faculty of Engineering, University of Tokyo, 1987.

[3] H. Fischer. Automatic differentiation of characterizing sequences. *Journal of Computational and Applied Mathematics*, 28:181–185, 1989.

[4] A. Griewank. *Mathematical Programming: On Automatic Differentiation in Mathematical Programming – Recent Developements and Applications*. Kluwer Academic Publishers, Boston, 1989.

[5] M. Liepelt and K. Schittkowski. PCOMP A Fortran Code for Automatic Differentiation. Report No. 254, DFG-Schwerpunktprogramm Anwendungsbezogene Optimierung und Steuerung, 1990.

[6] B. Christianson and L. Dixon. Reverse Accumulation of Jacobians in Optimal Control. Technical report, School of Information Sciences, Hatfield Campus, University of Hertfordshire, England, August 1992.

[7] R. Bulirsch. Die Mehrzielmethode zur numerischen Lösung von nichtlinearen Randwertproblemen und Aufgaben der optimalen Steuerung. In *Optimierungsverfahren — Software und technische Anwendungen*, R1.06 Dynamische Systeme, Carl-Cranz-Gesellschaft, Oberpfaffenhofen, Germany, 1971. Reprint: Department of Mathematics, Munich University of Technology, Germany (1993).

[8] A.E. Bryson and Y. Ho. *Applied Optimal Control*. Hemisphere Publishing Corporation, Halsted Press, New York, 1972.

[9] H.G. Bock. Numerische Behandlung von Zustandsbeschränkungen und Chebyshef-Steuerungsproblemen. In *Optimierungsverfahren — Software und*

technische Anwendungen, DR3.10 Systemdynamik und Regelung, Carl-Cranz-Gesellschaft, Oberpfaffenhofen, Germany, 1981.

[10] H.G. Bock. *Randwertproblemmethoden zur Parameteridentifizierung in Systemen nichtlinearer Differentialgleichungen.* Dissertation, Universität Bonn, 1985. Bonner Mathematische Schriften 183, Bonn, 1987.

[11] P. Hiltmann. *Numerische Lösung von Mehrpunkt-Randwertaufgaben und Aufgaben der optimalen Steuerung mit Steuerfunktionen über endlichdimensionalen Räumen.* Dissertation, Technische Universität München, 1990.

A Nonplanar Ascent of an Orbital Stage

A.1 Variables Declarations

$$\mathbf{x}^{\mathrm{T}} := (h, \gamma, V, m, \chi, \eta) \qquad \mathbf{u}^{\mathrm{T}} := (\varphi, \alpha_T, C_L, \delta_T) \qquad \mathbf{p}^{\mathrm{T}} := (t_f)$$

$$\lambda_{\mathbf{x}}^{\mathrm{T}} := (\lambda_h, \lambda_\gamma, \lambda_V, \lambda_m, \lambda_\chi, \lambda_\eta) \qquad \lambda_{\mathbf{u}}^{\mathrm{T}} := (\lambda_\varphi, \lambda_{\alpha_T}, \lambda_{C_L}, \lambda_{\delta_T}) \qquad \lambda_{\mathbf{p}}^{\mathrm{T}} := \left(\lambda_{t_f}\right)$$

$$\mathbf{y}^{\mathrm{T}} := \left(\mathbf{x}^{\mathrm{T}}, \mathbf{u}^{\mathrm{T}}, \mathbf{p}^{\mathrm{T}}\right) \qquad \lambda_{\mathbf{y}}^{\mathrm{T}} := \left(\lambda_{\mathbf{x}}^{\mathrm{T}}, \lambda_{\mathbf{u}}^{\mathrm{T}}, \lambda_{\mathbf{p}}^{\mathrm{T}}\right) \qquad \mathbf{z}^{\mathrm{T}} := \left(\mathbf{y}^{\mathrm{T}}, \lambda_{\mathbf{y}}^{\mathrm{T}}\right)$$

A.2 Symbolic Constants

$$g_0 := 9.80665 \ [\mathrm{m/s^2}] \qquad r_{Earth} := 6.371 * 10^6 \quad [\mathrm{m}] \qquad h_{ref} := 6659.2 \ [\mathrm{m}]$$

$$h_{Orbit} := 3 * 10^5 \quad [\mathrm{m}] \qquad \pi := 4 * \arctan(1) \qquad \omega := \frac{2*\pi}{24*3600} \ [\mathrm{1/s}]$$

Similarly, constants for $C_{D0}, K, \dot{m}_{max}, \rho_{ref}, S$ and

$$V_{Orbit} := \sqrt{\left(\frac{g_0 r_{Earth}^2}{r_{Earth} + h_{Orbit}}\right) - \omega * (r_{Earth} + h_{Orbit})}$$

are defined.

A.3 Modelling Functions

$$r := h + r_{Earth} \qquad g := g_0 * \left(\frac{r_{Earth}}{r}\right)^2 \qquad \rho := \rho_{ref} \exp\left(-\frac{h}{h_{ref}}\right)$$

$$C_D := C_{D0} + K C_L^2 \qquad C_{D,C_L} := \frac{\partial C_D}{\partial C_L} \qquad C_1 := \frac{\rho}{2} V S$$

The functions
$$\frac{L}{V}, D, I_{sp}, T, \alpha_n, \alpha_t, \dot{\gamma}_r, C_2$$
are modelled in a similar way.

A.4 Differential Equations for the State Variables

$$\frac{\mathrm{d}h}{\mathrm{d}t} := V \sin\gamma$$

$$\frac{\mathrm{d}\gamma}{\mathrm{d}t} := \alpha_n \cos\varphi + \dot\gamma_r - \frac{g\cos\gamma}{V} + 2\omega\cos\eta\cos\chi + \frac{C_2 r}{V}\left(\cos\gamma\cos\eta + \sin\gamma\sin\eta\sin\chi\right)$$

$$\frac{\mathrm{d}V}{\mathrm{d}t} := \alpha_t - g\sin\gamma + C_2 r\left(\sin\gamma\cos\eta - \cos\gamma\sin\eta\sin\chi\right)$$

$$\frac{\mathrm{d}m}{\mathrm{d}t} := -\delta_T \dot m_{\max}$$

$$\frac{\mathrm{d}\chi}{\mathrm{d}t} := \alpha_n \frac{\sin\varphi}{\cos\gamma} - \dot\gamma_r \cos\chi \tan\eta + 2\omega\left(\tan\gamma\cos\eta\sin\chi - \sin\eta\right) - \frac{C_2 \sin\eta\cos\chi}{\dot\gamma_r}$$

$$\frac{\mathrm{d}\eta}{\mathrm{d}t} := \dot\gamma_r \sin\chi$$

A.5 Performance Criterion of Mayer-Lagrange-Type

$$\mathcal{L} := 0 \qquad \Phi := m|_0 - m|_{t_f} \qquad \mathcal{J} := \Phi + \int_0^{t_f} \mathcal{L}\, \mathrm{d}t$$

A.6 Hamiltonian and Normalized Differential Equations

$$\mathcal{H}^* := \lambda_{\mathbf{x}}^{\mathrm{T}} \frac{\mathrm{d}\mathbf{x}}{\mathrm{d}t} \qquad \mathcal{H} := t_f \mathcal{H}^* \qquad \left(-\frac{\mathrm{d}\lambda_{\mathbf{y}}^{\mathrm{T}}}{\mathrm{d}\tau}, \frac{\mathrm{d}\mathbf{y}^{\mathrm{T}}}{\mathrm{d}\tau}\right) := \nabla\mathcal{H}$$

A.7 Modelling Functions Specifying Optimal Control

$$(\tan\alpha_T^*) := -\frac{\sqrt{\lambda_\chi^2 + (\lambda_\gamma \cos\gamma)^2}}{V\lambda_V \cos\gamma} \qquad\qquad \alpha_T^* := \arctan\left(\tan\alpha_T^*\right)$$

$$C_L^* := \frac{(\tan\alpha_T^*)}{2K}$$

$$\delta^* := \varsigma_1 + 0.17842\varsigma_3 + 0.0295\varsigma_4 + \varsigma_6 \qquad S_\delta := \frac{\mathrm{d}\lambda_{\delta_T}}{\mathrm{d}\tau}$$

$$\varphi^* := \arctan 2(-\lambda_\chi, -\lambda_\gamma \cos\gamma)$$

$$\varphi := \varphi^* \qquad \alpha_T := \alpha_T^* \qquad C_L := C_L^* \qquad \delta_T := \delta^*$$

A.8 Interior Point Conditions

$$\mathbf{z}|_{t_1^+} = \mathbf{z}|_{t_1^-} \quad \mathbf{z}|_{t_2^+} = \mathbf{z}|_{t_2^-} \quad \mathbf{z}|_{t_3^+} = \mathbf{z}|_{t_3^-} \quad \mathbf{z}|_{t_4^+} = \mathbf{z}|_{t_4^-} \quad \mathbf{z}|_{t_5^+} = \mathbf{z}|_{t_5^-} \quad \mathbf{z}|_{t_6^+} = \mathbf{z}|_{t_6^-}$$

$$S_\delta|_{t_1^-} = 0 \quad S_\delta|_{t_2^-} = 0 \quad S_\delta|_{t_3^-} = 0 \quad S_\delta|_{t_4^-} = 0 \quad S_\delta|_{t_5^-} = 0$$

A.9 Boundary Conditions

$$h|_0 = \text{hom} \;\; [\text{m}] \;\; | \; \tfrac{1}{3000} \qquad h|_{t_f} = h_{Orbit} \;\; | \; \tfrac{1}{300} \qquad \gamma|_0 = 0 \;\; [\text{rad}] \;\; | \; \tfrac{1}{0.0033}$$

$$\gamma|_{t_f} = 0 \quad [\text{rad}] \; | \; \tfrac{1}{0.0033} \qquad \lambda_m|_{t_f} = \tfrac{\partial \Phi}{\partial m|_{t_f}} \;\; | \; 500 \qquad \mathcal{H}|_0 = 0 \;\; [\text{kg}] \;\; | \; 1$$

The boundary conditions for $\tau|_{t_f}, V|_0, V|_{t_f}, m|_0, \chi|_{t_f}, \lambda_\chi|_0, \eta|_0, \eta|_{t_f}$ are prescribed in an analogous manner.

International Series of Numerical Mathematics, Vol. 115, © 1994 Birkhäuser Verlag Basel

OCCAL
A Mixed Symbolic-Numeric
Optimal Control CALculator

Rainer Schöpf*   and   Peter Deuflhard*

Problem

The numerical solution of optimal control problems by indirect methods (such as multiple shooting or collocation) usually requires a considerable amount of analytic calculation to establish a numerically tractable system. The reason for that is that certain steps in the analytical preparation of the calculation, which are simple in principle, may be very elaborate and can lead to rather complex expressions. Implementation of these into numerical code by hand is tiresome and error-prone.

In spite of the rather high standards of *numerical* algorithms for such problems (BOCK[1, 2, 3], BULIRSCH [4], DEUFLHARD/FIEDLER/KUNKEL [6]) the analytic calculations—often classified as simple in principle, but rather tedious in realistic examples—are nowadays mostly still done by hand—and thus prone to calculation errors. We present the package OCCAL [9] (mnemotechnically for Optimal Control CALculator), a system capable of automating this analytic processing to a reasonable extent by means of a modern symbolic manipulation language, intending to solve as many of the arising problems as possible.

Our combined symbolic-numeric system symbolically analyzes the given problem, using the Computer Algebra System REDUCE. This approach additionally permits to generate *optimized* and *correct* numerical subprograms for use with existing numerical solvers for boundary value problems. These are employed in the second step to obtain numerical solutions.

Let us introduce in short the notation we are going to use throughout this paper. We start from a vector of state variables $y(t)$, $y: [a, b] \to \mathbf{R}^n$ and a vector of control variables $u(t)$, $u: [a, b] \to \mathbf{R}^k$ which are subject to a system of ordinary differential equations:

$$y' = f(t, y, u)$$

with boundary conditions

$$r(y(a), y(b)) = 0, \quad r: \mathbf{R}^{2n} \to \mathbf{R}^n.$$

*Konrad-Zuse-Rechenzentrum für Informationstechnik Berlin, Heilbronner Str. 10, D-10779 Berlin

At the moment we restrict ourselves to problems where the functional to be minimized is an integral of a function depending on the state and control variables. We intend to enlarge this class in the future. The problem is the minimization of the integral

$$I[u] := \int_a^b f_0(y, u, t)dt \, .$$

Introducing the n adjoint variables $\lambda_i(t)$ we obtain the Hamiltonian

$$H(t, y, \lambda, u) := f_0(t, y, u) + \sum_{i=1}^n \lambda_i f_i(t, y, u)$$

that is to be minimized according to the minimum principle by Pontryagin. This leads to the canonical equations

$$y_i' = H_{\lambda_i} = f_i(t, y, u),$$

$$\lambda_i' = -H_{y_i}$$

$$= -\frac{\partial f_0}{\partial y_i}(t, y, u) - \sum_{j=1}^n \lambda_i \frac{\partial f_j}{\partial y_i}(t, y, u).$$

Together with the boundary conditions above, we now have a boundary value problem.

Overview of symbolic tasks

In this section we will present the essential features of the currently implemented system. For a start, we think that it is very important for the user of such a system to state his or her problem in a simple way. To relieve the user from the burden to write the input in a form that is directly understandable by the underlying computer algebra system REDUCE we have defined an easily understandable input format which is automatically translated into a sequence of REDUCE commands. This preprocessing step performs already a number of consistency checks, such as testing whether there is a differential equation for every dynamic variable, and so on. A first version of this preprocessor was implemented in perl, due to the widespread availability and excellent string manipulation facilities of this tool. A second version is being implemented in REDUCE's system programming environment RLISP.

The underlying computer algebra system REDUCE performs the symbolic step of the system: the problem at hand is analyzed and analytical computation is tried as far as possible. The generation of the adjoint equations is rather simple, since it involves only differentiation. Nevertheless it may lead to rather complicated expressions. Therefore implementation of other differentiation schemes such as automatic differentiation is under consideration.

To minimize the Hamiltonian it is necessary to solve the equation $H_u = 0$. This is more difficult since it can involve rather complicated systems of nonlinear equations. Fortunately, REDUCE is able to solve a large class of such systems analytically. It should be noted, however, that if it is not possible to solve these equations the resulting algebro-differential system has to be solved numerically.

Calculation of the switching function and the calculation and classification of singular control is again rather simple. The difficult part here is to determine which is the type of control: singular or bang-bang. This is, of course, not always possible beforehand. Nevertheless, our system will calculate the order of the singular control and try to check the Legendre-Clebsch condition. If it is not possible to determine the sign of the expression automatically, the system will interact with the user, asking him for more information about certain subexpressions to try to determine the overall sign.

The classification of state constraints can be important for the determination of the type of control. This part of the system is under development; we are currently working on a subsystem to handle inequalities.

Finally, the result of the symbolic calculations is used to automatically generate efficient numerical code for use with a boundary value problem solver.

The idea of automatic code generation is to release the user from the burden of typing in complicated arithmetic expressions. Instead, the system (here: REDUCE) produces a piece or even a complete program of numerical code (here: C) that is correct by its very definition.

In REDUCE we use the contributed package GENTRAN (which was developed at the University of Twente in the Netherlands). One of the advantages of this package is the possibility to add additional modules to generate code in another programming language.

Without further manipulation, the resulting code is generated directly from the expressions obtained in the previous phase, i.e. in a fully expanded form. This is, of course, non-optimal, except for really simple cases. Therefore several different optimization strategies can be applied.

We developed a very simple one that separates the calculation of certain common subexpressions like $\sqrt{x}$ or e^x, (so that their values are computed only once) and uses Horner's scheme to evaluate polynomials. In the case of sparse polynomials in many variables, however, using Horner's scheme is not straightforward, since it is not a priori clear how the variables should be ordered.

A more elaborate optimization scheme was developed and implemented by van Hulzen et al. [7]. It attempts to minimize the number of arithmetic operations. The optimizer performs heuristic searches on arithmetic expressions, detecting and extracting common subexpressions, and replacing them by temporary variable names. This package, called SCOPE (Source Code Optimizer) generates highly optimized numerical code at a level that is still difficult for many compilers. Too many of them just refuse to compile programs that contain very complex expressions.

The choice of optimization strategy depends mainly on the problem at hand. Complete optimization takes a considerable amount of computing time, but this may be payed off by the generation of more efficient numerical code. On the other hand, however, for certain applications such a complete optimization is overkill; in these cases the simpler scheme explained above is sufficient. For every new problem, it is therefore necessary to exploit all possibilities to arrive at an optimal solution.

The whole system is very robust with respect to user errors. The resulting code is syntactically correct and represents the analytical results.

To describe the code generation process more precisely it is first necessary to explain which subprograms are needed by the numerical solver of our choice.

Symbolic-numeric code

Up to now the OCCAL system has been using exclusively the numerical MUL-CON for solving the boundary value problem [5, 6]. MULCON stands for MULtiple shooting with adaptive numerical CONtinuation for one-parameter BVPs. We are working exclusively with C as programming language and intend to go to C++ in the future. MULCON was originally written in FORTRAN, but has been translated semi-automatically into C. A re-implementation of a whole collection of multiple shooting codes in C++ for various types of problems is currently under development at ZIB.

The type of problems solvable by means of the MULCON code are of the form

$$
\begin{aligned}
y' &= f(y; p) \\
r(y(a), \ldots, y(b); p) &= 0
\end{aligned}
$$

The associated multiple shooting Jacobian has the general block structure

$$
J_{ext} := \begin{bmatrix}
G_1 & -I & \cdot & & \cdot & P_1 \\
 & \cdot & \cdot & & \cdot & \cdot \\
\cdot & & \cdot & \cdot & & \cdot \\
\cdot & & \cdot & G_{m-1} & -I & P_{m-1} \\
R_a & & & & R_b & P_m
\end{bmatrix}
$$

in terms of the sub-matrices

$$
\begin{aligned}
G_j &:= \left. W(t_{j+1}, t_j) \right|_{y(t|x_j, p)} \\
R_a &:= \left. \frac{\partial r}{\partial y(a)} \right|_{y(a, p)} \\
R_b &:= \left. \frac{\partial r}{\partial y(b)} \right|_{y(b, p)}
\end{aligned}
$$

$$P_j \quad := \quad P(t_{j+1}, t_j)|_{y(t|x_j, p)}$$

$$P_m \quad := \quad \frac{\partial r}{\partial p}\bigg|_{y(t|x_j, p)}.$$

Here the matrices $W(t_{j+1}, t_j)$ are the *Wronskian matrices*

$$W(t, t_j) = \frac{\partial y(t)}{\partial y(t_j)}$$

which are the solutions of the *variational equation*

$$\frac{dW(t, t_j)}{dt} = f_y(y(t|x_j, p); p)W(t, t_j), \quad W(t_j, t_j) = I,$$

and the *sensitivity matrices* $P_j = P(t_{j+1}, t_j)$ are the solutions of the generalized variational equation

$$\frac{dP(t, t_j)}{dt} = f_y(y(t|x_j, p); p)P(t, t_j) + f_p(y(t|x_j, p); p)$$

with initial value

$$P(t_j, t_j) = 0.$$

The most important features of the numeric code are

- Rank-1 updates to Jacobian approximation.

- Computation of Moore-Penrose pseudoinverse J_{ext}^+ by QR decomposition.

- Rank-deficient Gauss-Newton method.

- Efficient steplength control.

We can now present the tasks that were implemented especially for the interface to the numerical solver MULCON: as can be seen from the above it is necessary to calculate the following entities:

- The boundary values r,

- the differential equations f,

- the derivative f_y, necessary to solve the variational equation,

- the derivatives of the boundary conditions R_a and R_b.

The symbolic calculation of the matrix entries is of some importance for the whole computation, since this avoids the (external) numerical differentiation of f and r, which might well lead to problems in the numerical calculation. Furthermore, it is conceivable to reorder the variables to take advantage of the possible block structure of R_a and R_b.

Examples

Turbo generator

To demonstrate our system on a simple example we'd like to present a not too complicated technical system, the so-called turbo generator. We don't want to discuss its technical background; let us note only that this is a model of a generator from a power plant which is to be controlled in a way to stay as close as possible to its stationary state if external perturbations are applied [12].

The necessary calculations can be done by hand without difficulty, the time for this can be estimated as one afternoon for someone who is familiar with this kind of problem, including the time to write the program.

In the following we present the input data provided to OCCAL, to give you a feeling of how it looks like. OCCAL needs about 4 seconds on a SUN 4 workstation to analyze the problem, work on it, and to generate the numerical subprogram.

```
COMMENT
                Turbo generator   Input file

NAME
turbo

TITLE
   T U R B O G E N E R A T O R

CONSTANTS
x10 := 0.60295
x20 := 0.0
x30 := 1.87243
x40 := 0.79778
alpha := 2.5
alpha2 := 1.0
alpha3 := 0.1
beta1 := 1.0
beta2 := 1.0
M := 0.04225
x1s := 0.60295
x2s := 0.00
x3s := 1.87253
x4s := 0.79778
s4 :=0.0
s5:=0.0
c:=0.0
A:=0.17
u1s := 0.80
u2s := 0.73962
kappad := 0.02535

FUNCTIONAL
alpha * ((x1 - x1s)**2 + (x4 - x4s)**2) +
 alpha2 * x2**2 +
 alpha3 * (x3-x3s)**2 +
 beta1 * (u1 - u1s)**2 + beta2 * (u2 - u2s)**2;
```

```
DYNAMIC_VARIABLES
x1
x2
x3
x4

CONTROL_VARIABLES
u1
u2

BOUNDARIES
0;
ts;

TIME_VARIABLE
t

DIFFERENTIAL_EQUATIONS
x1        x2 * x4;
x2        1/M *
          (u1 - s4 * x1 * x4 - s5 * x1 * x3 - kappad * x2);
x3        u2 - A * x3 + c * x4;
x4        - x1 * x2;

BOUNDARY_CONDITIONS
x1(0) - x10;
x2(0) - x20;
x3(0) - x30;
x4(0) - x40;
lamb_x1(ts);
lamb_x2(ts);
lamb_x3(ts);
lamb_x4(ts);
```

The notorious Reentry problem

As a more complicated example we'd like to present the Reentry problem from the book by Stoer and Bulirsch [10]. As you all know, this is a somewhat idealized problem of the reentry of a capsule into the earth's atmosphere.

In contrast to the previous example, this problem involves a large number of calculations. Furthermore, it serves to demonstrate one of the nontrivial computations performed by the OCCAL system: To determine the regular control a transcendental equation must be solved, yielding two solutions. Only one of these is of interest, as determined by the sign of the determinant H_{uu}. This sign is computed automatically, taking into account only the fact that one of the dynamic variables, representing the velocity v of the capsule over ground is always positive. This latter information has either to be supplied to the system beforehand, or during the interactive run, when it asks the user about the sign of v. In the case that it is not possible to decide which of the solutions of the equation $H_u = 0$ is to be selected, OCCAL will generate code for every of the solutions, and the necessary decision must then be done by the numeric step.

The time needed for the analytical calculations including the optimization of the resulting code is about 50 seconds on a Sun 4 workstation. The code optimization phase is able to reduce the size of the code by a factor of two.

Singular control for a two-link robot arm

This third and last example is meant to demonstrate the performance of OCCAL for a problem with singular control.

It is the problem of time-optimal movement of a two-link robot arm which moves frictionless in a horizontal plane [11, 8].

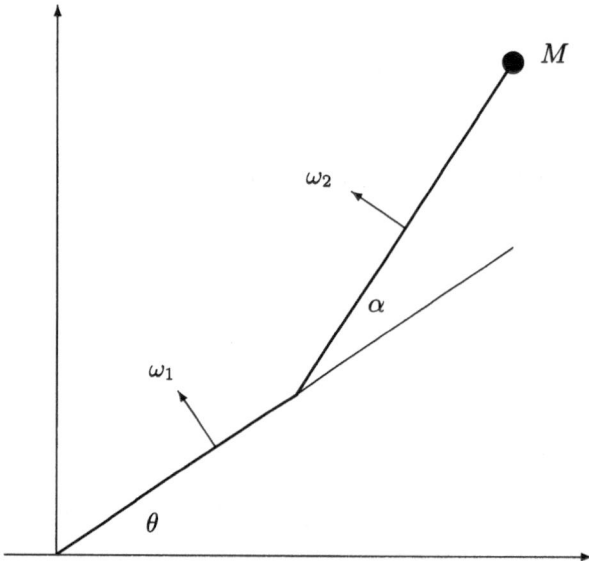

The control variables are the actuator torques U_1 and U_2 for the two links related to shoulder and elbow axis, respectively.

Starting from this information, OCCAL computes again the adjoint equations and the switching functions. Assuming the case of singular control (as opposed to bang-bang), it correctly determines the order of the singular control to be 1, as well as the singular control itself. The symbolic calculations, including the generation of the (optimized) C program, takes about 50 seconds of CPU time on our Sun 4.

Conclusions and Outlook

We have shown that our system OCCAL is a fast system for the treatment of nontrivial real-life optimal control problems. It offers a simple interface to work with and performs a number of operations fully automated, including the generation of optimized numeric code.

In the last months, work was done to allow a somewhat restricted treatment of singular sub-arcs. This is still under development. Another important development is the implementation of new multiple shooting codes at ZIB.

We'd like to conclude with the remark that it will probably not be possible in the near future to solve complex problems of this kind fully automatically without any support and interaction with the human user of the system.

But the state of the art in symbolic computing allows us to make many tedious calculations automatically, as well as a number of non-trivial ones. This means that people from applications can concentrate on the content of the problem rather than on the details of straightforward but tedious calculations.

References

[1] H. G. Bock: Randwertproblemmethoden zur Parameteridentifizierung in Systemen nichtlinearer Differentialgleichungen, Universität Bonn, Dissertation (1985).

[2] H. G. Bock: Numerical Solution of Nonlineaer Multipoint Boundary Value Problems with Applications to Optimal Control, GAMM, Copenhagen 1977.

[3] H. G. Bock: Numerische Berechnung zustandsbeschränkter optimaler Steuerungen mit der Mehrzielmethode, Carl-Cranz-Gesellschaft 1978.

[4] R. Bulirsch: Variationsrechnung und optimale Steuerung. Lectures given at the Universität zu Köln (1971). Unpublished.

[5] P. Deuflhard, B. Engquist (Ed.): Large Scale Scientific Computing, Birkhauser/Boston, Series "Progress In Scientific Computing", Vol. 7 (1987).

[6] P. Deuflhard, B. Fiedler, P. Kunkel: Efficient Numerical Pathfollowing Beyond Critical Points in ODE models, in: P. Deuflhard, B. Engquist (Ed.): Large Scale Scientific Computing, Birkhauser/Boston, Series "Progress In Scientific Computing", Vol. 7, pp. 97–113 (1987).

[7] J. A. van Hulzen: Code optimization of multivariate polynomial schemes: A pragmatic approach, Proceedings EUROCAL '83, Springer LNCS 162, pp. 286–300.

[8] H. J. Oberle: Numerical Computation of Singular Control Functions for a Two-Link Robot Arm, Preprint Universität Hamburg, Institut für angewandte Mathematik

[9] R. Schöpf, P. Deuflhard: OCCAL – A mixed symbolic-numeric Optimal Control CALculator, ZIB Preprint SC 91-13 (1991). Submitted for publication in: R. Liska and S. Steinberg (Eds.), Computer Algebra Applications in Science.

[10] J. Stoer, R. Bulirsch: Introduction to Numerical Analysis, Springer Verlag, Berlin 1980.

[11] A. Weinreb, A. E. Bryson: Minimum-Time Control of a Two-Link Robot Arm; presented at the 5th IFAC Workshop on Nonlinear Programming and Optimization, Capri, 1985.

[12] M. Wulkow: Vergleich numerischer Verfahren zur Berechnung der optimalen Steuerung eines Turbogenerators, Diploma Thesis, Univ. Münster, 1987.

5 Applications of Optimal Control

International Series of Numerical Mathematics, Vol. 115, © 1994 Birkhäuser Verlag Basel

A Robotic Satellite with Simplified Design

R. Callies*

Abstract

Modern techniques of optimization and control considerably increase the performance of small, robotic satellites. As an example, an advanced near-Earth mission is presented for such a spacecraft. Not only a point-mass model is considered, but the full rigid body dynamics of a highly realistic model spacecraft is taken into account. The arising problems are formulated mathematically as boundary-value problems for complex systems of highly non-linear differential equations. All scientific and technological constraints are exactly included as state and control constraints and interior point conditions. The numerical solution of the boundary-value problems is by a modified multiple shooting method. Problems of scaling and extremely small convergence areas require new solution techniques. For the first time the proof of mission feasibility is given. Moreover maximum thrust level is reduced by a factor of 4, cheap thrusters are used that are fixed to their positions, optimal momentum steering replaces conventional thrust vector steering – all this with a performance reduction of less than 2 per cent, but a dramatic decrease in mechanical complexity. Single stage design results in a further decrease of operation complexity and costs.

Introduction and Summary

With progress in microelectronics and -mechanics there is a renewed interest in small, robotic spacecrafts for precisely defined and limited scientific missions: either stand-alone missions (e.g. to asteroids and comets) or support missions that significantly improve the scientific output of a main mission by additional measurements. These space systems are competitive especially for certain types of payloads: highly developed scientific instruments with a large fraction of state-of-the-art microelectronics (e.g. multi-colour high-resolution CCD-cameras), miniaturized components and time-critical experiments.

The advantage of such spacecrafts is the rapid and often cheaper access to space. Scientific flexibility is increased, development and engineering cycles are relatively short. But this is true only as long as strict limitations of the overall size and mass are observed and system complexity is kept low. On the other hand, every decrease in system performance will result in a further decrease of the payload fraction, which is rather low anyhow.

---

*Department of Mathematics, Munich University of Technology, D-80290 Munich, Germany

In this paper it is shown that modern techniques of optimization and control considerably improve the efficiency of those small satellites. In addition, the optimization methods allow a simplification of the overall design by transferring complexity from the mechanical system – the "hardware" – to the build-in (electronic) intelligence and control. As for this, attention is mainly focused on the propulsion system.

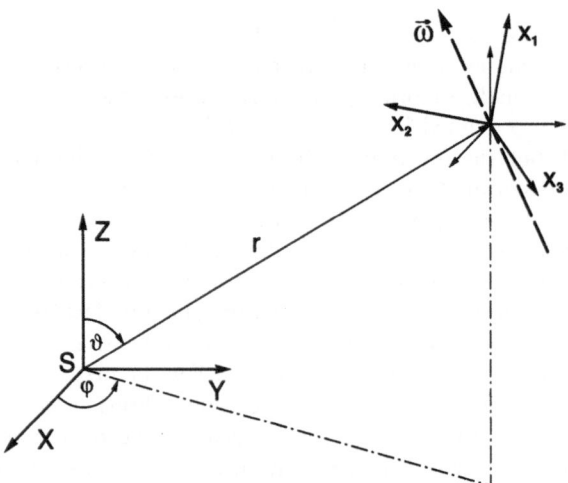

Fig. 1. Definition of the coordinate system: $\mathbf{C} := (r, \varphi, \vartheta)$ denotes the spherical inertial coordinate system, $\mathbf{x} := (x_1, x_2, x_3)$ the body-fixed centroidal coordinate system.

As an example, a typical small satellite for near-Earth operation is considered. The system is equipped with cheap liquid thrusters. Maximum thrust level could be reduced by a factor of about 4. Thrusters are fixed to their position, no gimbals, no momentum wheels or other moving parts are needed – clearly an important factor for reliability. Momentum steering is used instead of the conventional thrust vector steering. To achieve this, thruster positions are optimized with the center of mass of the satellite off thrust axes. This operation mode has been fully optimized for the first time. Single stage design results in a further decrease in operation complexity and costs. Due to the special construction of the satellite, dual-mode operation is possible: long time spin-stabilized, short time three-axes-stabilized with a pointing accuracy of better than 1°. All components on board are already found in commercial satellites.

To meet ESA restrictions the overall mass of the system is restricted to 250 kg, the outer diameter to 1260 mm and the total height of the satellite to 880 mm. This allows the spacecraft to be launched as a low cost piggy-back payload with the ARIANE launch vehicle. For this, the satellite is placed inside the adaptor between the two main payloads – typically two big commercial satellites.

The performance of this space robot is demonstrated for an advanced mission profile: Operating almost autonomously the spacecraft performs a double flyby at the asteroids 1989 JA and 2340 Hathor.

The Model System

The numerical calculations are performed in the spherical ecliptic coordinate system $\mathbf{C} := (r, \varphi, \vartheta)$ with the Sun S at its origin. $\mathbf{X} := (X, Y, Z)$ denotes the corresponding cartesian coordinate system, $\mathbf{x} := (x_1, x_2, x_3)$ the body-fixed centroidal coordinate system and $\vec{\omega}$ the vector of the instantaneous angular velocity (Fig. 1).

The spacecraft is an improved version of the INEO-spacecraft [1]. For the mathematical treatment a careful abstraction is made from the very realistic engineering system without losing relevant information: the model spacecraft has a cylindrical shape (phys. diameter 1260 mm, eff. diameter 1220 mm, phys. length 880 mm, eff. length 860 mm) and a uniform density. Initial mass is 244.7 kg, 132.0 of which are usable fuel. 33.4 kg of scientific instruments are carried. Hydrazinmonopropellant and N_2-cold gas thrusters are fixed to the circular bottom of the spacecraft (see Fig. 2).

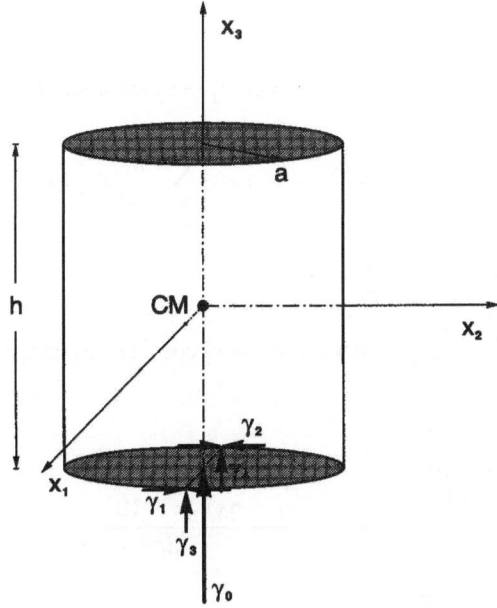

Fig. 2. Definition of the spacecraft model. CM denotes the center of mass, see text for further explanation.

The thrust direction of the Hydrazin main engine γ_0 (nominal thrust 10 N; eff. exhaust velocity: 2300 m/s, centered) and the two Hydrazin control engines

γ_1 and γ_2 (nominal thrust .5 N each, adjustable; eff. exhaust velocity 2300 m/s, diametrical position with a mutual distance of 1220 mm) is parallel to the body-fixed rotation axis x_3. For spin control two bidirectional cold gas thrusters γ_3 and γ_4 are used (nominal thrust .2 N each, adjustable; eff. exhaust velocity 700 m/s, diametrical position with a mutual distance of 1220 mm) with their axes anti-parallel to x_2.

In the coordinate system defined above the movement of the spacecraft with its six degrees of freedom is given by the following highly non-linear system of differential equations of the general form $\dot{x} = f(x, u, t)$ (x vector of state variables, u vector of control variables; t independent variable, here: time). Equations of motion of the rigid body are taken from [2], equations of motion of the center of mass are derived in [3]:

$$\dot{r} = v_r \qquad\qquad \dot{\varphi} = \frac{v_\varphi}{r \sin \vartheta} \qquad\qquad \dot{\vartheta} = \frac{v_\vartheta}{r}$$

$$\dot{m} = -\frac{\gamma_0}{v_{A0}} - \frac{2|\gamma_1|}{v_{A1}} - \frac{\gamma_3 + \gamma_4}{v_{A3}}$$

$$\dot{v}_r = F_r + \frac{v_\varphi^2 + v_\vartheta^2}{r} - \frac{\tilde{\gamma} M}{r^2} + \frac{\xi}{mr^2}$$
$$\qquad - \tilde{\gamma} \sum_{j=1}^{n} \frac{m_j}{s_j^3} [r - r_j \cos \vartheta \cos \vartheta_j - r_j \sin \vartheta \sin \vartheta_j \cos(\varphi - \varphi_j)]$$

$$\dot{v}_\varphi = F_\varphi - \frac{v_\varphi v_\vartheta}{r} - \frac{v_\varphi v_\vartheta}{r} \cdot \cot \vartheta - \tilde{\gamma} \sum_{j=1}^{n} \frac{m_j}{s_j^3} [r_j \sin \vartheta_j \sin(\varphi - \varphi_j)]$$

$$\dot{v}_\vartheta = F_\vartheta - \frac{v_r v_\vartheta}{r} + \frac{v_\varphi^2}{r} \cdot \cot \vartheta$$
$$\qquad - \tilde{\gamma} \sum_{j=1}^{n} \frac{m_j}{s_j^3} [r_j \sin \vartheta \cos \vartheta_j - r_j \cos \vartheta \sin \vartheta_j \cos(\varphi - \varphi_j)]$$

$$\dot{\omega}_1 = -\frac{3a^2 - h^2}{3a^2 + h^2} \omega_2 \omega_3$$

$$\dot{\omega}_2 = \frac{3a^2 - h^2}{3a^2 + h^2} \omega_1 \omega_3 + \frac{a \cdot (\gamma_4 - \gamma_3)}{m} \cdot \frac{12}{3a^2 + h^2}$$

$$\dot{\omega}_3 = \frac{2a \cdot \gamma_1}{m} \cdot \frac{2}{a^2}$$

$$\begin{pmatrix} \dot{\beta}_0 \\ \dot{\beta}_1 \\ \dot{\beta}_2 \\ \dot{\beta}_3 \end{pmatrix} = \frac{1}{2} \begin{pmatrix} 0 & -\omega_1 & -\omega_2 & -\omega_3 \\ \omega_1 & 0 & \omega_3 & -\omega_2 \\ \omega_2 & -\omega_3 & 0 & \omega_1 \\ \omega_3 & \omega_2 & -\omega_1 & 0 \end{pmatrix} \cdot \begin{pmatrix} \beta_0 \\ \beta_1 \\ \beta_2 \\ \beta_3 \end{pmatrix}$$

$\vec{r} := (r, \varphi, \vartheta)$ is the spacecraft's mass center location, $\vec{v} := (v_r, v_\varphi, v_\vartheta)$ its velocity and m the mass of the spacecraft. γ_i denotes the thrust magnitude of the respective engine and v_{Ai} its exhaust velocity.

$$\gamma_i := \begin{cases} F_{i0} \cdot \sin^2(\rho_i(t)) & i = 0, 3, 4 \\ F_{i0} \cdot (2\sin^2(\rho_i(t)) - 1) & i = 1, 2 \end{cases}$$

F_{i0} is the maximum thrust and $\rho_i(t)$ the thrust control function ($i = 0 \ldots 4$), a and h are effective lengths of the spacecraft (see Fig. 2).

$\vec{\omega}$ marks the instantaneous rotation axis of the spacecraft, the (redundant) Euler parameters $(\beta_0, \beta_1, \beta_2, \beta_3)$ describe the transformation between $\mathbf{x}$ and $\mathbf{X}$ in an unique way.

ξ denotes the solar pressure, $\tilde{\gamma}$ the gravity constant and M the mass of the Sun. The full complexity of gravitational forces of many other celestial bodies ($n = 202$) like planets, moons or planetoids is included: m_j is the mass of the j-th body and $s_j(t) := |\vec{r}(t) - \vec{r}_j(t)|$ its distance from the spacecraft.

The resulting force of the thrusters

$$\vec{F} = (F_r, F_\varphi, F_\vartheta)^T := \vec{F}_0 + \vec{F}_r + \vec{F}_3 + \vec{F}_4,$$

measured in $\mathbf{C}$, acts upon the center of mass of the spacecraft. The F_i are the solutions of the following equations

$$C(\beta_0, \beta_1, \beta_2, \beta_3) \cdot E(\varphi, \vartheta) \cdot \vec{F}_i = \vec{f}_i, \quad i \in \{0, r, 3, 4\}$$

with

$$E = E(\varphi, \vartheta) := \begin{pmatrix} \cos(\varphi)\sin(\vartheta) & -\sin(\varphi) & \cos(\varphi)\cos(\vartheta) \\ \sin(\varphi)\sin(\vartheta) & \cos(\varphi) & \sin(\varphi)\cos(\vartheta) \\ \cos(\vartheta) & 0 & -\sin(\vartheta) \end{pmatrix}$$

describing the transformation between the reference frames $\mathbf{X}$ and $\mathbf{C}$ and

$$C = C(\beta_0, \beta_1, \beta_2, \beta_3) :=$$

$$\begin{pmatrix} \beta_0^2 + \beta_1^2 - \beta_2^2 - \beta_3^2 & 2(\beta_1\beta_2 + \beta_0\beta_3) & 2(\beta_1\beta_3 - \beta_0\beta_2) \\ 2(\beta_1\beta_2 - \beta_0\beta_3) & \beta_0^2 - \beta_1^2 + \beta_2^2 - \beta_3^2 & 2(\beta_2\beta_3 + \beta_0\beta_1) \\ 2(\beta_1\beta_3 + \beta_0\beta_2) & 2(\beta_2\beta_3 - \beta_0\beta_1) & \beta_0^2 - \beta_1^2 - \beta_2^2 + \beta_3^2 \end{pmatrix}$$

describing the transformation between the coordinate systems $\mathbf{X}$ and $\mathbf{x}$. The $\vec{f}_i$ are defined by

$$\vec{f}_0 := \frac{\gamma_0}{v_{A0}} \cdot (\omega_2 h/2, -\omega_1 h/2, v_{A0})^T$$

$$\vec{f}_r := \frac{\gamma_1}{v_{A1}} \cdot (\omega_2 h, -\omega_1 h, 0)^T$$

$$\vec{f}_3 := \frac{\gamma_3}{v_{A3}} \cdot (\omega_2 h/2, -\omega_3 a - \omega_1 h/2, \omega_2 a + v_{A3})^T$$

$$\vec{f}_4 := \frac{\gamma_4}{v_{A3}} \cdot (0, 2\omega_3 a, -2\omega_2 a)^T + \vec{f}_3.$$

The following technical conditions have been considered in the state equations:

$$F_{01} = F_{02}, \; F_{03} = F_{04}, \; \rho_1(t) = \rho_2(t), \; v_{a1} = v_{a2}, \; v_{a3} = v_{a4}.$$

Boundary Conditions and Constraints

The launch situation at time t_0 is prescribed ($\vec{r}(t_0)$ given, $\vec{v}(t_0)$ given, $m(t_0) = 1$). Final time t_f is free, here the following condition holds: $|\vec{r}(t_f) - \vec{r}_{Hathor}(t_f)| \in [d_{H1}, d_{H2}]$, $d_{H1}, d_{H2} \in I\!R$ given. At the flyby at 1989 JA at the time t_F an interior point condition has to be fulfilled: $|\vec{r}(t_F) - \vec{r}_{1989JA}(t_F)| \in [d_{JA1}, d_{JA2}]$, d_{JA1}, $d_{JA2} \in I\!R$. For scientific reasons, d_{H2} and d_{JA2} are 175 km, whereas safety considerations yield d_{H2} and d_{JA2} equal 100 km.

A minimum distance between the spacecraft and every celestial body is necessary because of safety reasons: $|\vec{r}(t) - \vec{r}_j(t)| \geq d_j(> 0)$, $j = 1 \ldots n$, $\forall t \in [t_0, t_f]$. This leads to state constraints of 2nd order [4].

Maximization of the Payload

In a more abstract formulation the problem reads like this: Let us find a state function $x : [t_0, t_f] \longrightarrow I\!R^n$ and a control function $u : [t_0, t_f] \longrightarrow U \subset I\!R^m$, which minimize the functional

$$I(u) := -m(t_f)$$

subject to the conditions

$$
\begin{aligned}
\dot{x} &= f(x, u, t) = F(x, t) \cdot u \\
0 &= g(t_0, x(t_0)) \in I\!R^n \\
0 &= r(t_0, t_f, x(t_f)) \in I\!R^k, \; k < n \\
0 &= q_i(t_i, x(t_i), x(t_i^-), x(t_i^+)) \in I\!R^l, \; l < n, \; i = 1, \ldots K \\
0 &\leq C(x, t)
\end{aligned}
$$

t_0 denotes the initial time, t_f the final time and t_i an intermediate time with an interior point condition. With thrusters fixed to their position, only linear controls (the instantaneous thrust magnitude of the main engine and the attitude control engines) occur.

The so defined problem of optimal control theory is transformed in a well-known manner (see e.g. [4],[5]) into a multi-point boundary value problem. There the following system of coupled nonlinear differential equations results ($H := \lambda^T f$):

$$
\begin{aligned}
\dot{x} &= F(x, t) \cdot u \\
\dot{\lambda} &= -H_x(x, \lambda, u, t).
\end{aligned}
$$

In addition the Legendre-Clebsch condition has to be satisfied: $H_{uu}(x, \lambda, u, t)$ *pos. semidef.* The boundary conditions and interior point conditions are either prescribed a priori or obtained from the first variation of the extended functional. The solution of this boundary value problem satisfies the necessary conditions for an optimal solution.

The single components of a satellite are described by model functions and parameters that fit into this framework in a natural way. Only existing subsystems already space-qualified are taken into account. Scientific and technological constraints are included as state and control constraints and interior point conditions. The numerical solution of the boundary-value problems is by the multiple shooting method [6]–[8]. For the calculations a strongly modified version of the variant BOUNDSCO [8] is utilized. A high precision method is needed for the integration of the complicated systems of ordinary differential equations: extrapolation methods have been chosen for this purpose to solve the arising initial value problems [9]. For these complex problems arising from astronautics additional numerical techniques [3],[10] have been developed and applied.

Scaling problems and extremely small convergence areas required new solution techniques:

– *Interior points* T_i, $i = 1, \ldots, 14$, $T_i < T_j$ for $i < j$, $T_i \in [t_0, t_f]$ $\forall i$ are additionally introduced and allow the rescaling of the independent variable t in sections: $\zeta_i := e_{i,1}t + e_{i,0}$ $\forall t \in \,]T_{i-1}, T_i]$, $i = 1, \ldots, 15$ and $T_0 := t_0, T_{15} := t_f$. The scaling factors $e_{i,k}$ are determined in every iteration step of the multiple shooting method in that way, that $cond(M) \leq MIN * 1.5$; here M denotes the multiple shooting matrix [6] and MIN is an approximation of the minimum condition number that can be obtained by scaling operations.

– It is not before the last few iteration steps that the full problem is precisely calculated. Before that, the spiraling out maneuver and the interplanetary trajectory are handled separately (scaling problems arise due to different time and length scales), the rigid body motion and the motion of the center of mass are calculated sequentially: In the early steps of calculation the respective other parts are handled only approximately.

The Example Mission

As an example, a double flyby mission at the asteroids 1989 JA and 2340 Hathor has been designed as a reference mission (stimulated by [11]). Fig. 3 (Earth escape maneuver) and Fig. 4,5 (interplanetary flight) show a fully optimized flight trajectory from the Earth to the asteroid *2340 Hathor* via the asteroid *1989 JA*.

With an Ariane launch the spacecraft is delivered into an GTO-orbit with a perigee of 185 km and an apogee of 35786 km. Until now no other precise informations concerning launch are available. Therefore, a standard situation is defined: At launch time $T0$ the spacecraft has a distance of 125000 km from Earth, and the hyperbolic launch velocity $\vec{V}_{HL}$ lies in the plane of ecliptic and is parallel to the velocity vector of the Earth at $T0$ (*Definition* of launch geometry and $T0$!). L is the position of the spacecraft at $t = T0$. The 125000 km-boundary is artificially introduced, but has *no* mathematical meaning except the change of the coordinate system for technical reasons. The two coordinate systems are coupled

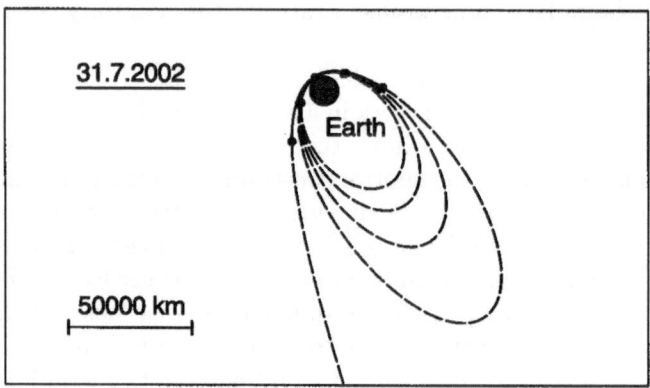

Fig. 3. Projection of the three-dimensional Earth escape
maneuver

by an interior point condition. V_{HL} is optimized; optimization of its direction will
further decrease fuel consumption [10]. The total final trajectory (spiraling-up and
interplanetary flight) was calculated without splitting-up or simplification.

Fig. 3 shows the projection of the optimal flight trajectory of the spacecraft
to the base plane of a geocentric coordinate system. Solid lines mark those parts of
the trajectory where thrusters are on, broken lines those with the thrust off. First
thrust arc starts at the perigee of the GTO-ellipse on July 27th, 2002. After 4
orbits and slightly more than 4 days the spacecraft has reached its optimal escape
velocity. Launch time $T0$ is July 31st, 2002; the spiraling up from the GTO-ellipse
takes 90.384 kg of Hydrazin. Optimal hyperbolic escape velocity is 1.221 km/s. On
the coast arcs the spacecraft is spin-stabilized with 5 rpm and the spacecraft main
axis x_3 has to be perpendicular to the plane of motion. Three axes stabilization is
necessary only while thrusters are in operation. The special design of the spacecraft
allows this fuel-saving dual-mode operation (spin- and three-axes-stabilization).

The interplanetary part of the journey is shown in Fig. 4 and Fig. 5. The orbits of Earth, *1989 JA* and *Hathor* are indicated by dashed-dotted lines. First deep
space maneuver T1 takes place on September 13th, 2002 (fuel consumption 7.811
kg), the second maneuver T2 takes place on March 23nd, 2003 (fuel consumption
26.727 kg). The spacecraft passes *1989 JA* on July 7th, 2003 (Fig. 4, point F in
Fig. 5) and reaches *Hathor* on September 28th, 2003. The positions of Earth, *1989
JA* and *Hathor* are plotted for the date given in the respective figure. Total fuel
consumption for this journey is 124.922 kg. On the coast arcs of the interplanetary flight the spacecraft is spin-stabilized with 5 rpm and the main axis x_3 has
to be perpendicular to the plane of ecliptic. The double flyby is mathematically
realized in an exact and elegant way by an interior point condition. The classical
formulation with a state constraint fails due to severe convergence problems.

Fig. 5 shows the three-dimensional plot of the complete and optimal inter-

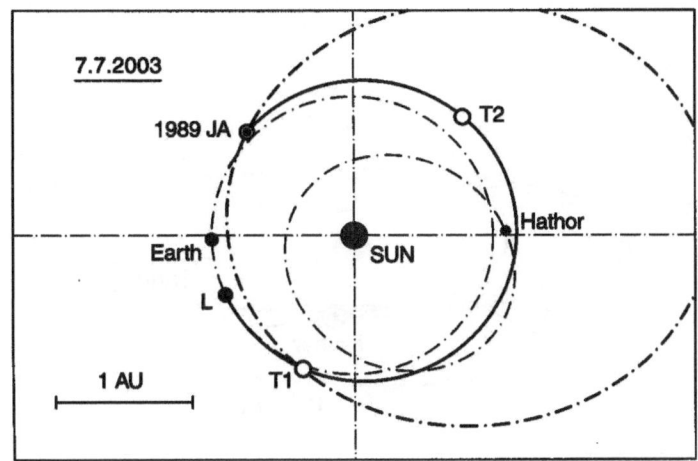

Fig. 4. Projection of the optimal flight trajectory to the plane of ecliptic for the 7.7.2003 (flyby at 1989 JA).

planetary flight trajectory.

Acknowledgement. The author is indebted to Prof. Dr. R. Bulirsch who always encouraged and supported this work. This research has been funded by the *Bavarian Consortium on High Performance Scientific Computing* and by the *DFG* in the *Special Research Center on Transatmospheric Vehicles (SFB 255)*.

References

[1] Iglseder, H., Arens-Fischer, W., Keller, H.U., Arnold, G., Callies, R., Fick, M., Glassmeier, K.H., Hirsch, H., Hoffmann, M., Rath, H.J., Kührt, E., Lorenz, E., Thomas, N., Wäsch, R., *INEO – Imaging of Near Earth Objects*. COSPAR-Paper 1-M.1.03, Washington (1992).

[2] Junkins, J.L., Turner, J.D., *Optimal Spacecraft Rotational Maneuvers*, Elsevier Science Publ. Comp., New York, 1986.

[3] Bulirsch, R., Callies, R., *Optimal trajectories for a multiple rendezvous mission to asteroids*. Acta Astronautica **26** (1992) 587–597.

[4] Bryson, A.E., Ho, Y.-C., *Applied Optimal Control*, Revised Printing, Hemisphere Publishing Corp., Washington D.C., 1975.

[5] Oberle, H.J., *Numerical Computation of Minimum-Fuel Space-Travel Problems by Multiple Shooting*, Report TUM-MATH-7635, Dept. of Mathematics, Munich Univ. of Technology, Germany, 1976.

[6] Bulirsch, R., *Die Mehrzielmethode zur numerischen Lösung von nichtlinearen Randwertproblemen und Aufgaben der optimalen Steuerung*, Report of the Carl-Cranz-Gesellschaft e.V., Oberpfaffenhofen, 1971. Reprint: Department of Mathe-

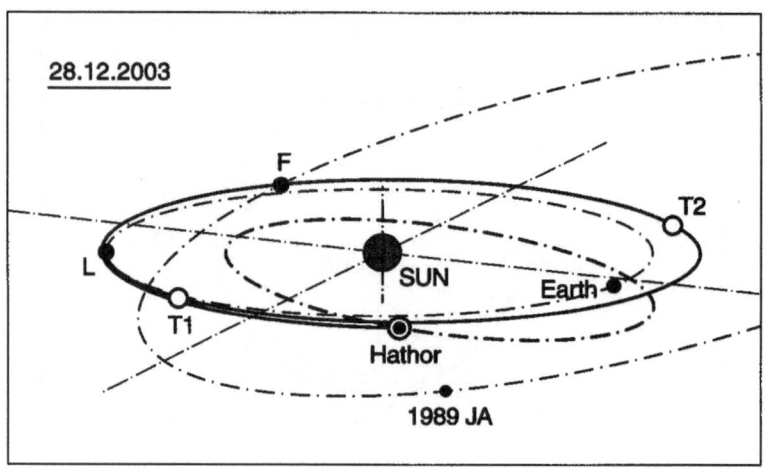

Fig. 5. Three-dimensional plot of the optimal double flyby
at 1989 JA and 2340 Hathor.

matics, Munich University of Technology, Germany (1993).

[7] Deuflhard, P., *Ein Newton-Verfahren bei fast singulärer Funktionalmatrix zur
Lösung von nichtlinearen Randwertaufgaben mit der Mehrzielmethode*, Thesis,
Cologne, 1972.

[8] Oberle, H.J., *Numerische Behandlung singulärer Steuerungen mit der Mehrzielmethode am Beispiel der Klimatisierung von Sonnenhäusern*, Thesis, Munich,
1977.

[9] Bulirsch, R., Stoer, J., *Numerical Treatment of Ordinary Differential Equations
by Extrapolation Methods*, Num. Math. **8** (1966) 1–13.

[10] Callies, R., *Optimale Flugbahnen einer Raumsonde mit Ionentriebwerken*,
Thesis, Munich, 1990.

[11] Penzo, P.A., *Mission Opportunities for Near-Earth Asteroids*, International
Conference on Near Earth Asteroids, San Juan Capistrano, 1991.

International Series of Numerical Mathematics, Vol. 115, © 1994 Birkhäuser Verlag Basel

Nonlinear Control under Constraints of a Biological System

D. Claude[*]   and   N. Nadjar[*]

Abstract

We propose a nonlinear adaptive control of a biological system involving the hormonal coupling between adreno-cortical hormones and vasopressin. Designing this control requires successive nonlinear parameter estimations subject to biological positivity constraints of the hormonal concentrations and mathematical constraints on the parameters to estimate.

Introduction

Biological systems are naturally nonlinear. Some of them may show homeostatic behaviours represented by dynamical systems with limit-cycles. This is the case of the adrenal-postpituitary system, showing stable physiological or pathological limit-cycles both with a circadian period since the system is forced by the "night-day" circadian synchronization.

The adrenal-postpituitary system results from the coupling between cortico-adrenal glands, which secrete cortisol and the neuro-postpituitary gland, which produces vasopressin. This system ensures an agonistic-antagonistic equilibration on which the cellular development depends: on the one hand, the antagonistic actions such that cell division and hydratation are inhibited by cortisol and stimulated by vasopressin. On the other hand, the agonistic actions mean that cortisol and vasopressin act in the same manner, seeing, for example, that both favour blood volume expansion. The adrenal-postpituitary system imbalances are responsible for some cases of cerebral œdemas or cerebral collapses [4], and play a role in the malignant cerebral tumors evolution [4].

A modelling of the agonistic-antagonistic equilibration of this biological system, has first been done by E. Bernard-Weil ([1], [2]) and has led to powerful bipolar therapies using corticoids and vasopressin ([1], [4]) whereas other classical therapies, defined by only the use of corticoids administration, failed.

The aim of this paper is to design a control law which is able, from a pathological state, to find the physiological behaviour again. Since the natural circadian periodicity of the endogenous hormones has to be kept, it is easy to show that this control law has necessarily the same periodicity. In this case, it is not generally possible to find the physiological rhythm again through a dynamical model

---

[*]Laboratoire des Signaux et Systèmes, C.N.R.S. – E.S.E., Plateau de Moulon, 91192 Gif-sur-Yvette Cedex, France

of the controlled pathological system and then one can be afraid not being able to find therapeutics for all the pathological cases. All the same, in order to solve the control problem, the locking thesis is presented: it allows system parameters to change, and thus suggests a nonlinear adaptive control. Designing this control requires successive nonlinear parameter estimations, using both global and local minimization strategies. Positivity constraints due to necessarily positive concentrations and constraints on the investigated control parameters are needed.

The paper is organized as follows. The agonistic-antagonistic model of the adrenal-postpituitary system is first presented in Section 2. We introduce a nonlinear adaptive control based on polynomial correctors in Section 3. The parameter estimation techniques under constraints are developed in Section 4. Conclusions are given in Section 5.

The adrenal-postpituitary system modelling

The notion of *agonistic-antagonistic equilibration*, by associating two basic principles in Biology, namely the equilibration and the agonistic-antagonism notion, allows us to give a particular structure to the nonlinear oscillators acting in some biological regulations. *Agonism* represents "parallel" actions and *antagonism* represents opposed actions.

Moreover, in a hormonal regulation, the biological system does not make any difference between endogenous and exogenous hormones.

So, we can consider two endogenous hormonal agents represented by two variables x and y, with positivity constraints because of the natural boundaries of hormonal concentration, and two exogenous hormones represented by X and Y. The dynamics of the biological system represented by the couple (x, y) is simultaneously controlled by the changes of both the antagonistic variable $z_1 = (x + X) - (y + Y)$ and the agonistic variable $z_2 = (x + X) + (y + Y) - m$, where m represents the reference level of the agonistic influence.

As a result of Bernard-Weil's work, the use of the agonistic-antagonistic equilibration principle allows us to propose, if there is no specific stimulus, the following state-space description of the biological system ([1], [2], [5]), by denoting $\mathcal{X} = x + X$ and $\mathcal{Y} = y + Y$ and thus, $z_1 = \mathcal{X} - \mathcal{Y}$ and $z_2 = \mathcal{X} + \mathcal{Y} - m$:

$$\begin{cases} \dot{x} &= \sum_{j=1}^{3}[k_j(u(z_1))^j + c_j(v(z_2) + S)^j] &= D_1(z_1, z_2) \\ \dot{y} &= \sum_{j=1}^{3}[k_j'(u(z_1))^j + c_j'(v(z_2) + S)^j] &= D_2(z_1, z_2) \end{cases} \tag{1}$$

$k_j, k_j', c_j, c_j', (j = 1, 2, 3)$, are the system parameters. The functions $u(.)$, $v(.)$ and $S(t)$, a synchronizer, usual in chronobiology, are defined, according to the adrenal-postpituitary system ([1], [2]) by:

$$\begin{cases} u(z_1) &= z_1 \\ v(z_2) &= m\ln(1 + z_2/m) \\ S(t) &= A\sin(\omega t + B) \end{cases} \tag{2}$$

When the control (X, Y) and the synchronizer S are zero, $(x, y) = (m/2, m/2)$ is a critical point of system (1).

The physiological system is represented by system (1) where the parameters are $k_j^\varphi, c_j^\varphi, k_j'^\varphi, c_j'^\varphi (j = 1, 2, 3)$. They determine the developments of D_1^φ and D_2^φ; the pathological system is represented by the same structure (1) but with different parameters values $k_j^\psi, c_j^\psi, k_j'^\psi, c_j'^\psi (j = 1, 2, 3)$, characterizing D_1^ψ and D_2^ψ.

The period of the synchronizer is considered to be circadian, and A, B, m, ω, as well as the physiological and the pathological parameters are assumed to be known. Their numerical values are given in [7].

Control

Introduction

The therapeutical strategy consists in searching the temporal rules to be followed by the exogenous hormones (X, Y) so that, after some transient period, system (1) ensures the control in order to get $(\mathcal{X}, \mathcal{Y})$ close to a physiological behaviour and fulfills the positivity constraints on the endogenous and exogenous hormones. The aim of the control is then to shift the pathological limit-cycle (x^ψ, y^ψ) towards the physiological limit-cycle (x^φ, y^φ) and thus to obtain $\mathcal{X} = x^\varphi$ and $\mathcal{Y} = y^\varphi$. Therefore, X and Y have a circadian periodicity since, biologically, x and y should have one. However, we can prove that no continuous control of circadian period allows us to reach the physiological limit-cycle again: the only alternative consists in reducing as much as possible a small residual limit-cycle [6]. One might then be afraid of rarely or never reaching a physiological behaviour. Yet, a feasible approach based on the locking thesis is presented.

In fact, in order to try and overcome this difficulty, we have to presume in the therapeutical act that the pathological system *parameters* may reach the physiological ones, by a reversible movement. This leads to introduce the locking concept [7].

This notion can be formulated in the following way: in some pathologies, when the controlled system of agonistic-antagonistic equilibration (1), which behaves according to the outputs z_1 and z_2, follows a periodic behaviour for a rather long time, it will copy it by adapting its parameters, now behaving as an uncontrolled system.

Indeed, the biological system does not make the difference between the endogenous and the exogenous hormones and thus it "ignores" that it is controlled.

Polynomial correctors

Through the usual control and decoupling techniques, a new state $(x, y, \mathcal{X}, \mathcal{Y})$ is defined and the pathological system (1), with the parameters values k_j^ψ, c_j^ψ, $k_j'^\psi$,

c'^ψ_j $(j = 1, 2, 3)$, can be rewritten [5]:

$$\begin{cases} \dot{x} &= D_1^\psi(z_1, z_2) \\ \dot{y} &= D_2^\psi(z_1, z_2) \\ \dot{\mathcal{X}} &= D_1^\varphi(z_1, z_2) + a_1 \\ \dot{\mathcal{Y}} &= D_2^\varphi(z_1, z_2) + a_2 \end{cases} \tag{3}$$

this system being submitted to the constraints $\mathcal{X} \geq x \geq 0$, $\mathcal{Y} \geq y \geq 0$.

The controller laws a_1 and a_2 depend on the bipolar control $X = \mathcal{X} - x$ and $Y = \mathcal{Y} - y$. The aim of the control is to minimize the difference between the pathological evolution and the physiological one.

Referring to [3] and [5], we propose the following controller defined by two *polynomial correctors* a_1 and a_2:

$$\begin{cases} a_1 &= \sum_{j=1}^3 \alpha_j [\mathcal{X} - x - \beta]^j \\ a_2 &= \sum_{j=1}^3 \alpha'_j [\mathcal{Y} - y - \beta']^j \end{cases} \tag{4}$$

α_j, α'_j, $(j = 1, 2, 3)$, β and β' are the controller parameters. The parameters β and β' are introduced to ensure satisfaction of the bounds constraints ([2], [3]).

Moreover, the possible solution $(a_1, a_2) = (0, 0)$ has to be rejected [6].

Nonlinear adaptive control

Possible locking of nonlinear oscillators suggests a nonlinear adaptive control process of imbalances, since pathological system parameters can change and since the physiological model is considered as a reference model.

The first step consists in searching the control to be applied to the pathological system in order to obtain a limit-cycle as close as possible to the physiological one. We estimate the eight parameters α_j, α'_j, β and β', $(j = 1, 2, 3)$, in (4), which define the first polynomial correctors (a_1, a_2) by minimizing a criterion function of the difference between the physiological limit-cycle and the controlled pathological one. We obtain a controlled system which is still far from the physiological system (see Fig. 3), and the resulting control (X, Y) is given in Figure 1.

According to the concept of agonistic-antagonistic equilibration, the system ignores the control. The locking thesis then applies: the nonlinear controlled system is considered to be an uncontrolled "new system", similar to (1), with $X = 0$ and $Y = 0$. The parameters k_j, c_j, k'_j and c'_j, $(j = 1, 2, 3)$, of this new system denoted by "identified 1" in Figure 5, are then estimated by minimizing a quadratic criterion, function of the distance between the behaviours of this uncontrolled system and the controlled pathological system. Figure 5 actually compares both systems.

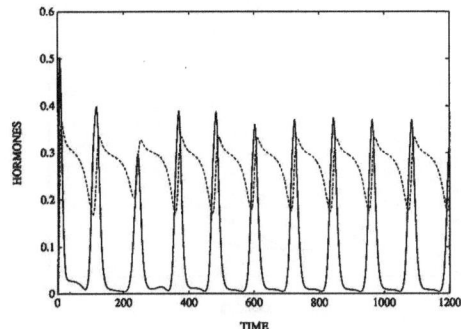

Figure 1: X(—) and Y(- -): control 1

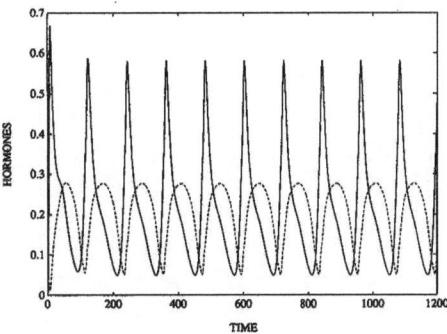

Figure 2: X(—) and Y(- -): control 2

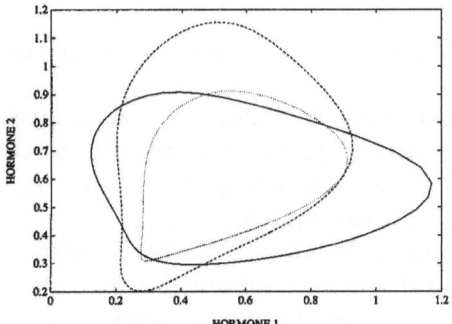

Figure 3: Limit-cycles: physio (—),
patho (- -) and controlled 1 (...)

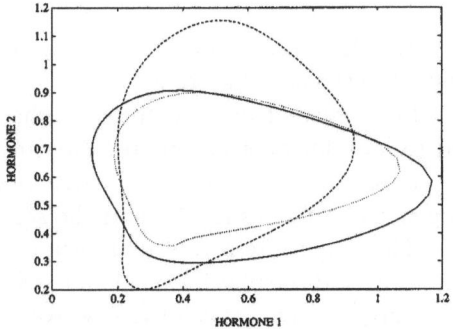

Figure 4: Limit-cycles: physio (—),
patho (- -) and controlled 2 (...)

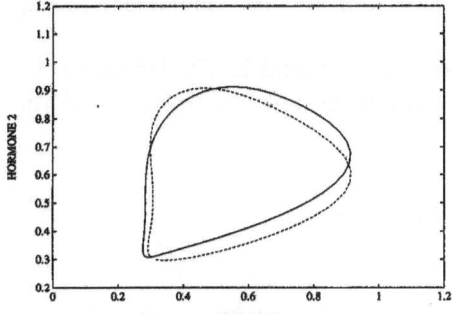

Figure 5: Limit-cycles: controlled 1 (—)
and identified 1 (- -)

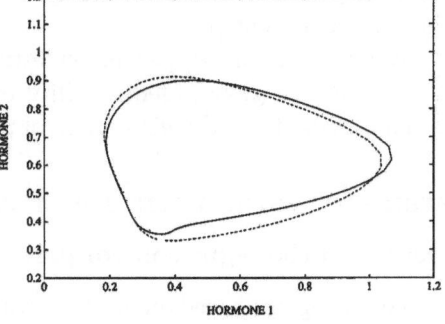

Figure 6: Limit-cycles: controlled 2 (—)
and identified 2 (- -)

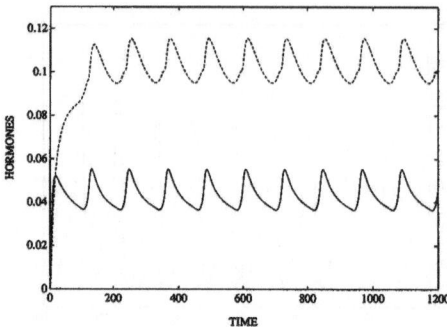

Figure 7: X(—) and Y(- -): control 5   Figure 8: Limit-cycles: physio (—),
patho (- -) and controlled 5 (...)

We then have another set of coefficients, which is no longer the set of the original pathological system.

Since system parameters have changed one has to remove the first control and to look for another one (denoted by "control 2" in Figure 2) in order to be even closer to the physiological behaviour. Figures 2 and 4, respectively, show the exogenous hormones (X, Y) and the second controlled system.

This procedure is repeated until the difference between the controlled system behaviour and the physiological one, is smaller than a fixed tolerance. After several iterations, a convergence of the twelve parameters towards the physiological ones is actually observed and, consequently, the control is gradually cancelled (in Figure 7, X and Y are almost close to zero).

Hence, the global control is finally a sequence of control laws coming close to zero, so that the successive systems show a convergence towards the physiological system. Notice that, when a "control i", $(i = 1, .., 5)$, is applied to an "identified system i", $(i = 1, .., 5)$, the resulting controlled system remains bounded, showing a very long transient [7].

System and control parameter estimates can be found in [7]. Moreover, we proved in [8] the global identifiability of the twelve system parameters and the local identifiability of the eight control parameters.

Parameter estimation under constraints

Research of the eight control parameters

We used a program based on an iterative random search strategy to determine the parameters that globally minimizes the chosen criterion [9]. The algorithm follows:

Step 1: Choose $\hat{\boldsymbol{\theta}}^0$; $k = 0$

Step 2: Compute $y^k = \hat{\boldsymbol{\theta}}^k + r^k$ where r^k is a random vector with a distribution to be determined.

Step 3: If $J(y^k) < J(\hat{\boldsymbol{\theta}}^k)$ then $\hat{\boldsymbol{\theta}}^{k+1} = y^k$ else $\hat{\boldsymbol{\theta}}^{k+1} = \hat{\boldsymbol{\theta}}^k$

Step 4: $k = k + 1$ and go to Step 2.

Suppose $\hat{\boldsymbol{\theta}}$ to be the global optimum such that the criterion J is smaller than a fixed tolerance. If $\hat{\boldsymbol{\theta}}^k$ is far from $\hat{\boldsymbol{\theta}}$, r^k should have a large variance, particularly in order to avoid local minima. If $\hat{\boldsymbol{\theta}}^k$ is close to $\hat{\boldsymbol{\theta}}$, r^k should have a small variance.

Therefore, the program principle alternates variance selection steps and, when a variance has been selected, it alternates computation steps.

This program requires neither special properties for the criterion function (quadratic, for instance) nor a priori knowledge on the initial parameters value (contrarily to a local one).

It needs an initial vector $\boldsymbol{\theta}^0$, representing the eight parameters, and chosen *equal to zero*, in our case, a space of research $[\boldsymbol{\theta}^{min}, \boldsymbol{\theta}^{max}]$ and a criterion function which can be calculated at each point $\boldsymbol{\theta}$:

First, we define a quadratic criterion J_1, given by :

$$J_1 = \frac{1}{2(N - 720)} \sum_{i=720}^{N} (\mathcal{X}^i - x_{physio}^i)^2 + (\mathcal{Y}^i - y_{physio}^i)^2$$

$(\mathcal{X}, \mathcal{Y})$ is defined in system (3) and (x_{physio}, y_{physio}) represents the physiological behaviour. N is the number of samples and i the current point ; the 720 first points correspond to the transient period and are not taken into account in the criterion calculation, since we are interested in the steady-state, here a limit-cycle. We then add to the criterion J_1 the essential penalties due to the positivity constraints, where (x, y) is also defined in system (3),

$$\text{If } x^i < 0 \quad : \quad J_1 = J_1 - 50x^i \text{ (respectively, for the variable } y)$$
$$\text{If } (\mathcal{X}^i - x^i) < 0 \quad : \quad J_1 = J_1 - 50(\mathcal{X}^i - x^i) \text{ (respectively, for the variable } Y)$$

Constraints on the eight control parameters

Moreover, in order to avoid obtaining an asymptotical steady-state instead of the desired limit-cycle, we studied the equilibrium points $(x^e, y^e, \mathcal{X}^e, \mathcal{Y}^e)$ of system (3)-(4) considered without the synchronizer $(S(t) = 0)$, where $(\mathcal{X}^e, \mathcal{Y}^e)$ always equals $(m/2, m/2)$. There exist nine possible equilibrium points and the linearized systems at these points can be easily found. Thus, using the Routh-Hurwitz criterion, we put some constraints on the real parts of the eigenvalues of each linearized system to exclude an asympotical convergence for the original system. This leads to constraints on the investigated parameters.

Equilibrium points $(x, y, \mathcal{X}, \mathcal{Y}) = (x^e, y^e, \mathcal{X}^e, \mathcal{Y}^e)$

By definition, the equilibrium points are such that $(\dot{x}, \dot{y}, \dot{\mathcal{X}}, \dot{\mathcal{Y}}) = (0, 0, 0, 0)$.

Moreover we take $\mathcal{X}^e = \mathcal{Y}^e = m/2$, since $(\mathcal{X}, \mathcal{Y})$ has to follow the physiological limit-cycle (x^φ, y^φ). That is:

$$\begin{cases} \alpha_1[m/2 - x^e - \beta] + \alpha_2[m/2 - x^e - \beta]^2 + \alpha_3[m/2 - x^e - \beta]^3 & = 0 \\ \alpha_1'[m/2 - y^e - \beta'] + \alpha_2'[m/2 - y^e - \beta']^2 + \alpha_3'[m/2 - y^e - \beta']^3 & = 0 \end{cases}$$

We obtain at most three real solutions for x^e and three real solutions for y^e easily obtained, noting that $(x^e, y^e) = (m/2 - \beta, m/2 - \beta')$ is an obvious solution.

Now, we will construct the linearized system of (3)-(4), where $S(t) = 0$, at each equilibrium point. However, to simplify the following calculations, we prefer considering the state vector (x, y, z_1, z_2) instead of $(x, y, \mathcal{X}, \mathcal{Y})$. Thus, the equilibrium points are $(x, y, z_1, z_2) = (x^e, y^e, 0, 0)$.

Linearization at $(x, y, z_1, z_2) = (x^e, y^e, 0, 0)$

$$\begin{cases} \dot{x} & = & k_1^\psi z_1 + c_1^\psi z_2 \\ \dot{y} & = & k'^\psi_1 z_1 + c'^\psi_1 z_2 \\ \dot{z}_1 & = & (k_1^\varphi - k'^\varphi_1)z_1 & + & (\sum_{j=1}^3 j\alpha_j[m/2 - x^e - \beta]^{j-1})((z_1 + z_2)/2 - x) \\ & & + (c_1^\varphi - c'^\varphi_1)z_2 & - & (\sum_{j=1}^3 j\alpha_j'[m/2 - y^e - \beta']^{j-1})((z_2 - z_1)/2 - y) \\ \dot{z}_2 & = & (k_1^\varphi + k'^\varphi_1)z_1 & + & (\sum_{j=1}^3 j\alpha_j[m/2 - x^e - \beta]^{j-1})((z_1 + z_2)/2 - x) \\ & & + (c_1^\varphi + c'^\varphi_1)z_2 & + & (\sum_{j=1}^3 j\alpha_j'[m/2 - y^e - \beta']^{j-1})((z_2 - z_1)/2 - y) \end{cases} \quad (5)$$

Therefore, the associated characteristic equation can be written:

$$P(\lambda) = \det \begin{bmatrix} -\lambda & 0 & k_1^\psi & c_1^\psi \\ 0 & -\lambda & k'^\psi_1 & c'^\psi_1 \\ -T & T' & a - \lambda & b \\ -T & -T' & c & d - \lambda \end{bmatrix} = 0$$

where

$$\begin{cases} T & = & \sum_{j=1}^3 j\alpha_j[m/2 - x^e - \beta]^{j-1} & \text{and } T' & = & \sum_{j=1}^3 j\alpha_j'[m/2 - y^e - \beta']^{j-1} \\ a & = & k_1^\varphi - k'^\varphi_1 + (T + T')/2 & \text{and } b & = & c_1^\varphi - c'^\varphi_1 + (T - T')/2 \\ c & = & k_1^\varphi + k'^\varphi_1 + (T - T')/2 & \text{and } d & = & c_1^\varphi + c'^\varphi_1 + (T + T')/2 \end{cases}$$

Thus,

$$P(\lambda) = \lambda^4 + A\lambda^3 + B\lambda^2 + C\lambda + D$$

with
$$\begin{cases} A & = & -a - d \\ B & = & ad - bc + T(k_1^\psi + c_1^\psi) - T'(k'^\psi_1 - c'^\psi_1) \\ C & = & T'[(d + b)k'^\psi_1 - (a + c)c'^\psi_1] - T[(d - b)k_1^\psi + (a - c)c_1^\psi] \\ D & = & 2TT'(k_1^\psi c'^\psi_1 - k'^\psi_1 c_1^\psi) \end{cases}$$

Use of the Routh-Hurwitz criterion

The Routh-Hurwitz criterion gives the necessary and sufficient conditions for all the zeros of $P(\lambda)$, and thus for all the eigenvalues of (5), in order to have negative real parts. That is:

$$A > 0 \text{ and } AB - C > 0 \text{ and } A(BC - AD) - C^2 > 0 \text{ and } D > 0$$

Therefore, in order not to obtain an asympotical steady-state instead of the desired limit-cycle, we demand to have at least one eigenvalue with a positive or zero real part. This condition is expressed by:

$$A \leq 0 \text{ or } AB - C \leq 0 \text{ or } A(BC - AD) - C^2 \leq 0 \text{ or } D \leq 0$$

The program tests these constraints and if they are not fulfilled, another set of parameters is immediately considered, thus preventing any unnecessary calculation: the explored space can be regarded as "smaller" and hence, the program converges faster towards the supposed global minimum of the criterion.

Lastly, in order to reduce the time of calculation again, a local minimization program, based on the Newton algorithm, has been coupled to the program of global minimization. This possibility was provided in the program: when the smallest variance of r^k has been selected several successive times, then the global optimization turns out to be a local one.

Estimation of the "identified" system parameters

In this case, a simple quadratic criterion J_2 becomes sufficient, without any specific constraints:

$$J_2 = \frac{1}{2(N - 720)} \sum_{i=720}^{N} \left(\mathcal{X}^i_{controlled} - x^i_{identified}\right)^2 + \left(\mathcal{Y}^i_{controlled} - y^i_{identified}\right)^2$$

We observe an easy convergence of the criterion J_2 towards the supposed global minimum and a program of local minimization, *initialized on the physiological parameters*, rapidly leads to the investigated minimum (see Fig. 5 and 7). The fifth identification exactly gives the physiological coefficients.

If we apply a sixth and last control to the physiological system, we obtain, as expected, that the parameters defining the control (a_1, a_2) are all zero, because of the initialization at zero of the minimization program: both X and Y end to tend to zero in the successive iterations (see Fig. 1, 2 and 7).

To estimate the physiological and pathological parameter values, the knowledge of the transient period in addition to the steady-state, is necessary.

Conclusions

Biology produces examples of nonlinear systems where control laws have to change limit-cycles. In this paper, we were interested in the adrenal-postpituitary system mathematically represented by a nonlinear system with a stable limit-cycle having a circadian periodicity, both in the physiological case as well as in the pathological case. Our goal was to find a control able to shift the pathological limit-cycle towards the physiological limit-cycle.

Since no continuous control with a circadian periodicity allowed us to reach the physiological limit-cycle again from a pathological state, it was therefore necessary to introduce the locking thesis, which allows system parameters to change, and then suggested a nonlinear adaptive control. In fact, our control problem can be formulated in terms of a model-reference adaptive control where the physiological model is actually the reference model and the pathological model parameters can change according to the locking principle. So control parameters have to be adapted to. Several successive nonlinear identifications were required using both global and local optimizations.

Acknowledgement. The authors are grateful to E. Bernard-Weil from "Fondation Rothschild", Paris, L. Pronzato and E. Walter from the "Laboratoire des Signaux et Systèmes", for their careful listening and their help.

References

[1] E. Bernard-Weil, Lack of response to a drug : a system theory approach, *Kybernetes* 14:25–30 (1985).

[2] E. Bernard-Weil, A general model for the simulation of balance, imbalance and control by agonistic-antagonistic biological couples, *Mathem. Modelling* 7:1587–1600 (1986).

[3] E. Bernard-Weil and D. Claude, Control in general or function models. Example of the model for the regulation of agonistic-antagonistic couples, *1er Congrès de Systémique, Lausanne*, 1989, pp. 473–482.

[4] E. Bernard-Weil et B. Pertuiset, Mathematical model for hormonal therapy (vasopressin, corticoids) in cerebral collapse and malignant tumors of the brain (36 cases), *Neurol. Res.* 5:19–35 (1983).

[5] D. Claude, On the agonistic-antagonistic equilibration and its control, *Analysis of controlled dynamical systems*, B. Bonnard, B. Bride, J.P. Gauthier, I. Kupka, eds, Birkhäuser, 1991, pp. 136–145.

[6] D. Claude, Shift of a limit-cycle in Biology and error-equation, *Math. Syst. Est. Contr.*, to appear.

[7] D. Claude and N. Nadjar, Adaptive polynomial control of adrenal-postpituitary imbalances, *ECC 91*, Grenoble, Hermès, Paris, 1991, pp. 797–802.

[8] D. Claude and N. Nadjar, Identifiability of a nonlinear model of the controlled adrenal-postpituitary system, *ECC 93*, to appear.

[9] L. Pronzato, E. Walter, A. Venot and J. F. Lebruchec, A general purpose global optimizer: implementation and applications, *Math. Comput. Simulat.* 26:412–422 (1984).

International Series of Numerical Mathematics, Vol. 115, © 1994 Birkhäuser Verlag Basel

An Object-Oriented Approach
to Optimally Describe and Specify
a SCADA System
Applied to a Power Network

Goran Ericsson*   Patrik Forsgren*   and   Erik Gyllensward*

Abstract

Total life-cycle costs for industrial control systems are high. A major reason is that most computer-based control systems have been developed step-by-step, in an ad-hoc fashion, and during a long period of time. An appropriate and uniform way of specifying, designing, and realizing such a system is lacking. A way to overcome these problems is by means of an overall systems' development concept which is described in this paper. The concept is based on an object-oriented approach, which is extended to be fully utilized for large SCADA systems. The same uniform concept can be used on different levels and during different phases of a SCADA[1] system project. The concept can be used, by means of existing computer technology, to obtain a more cost efficient system architecture. Such an architecture is a prerequisite for more advanced mathematical tools, such as optimal power flow algorithms, to be utilized to their best advantage. The ideas are highlighted by a case study, where a small part of a power network is considered.

Keywords. Control system analysis, Control system design, Control system, Industrial control, Power system control, Software engineering, Object-oriented analysis, Object-oriented design.

Introduction

Many research efforts today are put into obtaining good mathematical models and methods, such as optimal power flow calculation. These results are essential and serve as a bases for what is theoretically possible to implement in a control system. Often the engineering work stops after that point because "the rest is just implementation." However, this is not the case, as there are many delicate problems to tackle along the way. Some of them are discussed in this paper, and suggestions for improvement are addressed. Development of a complete control system can be divided into four domains (phases), the theory, the specification, the design, and the implementation domains. These are presented in Fig. 1.

---

*Royal Institute of Technology, KTH, Department of Industrial Control Systems, S-10044 Stockholm, Sweden

[1]Supervisory Control and Data Aquisition

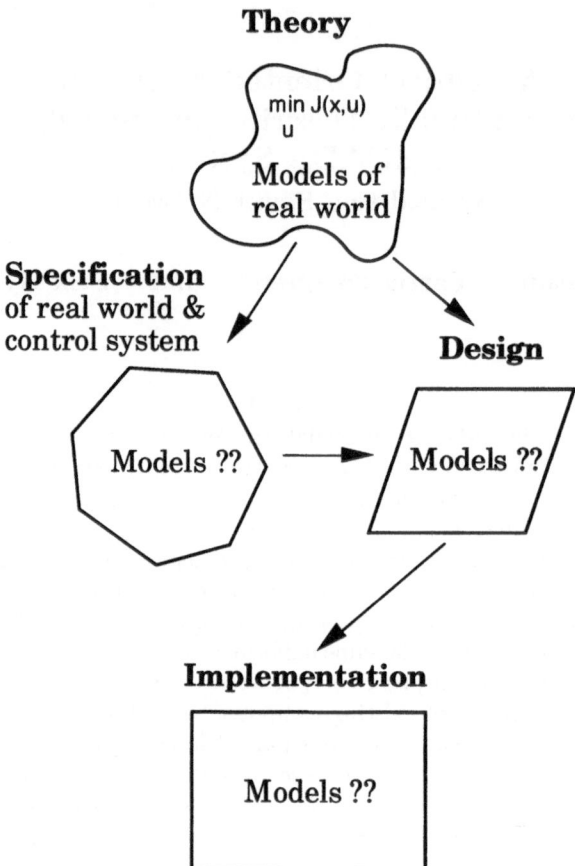

Figure 1: Development of a control system: theory, specification, design, and implementation domains

The theory domain carries formal and mathematical models of the real world, i.e., the process to be controlled and its corresponding control strategy. The process is a dynamic, complex, distributed large-scale real-time process, in this case a power network. Mathematical models commonly used are differential equations. This formal mathematical way of describing the process serves as input for the specification and design domains.

The specification domain consists of description of the process and the control system. It is more heuristic and possibly its form is verbal and non-mathematical. It is a description of what the human being (the user) and the process require from the control system.The design domain consists of a description of a system architecture that will satisfy the requirements listed in the specification. The architecture is described in terms of software components on higher and lower levels, such as building blocks, functionality, and routines. The language used may be similar to an implementation language but with a less precise syntax, for example, pseudo code.

Based on the design domain, the entire system is implemented, i.e., coded in a suitable programming language.

The research for improving mathematical tools has resulted in a great number of available models that can be used. However, it is well known among developers of complex and large-scale systems, that in the other three domains there is a shortage of corresponding models and methods. The existing models are different in the different domains, and often insufficient. As far as the authors know, no uniform model for all three domains exists. Thus, there is great risk that, when going from one domain to another, user requirements initially listed in the specification will not be met in the final implementation. Paradoxically, most research work is put into the mathematical part, whereas the highest costs are faced in the other three domains. Hence, more research effort must be carried out to attain a structured engineering approach within these domains, that supports transformation between them. A cost efficient systems architecture is necessary, in order to get optimal algorithms to work in an optimal way. This opens up new dimensions on optimal control such as "optimal life cycle costs" and "optimal performance during systems' total life time."

The aim of this paper is to present an overall modeling concept that can be used for describing and specifying large scale distributed SCADA systems in a natural and intuitive way. It can be applied on all levels, and it provides a uniform way of describing the entire system, during all phases (domains) of development. This paper mainly focuses on the three phases of specification, design, and implementation. The concept will most likely improve the cost efficiency in the specification, design, and implementation domains, and it is a necessary condition for more advanced control algorithms to be commercially available.

The modeling concept is based on an object-oriented approach, which is further discussed in section 2. The ideas are highlighted by a case study in section 3. The paper ends with a conclusion.

An Object-Oriented Approach

This section discusses why an object-oriented approach is a successful way to create a language that can describe the functionality in the real world domain. Object-oriented models can also be used during all phases of a project, i.e., not only in the real world domain but also in the design domain and in the implementation domain.

In order to fully utilize an object-oriented approach, the approach must be extended so that it can be used for describing and realizing not only smaller applications, but also large-scale distributed realtime control systems, such as SCADA systems. This is discussed further below, but first traditional and object-oriented systems' development are discussed.

Traditional and object-oriented systems' development

The traditional way to specify and implement a control system is to describe all functional requirements in a functional specification. The functionality is then mapped, via the design domain (phase), into the computer system where the different functions are implemented in a high level imperative language, for example, Fortran. Some parts may also be coded in Assembler, depending on performance requirements. In this case the system architecture is completely functionally oriented. Each module performs a certain function, for example, "status check." Because SCADA systems are large-scale and bound to be complex (Cegrell, 1986), development must be divided into different parts. This means that if the same functionality is used in different contexts and by different people, it is likely to be implemented differently. Thus a consequence is that the same functionality is likely to be non-reusable for other applications, where, for example "status check" is needed.

During the last five years a partly object-oriented approach has been used. Object-orientation has then been introduced on a low level in the specification domain, which in turn has been implemented in an object-oriented programming language, for example C++. However, the specification is functionally oriented on a high level. This implies that the overall system architecture will still be functionally oriented.

In this paper a fully object-oriented approach is emphasized. This means that object-orientation should be used on both higher and lower levels, and throughout all phases of a control system project. The aim is to provide a direct correspondence between objects in the different domains. An object specified in the real world will have its corresponding object in the design- and implementation domain respectively, since the same object model is used in all domains. The specification, design and implementation of a control system thus become completely object-oriented.

Advantages of Object-Orientation

A fundamental advantage of object-orientation is the way of introducing higher level abstractions, closer to the real world. This means that people with power system knowledge can describe the process and its requirements. These are then directly mapped into the implementation domain of the control system, without any major transformations. This is shown in Fig. 2, where there is a close coupling between the real world, the design, and the implementation.

This should be compared with traditional development, Fig. 3, where functional requirements are transformed into the design domain. The design solution, in turn, is then transformed into implementation. The risk is great that initial requirements listed in the specification will not be met in the implemented control system because the transformations are most likely not one-to-one; different models and ways of description exist in the different domains.

An object-oriented way to model and implement a computer-based control

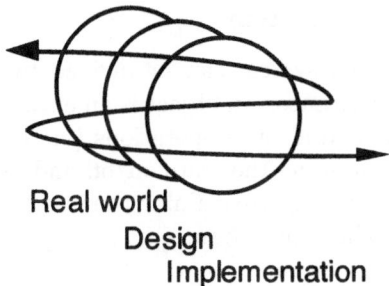

Real world
Design
Implementation

Figure 2: Object-oriented systems' development

system also has the following advantages over a traditional approach. According to Blair (1991) and Booch (1991), it becomes:

- easier to build reusable software components. Reusable components are a way of increasing the abstraction, because a user of a pre-defined component is interested only in the functionality, not the implementation.

- easier to reflect the real world in the implementation, by introducing higher abstractions; objects can contain other objects.

- easier to create well-defined interfaces. The implementation is separated from the interface, and it is hidden by encapsulation. This also makes it easier to build modular systems.

- possible to enforce evolutionary systems' development, since implementation may gradually be modified without having the interface changed.

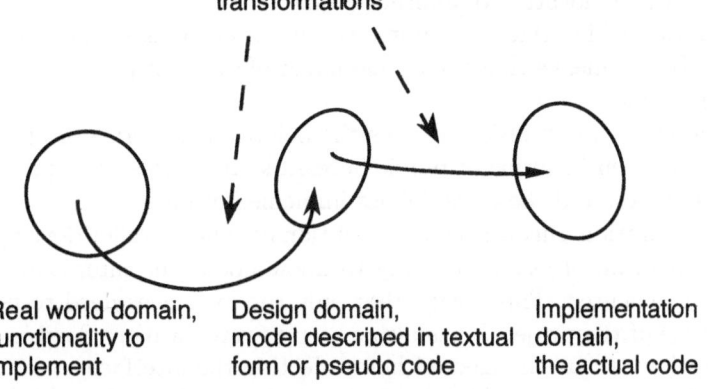

transformations

| Real world domain, functionality to implement | Design domain, model described in textual form or pseudo code | Implementation domain, the actual code |

Figure 3: Traditional systems' development

Extended Object-Oriented System

A conventional object-oriented system can be defined by three concepts (Blair, 1991): object, class, and inheritance. An object is an encapsulation of a number of data, which reflects a state, and a set of operations (methods). A simple example of an object is a breaker containing the state on/off and operations to switch the breaker on and off. A class is a template defining the interface and implementation. Objects are created from a class. Inheritance makes it possible to incorporate the behaviour of one class into a new one.

Conventional object-oriented languages support the concept of data abstraction, which makes it possible to break down the system into modules (objects). However, these languages do not hide the computer system well enough; the system designer has to deal with details on too low a level. Not only must he deal with interobject communication (Ellis, et. al., 1990; Lippman, 1990), he must also deal with interprocess communication. In order to hide the lower level, the conventional concept has to be extended, thus the concept of an extended object-oriented system is introduced.

The extended object-oriented system has to manage concepts not always covered by traditional object-oriented systems. These concepts, among other things, include:

- distributed communication,

- separation of interface and implementation,

- high level objects.

The Distributed Object Management System provides these features in the extended object-oriented system, as shown in Fig. 4.

Distributed communication has to handle object-to-object communication in a distributed and heterogeneous environment, i.e., an object does not have to be concerned with the location of another object.

Separation of interface and implementation encourages an evolutionary development. It also means that the development of different parts of the system can be done in parallel.

The interface of an object defines the behaviour in terms of operations and, in some cases, even in terms of public accessible attributes. The interface defines the type of object, and the class defines implementation.

This separation encourages an evolutionary and parallel development, because the user of an object needs only to know about the interface, not how the object is implemented. This means that only the interfaces need to be defined at an early stage of the project. It is also a step towards a more flexible system; the implementation can be changed without affecting the interface.

The aim is to preserve high level abstraction in the real world domain when the project proceeds to the design domain and the implementation domain. This is achieved by introducing the concept of high level objects, see Fig. 5.

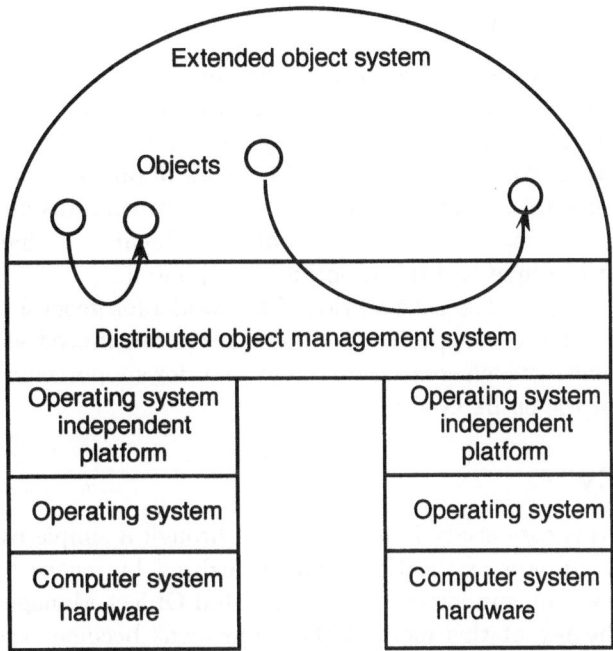

Figure 4: Extended object-oriented system

High level objects are able to contain other objects, but they are more then just "holders" of other objects; they can reflect some of the functionality defined by the contained objects. The high level object can also require some behaviour from the contained objects. For example, some high level objects deal with timing and require that contained objects fulfill certain timing constraints.

An example of a high level object is a transformer station, which contains objects of smaller granularity, such as breakers, disconnecting switches, transformer, and bus bars.

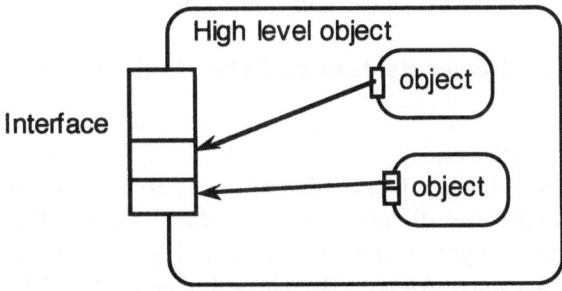

Figure 5: High level object

Evolutionary Approach

Reuse of software is one advantage of an object-oriented technique. This is often done by using class libraries. Inheritance is used to add new application-specific functionality. An improvement of the concept of class libraries can be achieved by introducing frameworks. A framework is a class library directed at a certain application, for example, power system control. It also contains an informal description, a policy, so, for example, if the state of a breaker is changed, how are other objects, for example, a display, notified or updated.

In the case study in the next section, frameworks for general control system functionality and for process specific functionality are introduced. Also, a tool framework is introduced, which contains different tools for supporting the description and realization of the entire system.

A Case Study

The purpose of this case study is to illustrate, through a simple example, how a control system for power applications can be realized by means of an extended object-oriented system, supported by a Distributed Object Management System.

Through the use of this method the main focus becomes the process and its requirements and not the computer realization. The process requirements are directly mapped into the computer environment, by an engineer with a good knowledge of the process, in this case, a good knowledge in power engineering.

Computer-related problems have very little influence on the application configuration of the control system, because they are basically hidden by means of the extended object-oriented system and the frameworks.

Direct-mapping reduces the risk of the process requirements being distorted, when mapping the real world domain into the design domain, and then the design domain into the realization domain. Furthermore, process objects can be found as high level objects in the computer system, and this, among other things, simplifies considerably maintenance and updating.

This study shows how required functionality for the supervision and control of a small part of a power system can be realized. Realization is based on the distributed object management system described in the previous section, with three frameworks placed on top of it, as shown in Fig. 6.

Description of the Power System and the Control System Functionality

The part of a power system used in this case study is shown in Fig. 7. It consists of a transformer station and part of a switch yard. The part of the switch yard consists of two power lines, L1 and L2, two bus bars, As and Bs, three breakers, B1, B2, and B3, and seven disconnecting switches, DS1 to DS7. The transformer station consists of three power lines, L3, L4, and L5, one three phase transformer, 130/24 kV, two bus bars, At and Bt, three breakers, B4, B5, and B6, and six disconnecting switches, DS8 to DS13.

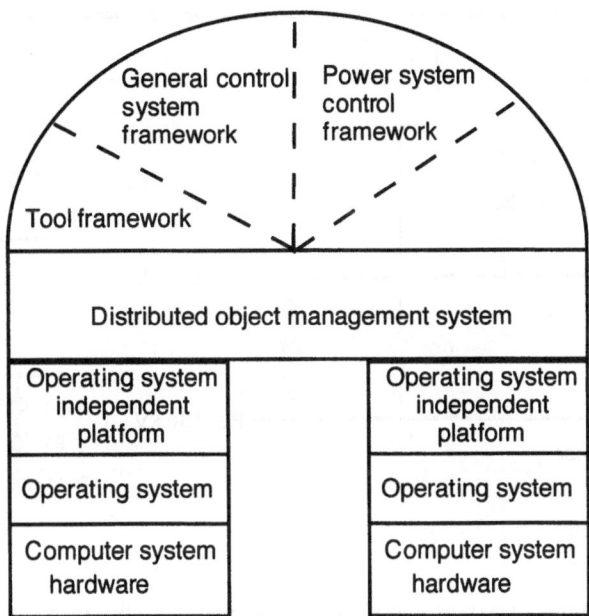

Figure 6: Three frameworks on top of the distributed object management system

The control system shall provide the following functionality for supervisory and control of the switch yard and the transformer station shown in Fig. 7. It shall provide single line diagrams, including breaker and disconnecting switch status, bus bar voltages, and active and reactive power in the power lines. From the single line diagrams, remote control of the breakers shall be possible by means of a two step procedure. An alarm shall be generated if the power flow in the power lines exceeds a pre-defined alarm limit. The alarm shall be presented on an alarm list.

Available Frameworks

On top of the distributed object management system, three frameworks have been implemented, as seen in Fig. 6: a general control system framework that provides classes with basic control system functionality, e.g., alarm lists, a power system control framework that provides classes with process specific functionality, e.g., breaker control, and a tool framework which supports the possibility to create, configure, and maintain the objects in the entire object system from different viewpoints. The process specific framework can be exchanged to adapt the control system to other processes, e.g., supervision and control of a district heating utility.

Even though the frameworks provide different functionalities their structures are similar to each other. They have a class hierarchy with single inheritance, in order to achieve a true hierarchy. Each subclass inherits all data and all methods from its superclass. The structure of the frameworks is made clear in Fig. 8, where

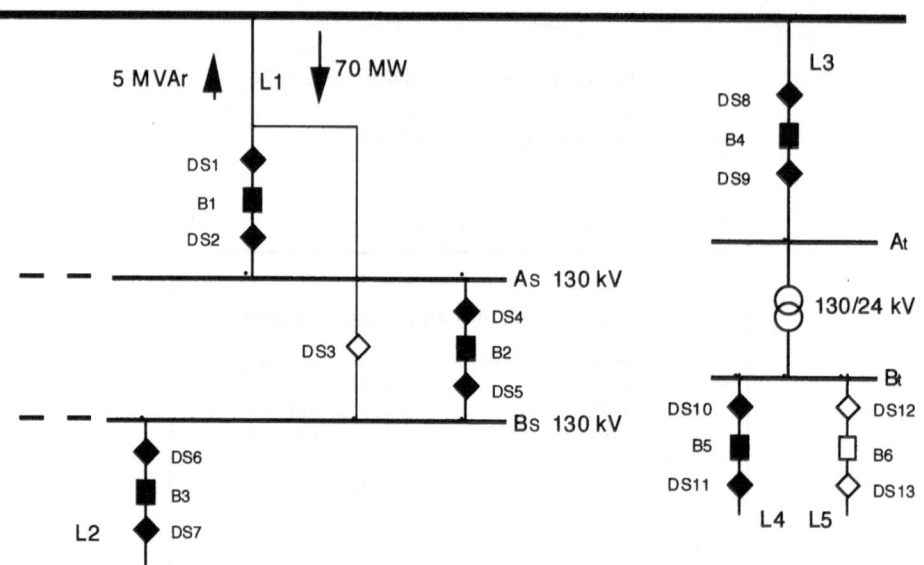

Figure 7: Part of a power system used in the case study

a part of the general control system framework, and a subset of the methods for the classes belonging to this part, are presented. However the type of data for the classes is not presented in the figure.

At higher levels in the hierarchy, classes are very general. The lower a class is in the hierarchy, the more specialized it is. A general control system class lies at the top of the class hierarchy for the control system framework. This class provides basic methods, e.g., create object, which all of the classes in the framework need. It is an abstract class which means that no objects are created from it, however its subclasses inherit methods and data from it. At the bottom of the hierarchy, for instance, a concrete class for creating alarm lists is found.

In a real case, the number of classes is large, and each class has often a large number of methods and data. It is therefore practical to regard these frameworks from different viewpoints, depending on what kind of information from them, or what kind of interaction with them, is required. From the operator's viewpoint, alarm list pictures, for example, are visible. From an application configuration view, on the other hand, the information needed to create alarm lists is available.

Each viewpoint is supported by a tool in the tool framework. The application configuration tool supports, for instance, the application configuration viewpoint. It can be used to configure both the control system functions provided by the general control system framework and the power system control framework.

The main purpose of the application configuration tool is to support adaption of the control system to a certain process. This is done by mapping the process and its requirements into the computer system. Since such a configuration should

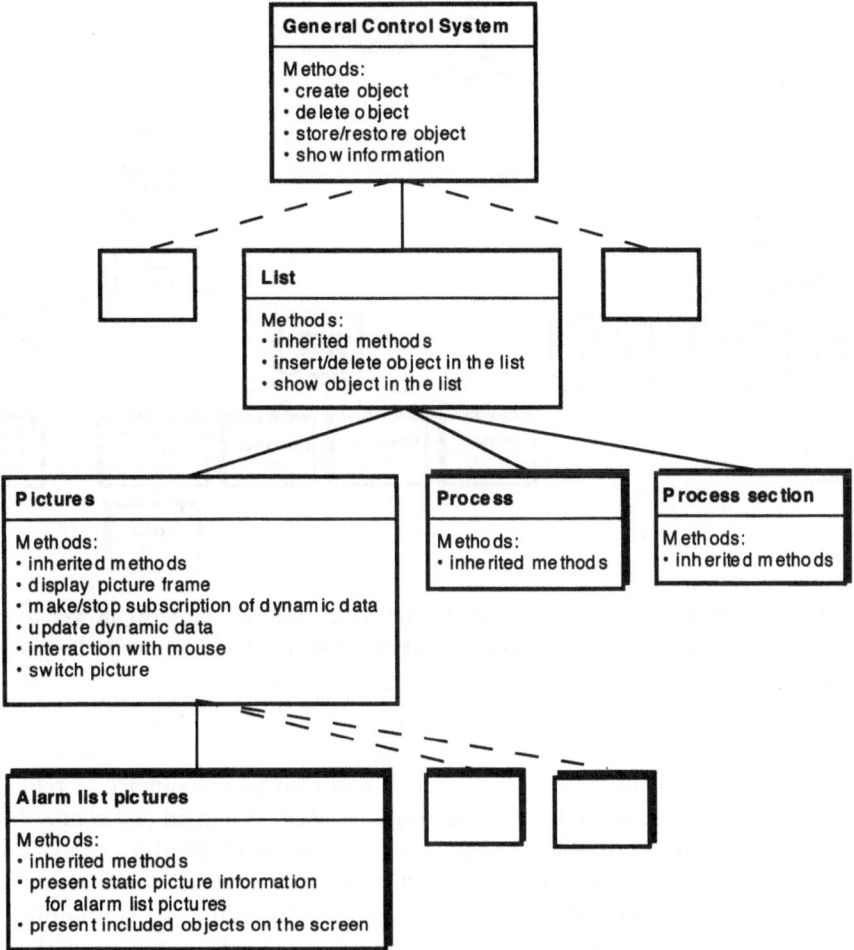

Figure 8: A part of the general control system framework

be carried out by people with the best knowledge of the process, a minimal of computer knowledge should be necessary. Roughly, such an adaption is achieved in the following way.

With support from the application configuration tool, the configurator sorts through the frameworks looking for high level classes, e.g., transformers, breakers, and alarm lists, which can fulfill the required control system functionality. The classes provide a show information method, where they present themselves for the configurator. When the configurator has found a suitable class he creates the necessary number of objects from it. When an object is created, a memory area is allocated for its data. Often the data has default values which can be changed to adapt the object to fulfill certain requirements. Such an adaption can be very simple, such as setting a parameter to choose a certain symbol for presentation

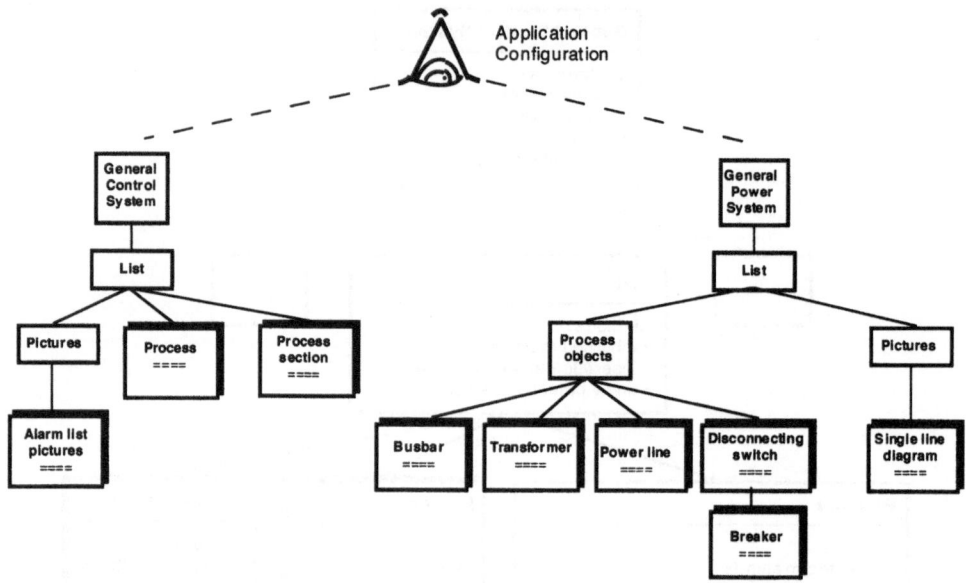

Figure 9: The general control system framework and the general power system framework, from the application configuration viewpoint. Shadowed classes have methods which are visible from this viewpoint

of a process object on the screen. However, it can also be a sequence of relatively extensive steps, as when a single line diagram is created. When all necessary objects have been created, it is possible to end the application configuration session, start the control system, and supervise and control the process from an operator's point of view by use of the operator tool. It is possible to return to the application configuration view and modify the control system s functionality or insert new functionalities without regenerating the system.

If a requisite class is missing, it can be created by refining and specializing an existing class from some of the frameworks. The new class inherits all methods and all data from the existing class, and additional methods and data are added. From the new class, an object can then be created. This too, can be supported by the tool framework. One way of realization is to provide a high-level language which makes it possible for the configurator to create new classes based on specialization of existing classes (Borning, et. al., 1979; Ingalls, et. al., 1988). It must be imperative that this is easy to do, without certain computer knowledge. Another way of realizing the new class is to provide a high level specification language which makes it possible for the configurator to define the required new functionality in a clear-cut way. The class is then created by a system engineer with knowledge of the realization domain of the system, through specializing of an existing class.

To make it possible to adapt a control system to a certain process in the way described above, the following requirements, among others, are placed on the application configuration tool:

- The tool must provide a link between the configurator and the classes, making it possible to both search for a certain class, get information about it, and define and create objects from it.

- Information filtering must be provided. When looking at frameworks, methods not necessary for application configuration should be filtered out. The unnecessary description of the type of data which belongs to each class should also be filtered out.

- When looking at a certain class, both its inherited methods and its own methods which have an application configuration view should be explicitly presented.

- The man-machine interface must be easy to use.

Information filtering is illustrated in Fig. 9. For some classes, all methods are viewed by the application configuration tool, whereas for others just a few or none are viewed. The shadowed classes have methods which are visible from the application configuration viewpoint.

Application Configuration

The configuration of the control system emanates from the description of the part of the power system to be controlled in section 3.1 and from the specification of the functionality of the control system in section 3.2. The process and its requirements are mapped into the computer system by means of the application configuration tool, the general control system framework, and the power system control framework. It is supported by the application configuration tool, to create objects from the classes which are shadowed in Fig. 9. It is not necessary, nor even possible, to create objects from the other classes during the application configuration session. When looking at a certain class, e.g., alarm list class, with support from the application configuration tool, a view similar to the one shown in Tab. 1 is presented.

The objects listed in Tab. 2 must be created to fulfill the requirements stated in the specification of the control system s functionality described in section 3.2.

One object from the process class is created. The object represents the entire power system to be controlled and manages the process section objects.

Two objects from the process section class are created. They represent the considered transformer station and the chosen part of the switch yard. Each of them manage the process objects, e.g., breakers and bus bars, which they include.

One alarm list is created. It shows the alarms generated when the power flow in the power lines exceed pre-defined alarm limits.

Three single line diagram objects are created, showing the entire power system, the transformer station, and the chosen part of the switch yard. Each single

> Classname: Alarm List
> *Declaration of methods*
> • Create object
> • Delete object
> • Update object
> • Show information about object

Table 1: The Application Configuration View of the Class Alarm List

line diagram object consists of a static and a dynamic part. The static part is the static picture drawn on the screen. The dynamical data is linked to this picture and reflects the status of the breakers and disconnecting switches, bus bar voltages, and active and reactive power in the power lines. When a change occurs in the physical process, this is reflected through the dynamical part of the single line diagram object. Information of the status can in this way be shown to the operator. The single line diagram object for the entire process shows overview information about the entire power system, while single line diagram objects for the switch yard and the transformer station consist of more detailed information.

class	object names
Process	entire power system
Process section	switch yard transformer station
Alarm list	alarm list for power flow alarms
Single line diagram	single line - entire power system single line - switch yard single line - transformer station
Bus bar	2 in the switch yard, As & Bs 2 in the transformer station, At & Bt
Transformer	transformer, 130/24 kV
Power Line	power lines L1-L5
Disconnecting switch	7 in the switch yard, DS1-7 6 in the transformer station, DS8-13
Breaker	3 in the switch yard, B1-3 3 in the transformer station, B4-6

Table 2: Created Objects in the Case Study

Four bus bar objects, two in the transformer station and two in the switch yard, are created. They contain information about bus bar voltages in the power system.

One transformer object is created from the transformer class. The object is defined by a number of factors, for example, the ratio of voltage transformation, 130/24 kV.

Five power line objects, corresponding to L1-L5 in Fig. 7, are created. Alarm limits for the power flow in the power lines are defined.

Thirteen disconnecting switch objects are created, seven in the switch yard, DS1-7, and six in the transformer station, DS8-13.

Six breaker objects are also created, B1-3 in the switch yard, and B4-6 in the transformer station.

Provided that all hardware is connected in the proper manner, it should now be possible to quit the application configuration tool, start the control system, and begin to supervise and control the process from an operator's point of view, according to the requirements set up in section 3.2.

Conclusions

In this paper it has been stressed that uniform models for specification, design, and implementation of large-scale, computer-based real-time industrial control systems are needed. These models are necessary to be able to provide a cost efficient system architecture, which in turn is a prerequisite for more advanced mathematical tools, such as optimal power flow algorithms, to be utilized to their best advantage. An object-oriented approach has been shown to be advantageous for design and implementation of medium scale applications and corresponding control systems, and for delimited parts of large-scale systems. But a conventional object-oriented approach must be extended to meet the requirements in a distributed environment, for example, interaction between objects that are located on different computer nodes. In order to fully reach maximum utilization, object-orientation must also be used for specification of such distributed complex systems. The ideas have been highlighted through a case study. Based upon this the following conclusions can be drawn.

- Uniform object models can be used during all development phases of a control system project: specification, design, and implementation. This implies that a real world process object will be found as a well-defined high level object in the software system. The process requirements from the specification are directly mapped into the computer implementation. This is readily done by an engineer with a good process knowledge, in this case power engineering, and with a minimum of computer related problems. This will make maintenance and further development of the entire control system easier.

- The presented object-oriented approach can be used on different levels when describing a control system. On a high level, the overall functionality can be specified. On a low level, the system can be realized by means of a distributed object management system and a number of frameworks on top of it, as shown in the case study. This considerably reduces the risk for distortion of the requirements on the control system when moving between the different development phases.

- It is probable that costs for specification, design, and implementation of SCADA systems for large industrial applications, for example, power systems, can be lowered considerably by means of a further developed object-oriented approach, especially if it is used on all levels of a SCADA system project.

The concept has been applied to a part of a power network, but it is general enough to be applied to other large-scale processes with real-time requirements, such as process industry, manufacturing industry, and telecommunication networks.

As this paper describes a novel development concept for control systems, several suggestions for future work can be made. The most important may be to further develop the object model and supporting methods that can be used during the specification, design, and implementation phases. To find strict guidelines for the transformation between the different phases is another suggestion. Last, but not least, the ideas shall be implemented.

Acknowledgement. The authors would like to thank the advisor of the project, Professor Torsten Cegrell, for his assistance, valuable discussions, and encouragement during this work.

References

Blair, G. (1991). Object-Oriented Languages, Systems and Applications. Pitman Publishing.

Booch, G. (1991). Object-oriented Design with Applications. Benjamin/Cummings Publishing Company, Inc.

Cegrell, T. (1986). Power System Control – Technology. Prentice Hall International.

Ellis, M. and Stroustrup, B. (1990). The Annotated C++ Reference Manual. Addison Wesley.

Lippman, S. (1990). C++ Primer. Addison Wesley.

Borning, A.H. (1979). ThingLab, A Constraint-Oriented Simulation Laboratory, Tech. Report SSL-79-3, Xerox Palo Alto Research Center, Palo Alto, CA.

Ingalls, D., et. al. (1988). Fabrik – A Visual Programming Environment, OOPSLA.

International Series of Numerical Mathematics, Vol. 115, © 1994 Birkhäuser Verlag Basel

Near-Optimal Flight Trajectories
Generated by Neural Networks

Bernt Järmark*   and   Henrick Bengtsson*

Abstract

A particular rendezvous mission is modeled, which is difficult to perform in practice by pilots. It serves as a demonstration of using neural networks to generate control laws. The problem emphasizes the need for a control law since it is non-linear and has a sensitive element included. A particular formulation of the optimal criteria is made. The optimization method used must be efficient and cheap to use as a large set of optimal solutions is needed as learning material for the neural networks. The worst problem with neural network is to show it all information needed for control within the whole state space. The application in this paper shows some of the difficulties found during our experimental sessions.

Introduction

Optimal flight trajectories are not possible to generate in closed-loop form concerning realistic aircraft and missions. Several sub-optimal and near-optimal approaches like references 1, 2 have been considered in the literature. Off-line (open loop form) studies can be achieved for realistic models of both the aircraft and its mission, cf. reference 3.

A rather new approach to generate near-optimal flight paths is to use neural networks. In reference 4 optimal climb paths have been the issue and the neural network seems to be a good tool for control within the entirely flight envelope. As an initial demonstration a maximum energy climb problem was chosen. Also, the problem of reaching a launch envelope (includes catching up a distance) was tested and seems to work satisfactory too. The motion was in the vertical plane with four states and the load factor as the control.

The present study, which was done as a master thesis, ref. 5, will expand to cover three dimensional motion and a particular mission which is known to be difficult to perform in practice. The mission considered is to repel an intruder as quickly as possible. It contains a straight flight path as well as a sharp turn at the end. This means we have a large variation in the dynamics and it's doubtful if sub-optimal methods like singular perturbation technique would work.

---

*Saab Aircraft Division, Saab-Scania AB, S-58188 Linköping, Sweden

There is a must of having an efficient optimization algorithm in order to generate a large set of optimal trajectories as learning material. The fact that we need optimal ones is obvious as we want to generate a near-optimal network. Also, a distinct evaluation of the neural net performance can then be achieved.

The purpose with the paper is to demonstrate a particular formulation of the optimal criteria in order to model the rendezvous problem. Then the neural network application and the basic problems with it will be discussed and demonstrated in both two and three dimensional cases.

Model Formulation

Mathematical models of both the system and the mission are required. Particularly the latter one will be accentuated in this paper in order to really describe the mission as correct as possible.

Aircraft Model

The equations of aircraft motion are the point mass model,

$$\dot{x} = v \cos \chi \cos \gamma, \tag{1}$$

$$\dot{y} = v \sin \chi \cos \gamma, \tag{2}$$

$$\dot{z} = v \sin \gamma, \tag{3}$$

$$\dot{\chi} = \frac{g_0}{v} \frac{u_h}{\cos \gamma}, \tag{4}$$

$$\dot{\gamma} = \frac{g_0}{v}(u_v - \cos \gamma), \tag{5}$$

$$\dot{v} = \frac{T(z, v) - D(z, v, n)}{m} - g_0 \sin \gamma, \tag{6}$$

$$\dot{m} = -F(z, v), \tag{7}$$

where x, y and z are earth fixed coordinates (z is the altitude), χ is the course angle, γ is the climb angle, v is the velocity, m is the mass and g_0 is the acceleration due to gravity. The model of the aircraft is described by non-linear physical components representing thrust (T), drag (D), and fuel flow (F). These functions will model the physics to an arbitrary level of accuracy as discussed in ref. 3. In this paper a generic model is used, with a wing area of 35 m^2, initial mass of 10000 kg and an engine similar to F404.

The controls are represented by the vertical, u_v, and horizontal, u_h, components of the load factor n. Consequently we have

$$n = \sqrt{u_v^2 + u_h^2}. \tag{8}$$

This has shown to be convenient instead of using load factor and bank angle as control inputs. The load factor is limited to 6 g's.

The drag can be split into two components: the zero lift drag and the induced drag, with indexes '0' and 'i' respectively.

$$D(z, v, n) = D_0(z, v) + D_i(z, v, n). \tag{9}$$

The zero lift drag increases with velocity and decreases with altitude, while the induced drag increases with altitude and load factor and decreases with velocity. These effects are notable in the optimal solutions.

Mission Formulation

The scenario we have chosen is to repel an intruder, which is a special case of rendezvous. It is as follows: An unknown aircraft is approaching and our goal is to perform a rendezvous (i.e. equal aircraft states) as far out as possible. The geometry is closing up very fast which makes the final turn, in the order of 90 to 180 degrees, very critical.

Figure 1 shows one typical example of an optimal trajectory in our application. The y vs. x curve acts as described with a rapid turn into the wanted course and after a long segment of straight flying it ends up with a quick turn. The other curve describes the optimal vertical motion and its behaviour is a result of the drag components in relation to altitude, speed and load factor. The horizontal and vertical load factors are exemplified in figure 2.

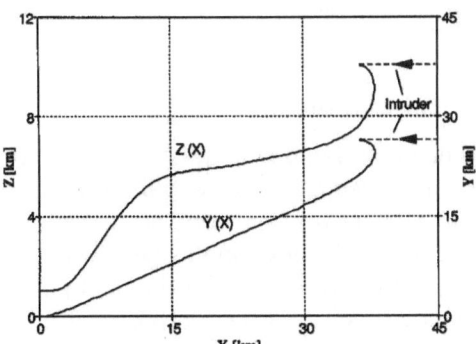

Figure 1: Rendezvous Trajectory

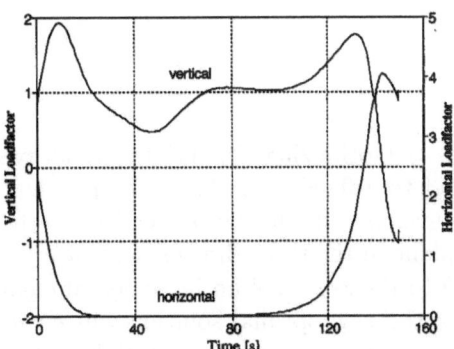

Figure 2: Example of Load Factors

Mathematical Model of the Mission

The total cost function V to be minimized weights the final states (indicated by subindex 'f') of the altitude, the velocity, the position in the horizontal plane, the climb angle and the final heading to coincide with the intruder.

$$V(X_0) = -\Delta V + k_\chi (\chi - \chi_f)^2 + k_\gamma (\gamma - \gamma_f)^2 + k_z (z - z_f)^2 + k_v (v - v_f)^2. \tag{10}$$

In order to meet the requirement of making the rendezvous as far out as possible, the particular component, ΔV, in the cost function has to be maximized. This is the driving term in the horizontal plane. The function is,

$$\Delta V = x(t_f)\cos\theta + y(t_f)\sin\theta, \tag{11}$$

where θ is an auxiliary angle controlling the rendezvous point.

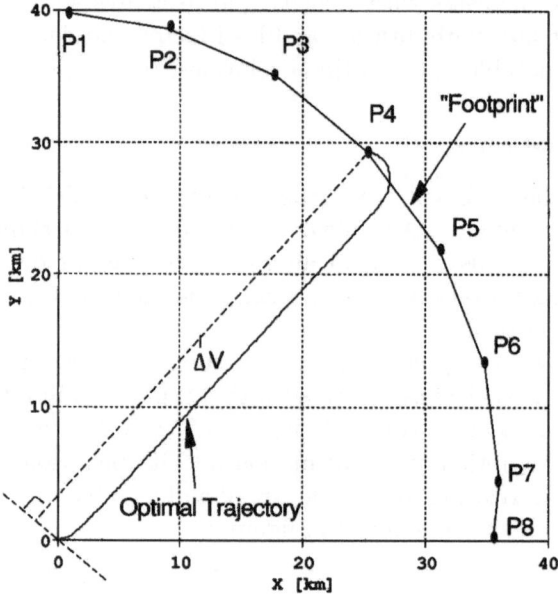

Figure 3: Example of a Footprint

Optimizing Eq.(11) the rendezvous will be as far out as possible and θ is set to a fixed value for each optimization run. Sweeping the θ we will obtain an envelope, like in figure , called the footprint (the projection into the horizontal plane in the three dimensional case). The normal to the envelope makes the angle θ to the x-axis. If an incremental change in the final time is made the final point of the new optimal solution will move in the direction of θ, i.e. perpendicular to the footprint. The straight flight path, when it exists, is in the direction of θ. This can be proven by a closer look at the adjoint equations found in the optimization algorithm. This also confirm the fact that the normal of the footprint is pointing in the θ direction as an incremental change in time will prolong this straight path. The interpretation of the optimal ΔV in Eq.(11) is that it represents a distance to a line through origin with the slope to the y-axis by θ, according to figure 3. By the choice of θ and t_f we will control the final x and y values. The purpose of this approach is to obtain a good and robust convergence in the optimization scheme as the adjoints for x and y are constants. Also, this is a convenient way of producing the footprint.

A set of optimal trajectories for various θ and t_f is produced. Afterwards, in the learning phase, we apply the intruders which satisfy these final states (x_f, y_f, v_f).

The optimization program used is an extension to three dimensions of the program behind the work in reference 4. It is run on a 386-PC in interactive mode where a new optimal problem can be solved fast (and cheap).

Neural Network

The neural network technique, which has been given much attention during the last years, is often built up on empiric experience. Literature dealing with neural network is very sparse. Reference 6 handles general theories from biological memories to some applications with different models of neural networks. A more direct description of modern neural network computing could be found in reference 7. The program package used is the NeuralWorks Professional II/Plus. In the manuals (reference 8) some theory is discussed and particularly the back propagation algorithm is brought up.

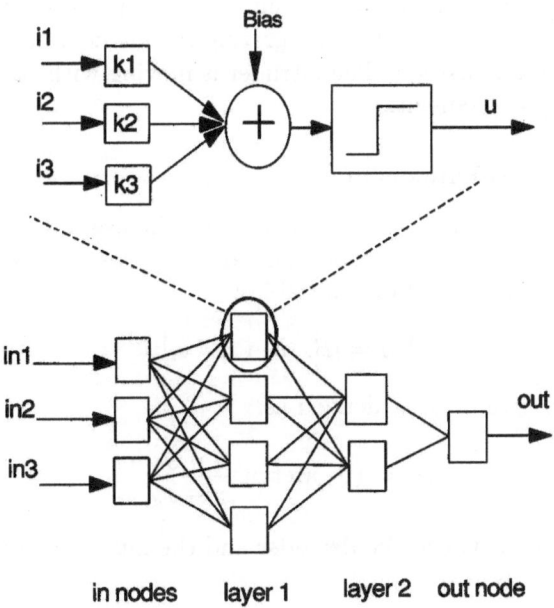

Figure 4: Structure of a Neural Net

Roughly speaking a neural network is a systematic non-linear curve fitting technique. The use of neural network has two phases:

1. The learning phase where the net models the behaviour of some given optimal trajectories.

2. The application phase where the net is used to generate controls.

Network Theory

A neural network is a complex structure built up of small elements called neurons or processing units. A neuron gets a number of weighted input signals which are added to a bias signal and processed through a "step function" before output. In practice the "step function" needs to be differentiable. In our application we used a tangent hyperbolicus function to represent the step.

The net architecture used here is the most common, consisting of one input layer, one to three hidden layers and an output layer as shown in figure . The net is fully connected which means that each neuron in one layer is connected to all neurons in the following layer and so on. It can be seen as an nonlinear function $g(X)$.

The structures used in this paper are the best ones obtained in reference 5, where several compositions were tested.

Results

Before the net learning we had to add the intruder equations to our system. The intruder trajectories were produced by integrating the intruder equations backwards from the obtained final position. The intruder is moving with constant velocity at fixed altitude and y-component.

Experiments in Two Dimensions

In order to illustrate the rendezvous problem and the use of Eq.(10) introductory studies in two dimensions were made. The altitude was constantly set to 1 km. The state vector X_N consisted of 5 variables:

$$X_N = [B_r, R, \chi, v_d, v_i], \tag{12}$$

where B_r is the relative bearing defined as:

$$B_r = \chi - \arctan \frac{y_i - y_d}{x_i - x_d}, \tag{13}$$

and R is the distance between the defender and the intruder (subindex 'd' and 'i' respectively) defined as:

$$R = \sqrt{(y_i - y_d)^2 + (x_i - x_d)^2}. \tag{14}$$

Our load factor is:

$$n = \sqrt{1 + u_h}. \tag{15}$$

In this experiment we used a learning material consisting of 15 different optimal trajectories with $t_f = [75, 120]$ seconds and the step length between two samples set to 3 seconds. Seven of the trajectories were spread out to cover straight forward incoming intruders in the first quadrant. The remaining eight trajectories were used to prevent us from getting a static error before and while we are flying in straight course. (This represents a new trajectory from the actual point from where the net has no information). The problem is that the net has not learnt how to compensate an error in this segment because of the constant load factor in the optimal case. The artifice is to show the net trajectories with higher initial velocities which will fill up the lack of data in our subspace.

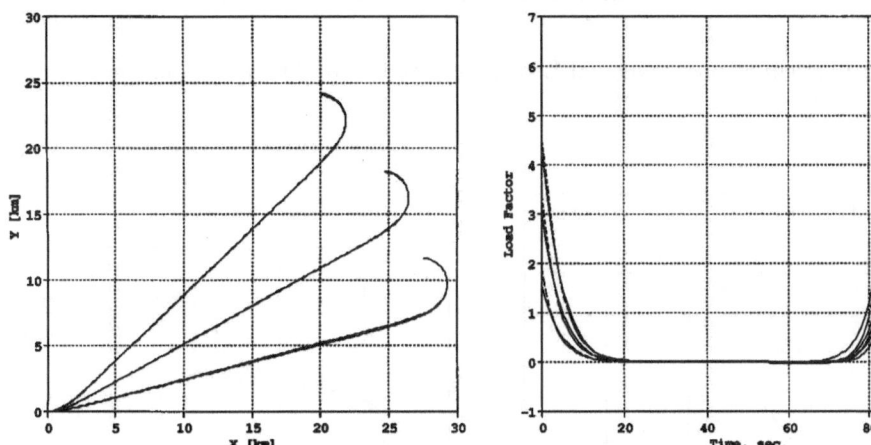

Fig. 5: Trajectories in Horizontal Plane   Figure 6: Load Factor vs. Time

For evaluation we used trajectories not previously seen by the net of duration $t_f = 100$ seconds. The result of the two-dimensional case is illustrated in figures 5 and 6 where the optimal trajectories are represented by solid lines and the net generated by dashed ones. The load factor constraint (6 g's) results at the abrupt change at the end of the plot in figure 6.

Experiments in Three Dimensions

Altitude is the third dimension and the final state vector is:

$$X_N = [z, B_h, B_r, R, \chi, v_d, v_i]. \tag{16}$$

We made an estimate for the needed learning material in three dimensions and found that we have to increase it as shown in table 1.

The amount of learning material will increase even more if we consider to model a full three dimensional system where a couple of other state variables are added.

Variable	Number	Motivation
initial altitude z_0	5	early experiments in the vertical plane
initial velocity v_0	2	experiments in the horizontal plane
course angle χ	5	experiments in the horizontal plane
final time t_f	≥ 2	experiments in the horizontal plane
final altitude z_f	≥ 2	estimated
intruder velocity v_i	2	experiments in the horizontal plane
number of trajectories	$\Pi 400$	multiplication

Table 1: Estimation of Needed Learning Material in Three dimensions

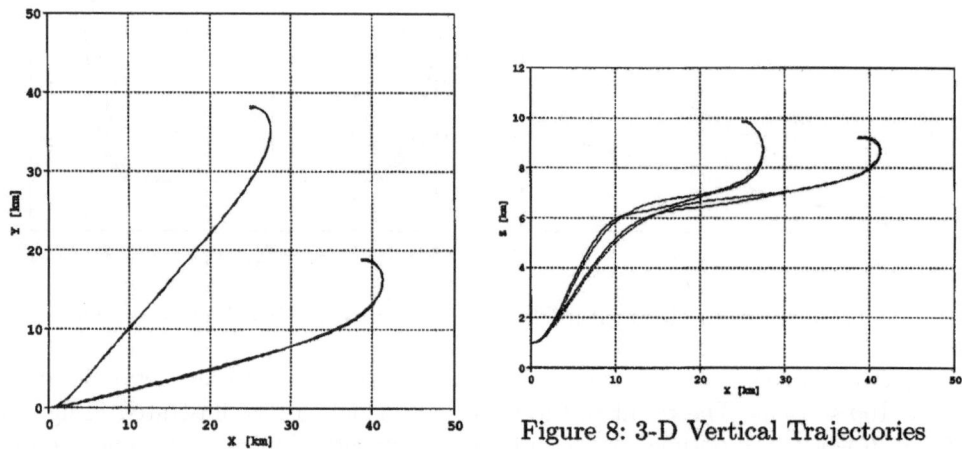

Figure 8: 3-D Vertical Trajectories

Figure 7: 3-D Horizontal Trajectories

In order to limit the amount of calculations we restricted our state space as follows:

- Course angle χ is limited to the interval -15 – 60 degrees.

- Initial altitude z_0 is constantly set to 1 km.

- Final altitude z_f is limited to the interval 8 – 10 km.

The learning material produced, consisted of 124 different trajectories and they were all sampled with 6 seconds interval. The large number of trajectories in the learning material reduces the importance of each single sample which motivates the increased interval between each sample. Like the two-dimensional case we showed the net a full set of trajectories from the initial altitude at 1 km, in this case 41. The rest was used to prevent static errors in both altitude and horizontal position.

Those trajectories were generated with initial altitudes varying between 1 and 10 km and with two different initial velocities. The quick end phase is very sensitive where velocity depends on climb angle and vice versa in accordance to Eq.(5) and (6).

The three dimensional experiment results are presented in figures 7, 8, 9 and 10 where the optimal trajectories and controls are compared with the neural network produced ones. The error in vertical movement can be explained with the fact that we could change altitude in behalf of velocity and get another optimal climb path. We also notice that the net control has a large gain at the end which results in a heavy control in order to correct errors. This might be removed by more data in this region. The changes in geometry is rapid close to the rendezvous point.

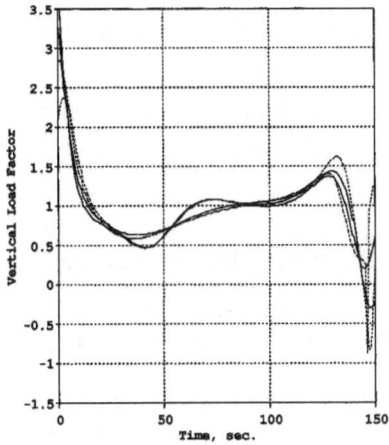
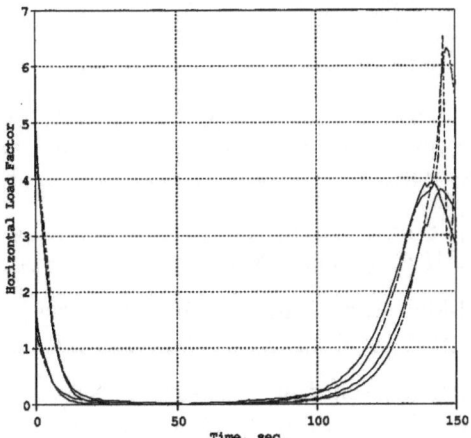

Figure 9: Horizontal Load Factor   Figure 10: Vertical Load Factor

Conclusions

The net used in the two dimensional experiments was built up with 5 nodes in the input layer, 2 hidden layers with 10 and 5 nodes respectively and one output node. The amount of nodes can be decreased with about 4-5 nodes to get an optimal configuration. The evaluation of the two dimensional net shown errors well inside given specifications based on a practical level of accuracy (see reference 5).

In three dimensions we used a net built up with 8 nodes in the input layer, 2 hidden layers with 15 and 10 nodes respectively and 2 nodes in the output layer. In most cases the net generated three dimensional trajectories were close to the optimal ones, but there is sometimes a discrepancy caused by insufficient control of χ at the end. This might not spoil the use of neural net as in practice we must use another controller (e.g. pilot) in the end phase.

Summary

The neural network has proven to be a good device for generating near-optimal flight trajectories for a realistic problem. For our demonstration a limited subspace was considered. It indicated the need for a large set of learning material which seems to be the limitation of using neural networks. The amount might rise quickly with the number of states in the system. The optimization has to be performed in an efficient way in order to generate a large set of learning material. This might be a weakness if not having a good optimization method available. The method and the problem formulation used in our work seems to be a good approach.

References

[1] Shinar, J., Well, K. and Järmark, B.: Near-Optimal Feedback Control for Three-Dimensional Interceptions, *ICAS paper No 86-5.1.3*. Presented in London, September 7-12, 1986.

[2] Järmark, Bernt: On Closed-Loop Controls in Pursuit-Evasion, *Journal of Computers and Mathematics with Applications*, (Edits. Y. Yavin and M. Pachter). Vol. 13, 1987.

[3] Järmark, Bernt: Various Optimal Climb Profiles, *AIAA paper No. 91-2859-CP*, *AFM Conference*, New Orleans, Louisiana, August 12-14, 1991, pp. 124-130.

[4] McKelvey, Tomas: Neural Networks Applied to Optimal Flight Control, IFAC-IFIP-IMACS Int. Symp. on AI in real control, Delft, Netherlands June 1992.

[5] Bengtsson, Henrick: Neurala Nätverk tillämpade pårendezvousproblemet, Master Thesis, LiTHISY-EX-1198, Department of Electrical Engineering, Division of Automatic Control Linköping, June 1992, (in Swedish).

[6] Kohonen, Teuvo: Self-Organization and Associative Memory, Springer Verlag, 1989.

[7] Lippmann, P. Richard: An Introduction to Computing with Neural Nets, *IEEE*, *ASSP Magazine*, April 1987.

[8] NeuralWare Inc.: Neural Computing, manual to program package NeuralWorks Professional II/PLUS, 1991.

International Series of Numerical Mathematics, Vol. 115, © 1994 Birkhäuser Verlag Basel

Performance of a Feedback Method with Respect to Changes in the Air-Density during the Ascent of a Two-Stage-To-Orbit Vehicle

B. Kugelmann*

Introduction

In recent years there has been a lot of research with respect to the so-called Sänger model: the corresponding aerospace system consists of two hypersonic airplanes, where the smaller one initially is fastened on top of the lower vehicle. This compound is to be launched horizontally by means of the propulsion system of the lower airplane which takes the upper part to a certain altitude. At this point the two stages separate, the lower stage returning to the ground, while the upper stage continues its ascent to an Earth-orbit, capable of a later re-entry manœuvre similar to the Space-Shuttle orbiter. For more details see Kuczera et al. [5]. The research efforts devoted to this Two-Stage-To-Orbit (TSTO) system were concentrated both on the propulsion system and the trajectory optimization (Bulirsch,Chudej [2]). In the latter paper a-priori solutions or nominal trajectories for prescribed physical data have been given.

The present paper is concerned with another aspect of the trajectory optimization, that is the robust and optimal guidance of the vehicles under the influence of disturbances, which cannot be included in the mathematical model of the system and therefore cannot be taken into account when an a-priori solution of the optimal trajectory is computed. These disturbances are due to the fact, that the mathematical model used to compute the nominal solutions is just an approximate description of the intricate behavior of the system in reality. For example, for the aforementioned aerospace system, perturbations of the air density have a considerable influence on the solution, but obviously they cannot be included in the mathematical model for the a-priori computations. Therefore an in-flight guidance method has to be applied, that guarantees the observance of the mission targets and, if possible, the optimality of the perturbed flight trajectory as well. In [6] such a feedback algorithm is described, which computes an update for the control variables of the system in minimal time, based on the measured deviation of the actual path from the precalculated nominal path. In the present paper this method is to be applied to the problem of unexpected changes in the air density during the Sänger ascent.

*Department of Mathematics, Munich University of Technology, D-80290 Munich, Germany

Mathematical Model and Nominal Solution

The object of the mathematical optimization is to maximize the payload, that is
the mass which can be taken into a circular Earth-orbit by the upper stage:

$$\max_{u,\mu,\epsilon,b} m(t_f), \tag{1}$$

where $m(t_f)$ represents the final mass at the moment t_f when the upper stage
enters the circular orbit. The control functions are angle of attack $u(t)$, lateral
inclination angle $\mu(t)$, thrust angle of attack $\epsilon(t)$ and mass flow $b(t)$. Time t is
the independent variable. Additional control parameters, which may be varied in
order to increase the payload, are the separation time t_s and the final time t_f. The
motion of the dynamic system is described by a point mass model with six state
variables. The state variables satisfy the following nonlinear system of differential
equations:

$$\dot{v} = \frac{1}{m}[T(h;b)\cos\epsilon - D(v,h;u)] - g(h)\sin\gamma +$$
$$+\omega^2 R\cos\Lambda(\sin\gamma\cos\Lambda - \cos\gamma\sin\chi\sin\Lambda)$$

$$\dot{\gamma} = \frac{1}{mv}[T(h;b)\sin\epsilon + L(v,h;u)]\cos\mu - \left(\frac{g(h)}{v} - \frac{v}{R}\right)\cos\gamma +$$
$$+2\omega\cos\chi\cos\Lambda + \omega^2\cos\Lambda\frac{R}{v}(\sin\gamma\sin\chi\sin\Lambda + \cos\gamma\cos\Lambda)$$

$$\dot{\chi} = \frac{1}{mv\cos\gamma}[T(h;b)\sin\epsilon + L(v,h;u)]\sin\mu - \frac{v}{R}\cos\gamma\cos\chi\tan\Lambda + \tag{2}$$
$$+2\omega(\sin\chi\cos\Lambda\tan\gamma - \sin\Lambda) - \omega^2\cos\Lambda\sin\Lambda\cos\chi\frac{R}{v\cos\gamma}$$

$$\dot{h} = v\sin\gamma$$

$$\dot{m} = -b$$

$$\dot{\Lambda} = \frac{v}{R}\cos\gamma\sin\chi$$

The state variables and the corresponding boundary conditions are:

v	velocity	$v(0) = 0.275$ [km/sec]	
γ	path inclination	$\gamma(0) = 0$ [°]	$\gamma(t_f) = 0$ [°]
χ	azimuth inclination	$\chi(0) = 0$ [°]	
h	altitude	$h(0) = 0$ [km]	$h(t_f) = 500$ [km]
m	mass	$m(0) = 184$ [10^3kg]	$m(t_f)$ free
Λ	geographical latitude	$\Lambda(0) = 0$ [°]	

There is one additional end condition which guarantees that the trajectory connects
smoothly to the final circular orbit. This condition combines the final values of v, χ
and Λ. The corresponding equation as well as the complicated formulas for thrust
$T(h;b)$, aerodynamical lift $L(v,h;u)$ and drag $D(v,h;u)$ are discussed in great

detail in Chudej et al. [3]. The mass at an arbitrary time is considered to be the amount of mass connected with the upper stage. Therefore there is a jump in the mass function at the separation point. From a physical point of view as well as from the point of view of the optimization, there have to be bounds for the mass flow. These bounds are chosen differently for the different engines of lower and upper stage. The calculations in the present paper have been done for rocket-propelled engines in both stages, while the optimization problem has also been solved for airbreathing systems (see Bulirsch, Chudej [1]). The aerodynamic model includes a Mach-dependent quadratic polar together with a widely used standard atmosphere (NASA [8], Shau [9]). The terms in (2) that contain the angular velocity ω of the Earth are due to the Coriolis effect. R is the distance of the point mass from the center of gravity and $g(h)$ describes the gravitational acceleration.

From the mathematical point of view (1) together with (2) represents an optimal control problem. By using the necessary conditions of the calculus of variations such a problem can be reduced to a multipoint boundary value problem, which in turn can be solved numerically. For the Sänger ascent this has been done in [2] by using a multiple shooting algorithm. The computations yielded the following series of different ascent phases, mathematically called phases of the switching structure:

1st phase	$[0, t_s]$	both stages connected; maximal thrust of lower engine
2nd phase	$[t_s, t_1]$	upper stage only with maximal thrust
3rd phase	$[t_1, t_2]$	upper stage only without thrust
4th phase	$[t_2, t_f]$	upper stage only with maximal thrust

The total ascent time turned out to be 3170 seconds and the optimal staging point occurred 134 seconds after take-off. The payload was 16044 kg for a launching site at the equator.

The resulting solution will be called nominal solution or nominal trajectory and the corresponding variables will be characterized by the subscript 0:

$$(v_0, \gamma_0, \chi_0, h_0, m_0, \Lambda_0) = x_0(t) \in \Re^6 \qquad \text{nominal state,}$$

$$(u_0, \mu_0, \epsilon_0, b_0) = w_0(t) \in \Re^4 \qquad \text{nominal control}$$

and

$$t_{s_0}, t_{1_0}, t_{2_0}, t_{f_0} \qquad \text{nominal switching points.}$$

Guidance Problem and Feedback Control

Unfortunately it is not feasible to use the precalculated optimal control w_0 to guide the dynamic system in actual flight. This is due to the fact that the mathematical model is only an approximation of the real physical situation: especially the aerodynamic forces like lift and drag are subject to constant changes and therefore they cannot be represented exactly by mathematical formulas in the form $L(v, h; u)$ or

$D(v, h; u)$ respectively. But also the thrust cannot be described exactly as a function of altitude and massflow, because the appropriate relationship is not only too complicated for a mathematical formulation but also even unknown in some of its aspects. As a matter of fact thrust, lift and drag functions will always be chosen in order to fit some measured data as precise as possible and are not based on a physical law. Beyond that there might be additional influences (wind,...) which are not included in the model. The resulting inconsistencies between reality and mathematical model will inevitably cause perturbations from the precalculated nominal path when only the nominal control is used for in-flight guidance. Therefore an update mechanism has to be used which computes in minimal time a correction δw for the control, based on measured deviations δx in the state variables. Additionally the new, actual control $w_0 + \delta w$ should yield an actual trajectory which still is optimal, at least to first order, in the sense of (1) and which satisfies the given end conditions. At first sight there are three different ways to compute the actual control of the perturbed problem:

(i) The same procedure as for the computation of the nominal solution can be applied: The optimal control problem is reduced to a boundary value problem which is solved by a multiple shooting algorithm. In this case the information about the nominal solution, which has been pre-computed, is used only as an initial guess for the multiple shooting iteration. The disadvantage of this method is, that the computation of the solution normally takes too much time in order to be applicable to a rapidly changing dynamic process. On the other hand the basic idea for the feedback algorithms in (ii) and (iii) is already included here.

(ii) Again the optimal control problem is reduced to a nonlinear boundary value problem. This boundary value problem is linearized and the resulting linear boundary value problem can be solved directly. This is equivalent to doing one iteration step for the nonlinear boundary value problem starting at the nominal solution. Because of the neighborhood of actual and nominal trajectory this single step is sufficient in order to get a good approximation of the actual solution. For more details see Krämer-Eis, Bock [4].

(iii) Here the necessary conditions of the optimal control problem and the boundary conditions are linearized about the nominal solution. This yields a linear multipoint boundary value problem for the deviations $\delta x, \delta w$ and the deviations $\delta \lambda$ of the so-called adjoint variables $\lambda \in \Re^6$. Again one iteration step is sufficient to get the solution of this linear problem.

Since the third method will be used in this paper to compute a feedback control for the Sänger vehicle, it is sketched here for a general optimal control problem of the form (Mayer type): Maximize, for a given function φ,

$$\max_w \varphi(x(t_f), t_f)$$

under the constraint, with given functions f,

$$\dot{x} = f(x, w) \tag{3}$$

and some given boundary conditions. From the calculus of variations it is known, that there exist adjoint functions $\lambda(t)$ so that, defining the *Hamiltonian*, $H = \lambda^T f(x, w)$, the optimal solution satisfies the following equations:

$$\dot{\lambda} = -H_x(x, w, \lambda) \tag{4}$$
$$0 = H_w(x, w, \lambda) \tag{5}$$

The equations (3), (4) and (5) hold for the nominal solution as well as for the actual solution. By formulating the perturbed problem as a variation of the nominal solution

$$x = x_0 + \delta x$$
$$\lambda = \lambda_0 + \delta\lambda$$
$$w = w_0 + \delta w$$

and by linearizing the equations (3), (4), (5) for the perturbed problem along the nominal solution one obtains the following equalities to first order:

$$\dot{\delta x} \doteq f_x(x_0, w_0)\delta x + f_w(x_0, w_0)\delta w \tag{6}$$
$$\dot{\delta\lambda} \doteq -H_{xx}(x_0, w_0, \lambda_0)\delta x - H_{xw}(x_0, w_0, \lambda_0)\delta w - H_{x\lambda}(x_0, w_0, \lambda_0)\delta\lambda \tag{7}$$
$$0 \doteq H_{wx}(x_0, w_0, \lambda_0)\delta x + H_{ww}(x_0, w_0, \lambda_0)\delta w + H_{w\lambda}(x_0, w_0, \lambda_0)\delta\lambda \tag{8}$$

For a regular H_{ww} equation (8) can be rewritten

$$\dot{\delta w} \doteq -H_{ww}^{-1}H_{wx}\delta x - H_{ww}^{-1}H_{w\lambda}\delta\lambda \tag{9}$$

and by insertion of this expression for δw into the equations (6),(7) one obtains a linear boundary value problem for the variations δx and $\delta\lambda$ of the form:

$$\begin{pmatrix} \dot{\delta x} \\ \dot{\delta\lambda} \end{pmatrix} = T_0(t) \begin{pmatrix} \delta x \\ \delta\lambda \end{pmatrix}$$

where the coefficient matrix T_0 depends only on the data of the nominal solution. Because of the linearity of the differential equation there is also a linear relation between the solutions at two different time points t and t_c:

$$\begin{pmatrix} \delta x(t) \\ \delta\lambda(t) \end{pmatrix} = Y_0(t_c, t) \begin{pmatrix} \delta x(t_c) \\ \delta\lambda(t_c) \end{pmatrix} \tag{10}$$

where the transition matrix $Y_0(t_c, t)$ is computable as soon as the coefficient matrix T_0 and the pair (t, t_c) are known. Assuming that at point t_c a deviation s_c in the state variables has been measured

$$\delta x(t_c) = s_c$$

then the update $\delta w(t_c)$ for the control can be computed by the following steps: First of all relation (10) for $t = t_{f_0}$ is applied to the linearized version of the end conditions:

$$B_0 \left(\begin{array}{c} \delta x(t_{f_0}) \\ \delta \lambda(t_{f_0}) \end{array} \right) = 0$$

which yields the following linear system of equations:

$$B_0 Y_0(t_{f_0}, t_c) \left(\begin{array}{c} s_c \\ \delta \lambda(t_c) \end{array} \right) = 0 \tag{11}$$

In this equation B_0 and Y_0 depend only on the nominal solution and the correction point t_c, and s_c is the result of the measurement of the state variables. Therefore the unknown variable in this system is $\delta \lambda(t_c)$, a fact which suggests the following partitioning of the matrix:

$$B_0 Y_0(t_{f_0}, t_c) = (E_{1_0}, E_{2_0})$$

so that the linear system (11) can be rewritten as

$$E_{1_0} s_c + E_{2_0} \delta \lambda(t_c) = 0$$

or

$$\delta \lambda(t_c) = -E_{2_0}^{-1} E_{1_0} s_c$$

Finally this expression for $\delta \lambda(t_c)$ is inserted in (9) in order to obtain a linear feedback law:

$$\delta w(t_c) = -H_{ww}^{-1} [H_{wx} - H_{w\lambda} E_{2_0}^{-1} E_{1_0}] s_c \tag{12}$$

or

$$\delta w(t_c) =: G(t_c) s_c. \tag{13}$$

Looking at the derivation of this feedback law it is obvious that the gain matrix $G(t_c)$ depends exclusively on the nominal trajectory and the correction point t_c. Therefore this matrix can be pre-computed if the correction point is fixed in advance. On the other hand (13) has been derived by a linearization of the necessary conditions for an optimal control problem. It is known that the corresponding linearization error will increase exponentially with respect of the time distance from the correction point. For the feedback algorithm this means that the feedback law has to be applied repeatedly in order to dampen the linearization error. The resulting feedback scheme is divided into two parts:

- A number of tasks can be performed before the actual event takes place, i.e. before the Sänger vehicle takes off: the nominal solution is computed, the nominal state and the nominal control are stored and the gain matrices $G(t_c)$ are computed for a number of sample points.

- During the actual process, i.e. the flight of the vehicle, the amount of computational work in order to update the control is minimal: after having measured the actual state at t_c, the deviation s_c from the nominal trajectory is computed by comparison with the stored data. If the gain matrix G has not been pre-computed at this point t_c the corresponding matrix is linearly approximated by means of the gain matrices of the two neighboring sample points. Then the feedback law is applied to s_c, which requires the multiplication of the small matrix G with the vector s_c of the perturbations. Finally the correction δw for the control is added to the stored nominal control.

Because of the simplicity of the feedback law the time for the computation of one control update is negligiblé. Therefore in practice the number of correction steps depends mainly on the number of measurements that are available during the ascent. It can be shown that the resulting actual trajectory is optimal to first order. On the other hand the algorithm cannot handle perturbations for which the exact solution has a switching structure which is different from the switching structure of the nominal solution. The same algorithm can be applied to problems which have inequality constraints for the state and/or control variables. For more details see [6].

A-posteriori Tests

The algorithm has been tested for the Sänger problem by doing some numerical simulations. This was done by introducing certain perturbations to the system and by numerically integrating the resulting perturbed system (2) with the updated control. The feedback scheme is said to be successful if the computed actual trajectory satisfies the prescribed end conditions within the following bounds:

$$|\gamma(t_f)| < 0.01 \, [^\circ]$$

$$|h(t_f) - 500| < 1 \, [\text{km}]$$

and the third end condition which combines $v(t_f), \chi(t_f)$ and $\Lambda(t_f)$ has to be satisfied within a relative precision of 10^{-4}.

There are two different ways to prescribe perturbations: the first is to check whether the feedback law is successful if there is a deviation s at time t but no further perturbation for the rest of the trajectory. That means that for the rest of the flight path from t to t_f the system is to behave as it is modelled in (2). This kind of a solitary perturbation is investigated in [7] where the corresponding controllability regions are described. The second possibility is that a systematic error is simulated. That means that some data in (2) or even some functional dependency like the formula for the lift or the drag are changed. It is known that the air density is a constantly changing parameter for aerodynamic processes while on the other hand it has a great influence on the behaviour of the system. Therefore changes in the air density have been chosen in order to test the performance of the feedback scheme. The nominal model for the air density in (2) is:

$$\rho(h) = \rho_0 e^{-h/h_{scal}}$$

with given values ρ_0 and h_{scal}. While there are many different ways to vary this formula, the perturbation that has been investigated in this paper is the following: the coefficient ρ_0 is changed for the whole flight by a constant factor:

$$\rho_{0,actual} = c\rho_0.$$

The simulation of the perturbed flight was carried out as follows: the system of differential equations (2) was numerically integrated starting with the exact initial values of the nominal solution but with the perturbed air density. At each correction point the control is updated in order to provide the actual control which is valid for the numerical integration to the next correction point.

As it is explained and justified in [7] it is not necessary to implement the feedback scheme exactly as it is described in the previous section but it is possible to check the performance of this scheme by using a reduced multiple shooting method. This method is easier to implement while it lacks the efficiency of the feedback law (13). But the aim of this paper is to compute the range of perturbations in the air density, that can be corrected successfully, therefore efficiency is of no importance.

For the numerical simulation a grid of 100 equidistant correction points has been chosen and for this case the range for the perturbation parameter c was

$$0.91 \leq c \leq 1.10 \tag{14}$$

which means that the feedback scheme still guarantees a flight trajectory that satisfies the end conditions if the air density is 9% lower respectively 10% higher than the nominal density. The change in the performance index during the flight is shown for these two cases in the following table:

time	c=0.91	c=1.10
0.0	16044	16044
31.7	16061	16006
95.1	16098	15972
126.8	16122	15950
158.5	16123	15948
$\vdots$	$\vdots$	$\vdots$
3169.3	16124	15948

This has to be interpreted as follows: if, for example, the air density is disturbed by the factor $c = 0.91$ for the first 31.7 seconds only and takes the nominal value for the rest of the flight, then the feedback scheme yields an actual trajectory which has a final mass of 16061 [kg]. If the factor $c = 1.10$ is valid for the entire ascent, the actual payload is 15948 [kg]. Figures 1 and 2 show the different histories for the altitude and the path inclination for the case when the perturbation is valid during the entire ascent. The dotted line represents the nominal trajectory, the solid line represents the corrected path for the perturbed problem and the dashed line shows the result when the nominal control is used to guide the system under perturbations. Figure 1 shows the altitude history for the entire ascent in the case when the perturbed air density is 10% higher than it was assumed for the nominal solution. The difference between nominal and corrected trajectory can hardly be seen in the picture. Figure 2 shows the history of the path inclination for the first part of the ascent, where the influence of the air density is most intense. In figure 2 the actual air density is 9% lower than the nominal density.

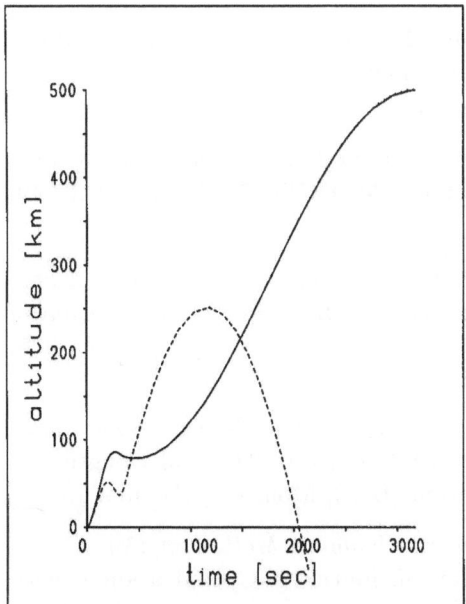 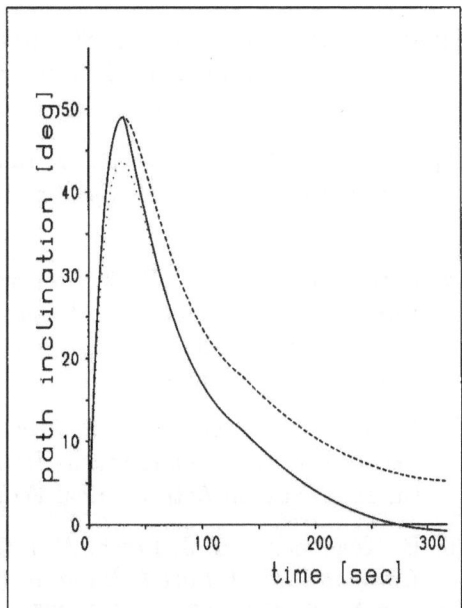

Figure 1: altitude histories for perturbation factor $c = 0.91$

Figure 2: path inclination histories for perturbation factor $c = 1.10$

————: trajectory for perturbed problem using the updated control

— —: trajectory for the perturbed problem using the nominal control

· · · · · ·: nominal trajectory, i.e. nominal problem and nominal control

The algorithm has been developed and the results have been obtained under the assumption that during the flight it is not possible to measure the deviation of the air density directly. From the technical point of view this seems to be correct but theoretically it is also interesting to investigate the range of correctable deviations if the changes in the air density are included in the feedback law (compare (13)):

$$\delta w(t_c) = \tilde{G}(t_c) \left(\begin{array}{c} s_c \\ \delta\rho(t_c) \end{array} \right)$$

A simulation of the corresponding feedback scheme yielded the following range for the perturbation parameter c (see (14)):

$$0.72 \leq c \leq 1.28.$$

References

[1] R. Bulirsch, K. Chudej. *Ascent Optimization of an Airbreathing Space Vehicle*, Proceedings of AIAA Guidance, Navigation and Control Conference, pp. 520–528, New Orleans, Louisiana, 1991.

[2] R. Bulirsch, K. Chudej. *Staging and Ascent Optimization of a Dual-Stage Space Transporter*, Zeitschrift für Flugwissenschaften und Weltraumforschung, pp. 143–151, Springer Verlag, 1992.

[3] K. Chudej, R. Bulirsch, K. D. Reinsch. *Optimal Ascent and Staging of a Two-Stage Space Vehicle System*, Jahrbuch der DGLR 1990 I, pp. 243–249, 1990.

[4] P. Kraemer-Eis, H. G. Bock. *Numerical Treatment of State and Control Constraints in the Computation of Optimal Feedback Laws for Nonlinear Control Problems*, Proceedings of "Large Scale Scientific Computing", pp. 287–306, Oberwolfach, Germany, 1985.

[5] H. Kuczera, P. Krammer, P. Sacher. *Sänger and the German Hypersonics Technology Programme*, Status Report 1991, IAF-91-198, 42nd Congress of the International Astronautical Federation (IAF), Montreal, Canada, 1991.

[6] B. Kugelmann, H. J. Pesch. *New General Guidance Method in Constrained Optimal Control, Part 1: Numerical Method*, Journal of Optimization Theory and Applications 67, pp. 421–435, 1990.

[7] B. Kugelmann, H. J. Pesch. *Controllability Investigations of a Two-Stage-To-Orbit Vehicle* Proceedings of "Optimalsteuerung und Variationsrechnung — Optimal Control", 1991, Oberwolfach, Germany, to appear in Birkhäuser.

[8] NASA. *U.S. Standard Atmosphere*, Washington D.C., 1962.

[9] G. C. Shau. *Der Einfluß flugmechanischer Parameter auf die Aufstiegsbahn von horizontal startenden Raumtransportern bei gleichzeitiger Bahn- und Stufungsoptimierung*, Dissertation, Report DLR-FB 73-86, Department of Mechanical and Electrical Engineering, University of Technology, Braunschweig, Germany, 1973.

International Series of Numerical Mathematics, Vol. 115, © 1994 Birkhäuser Verlag Basel

Linear Optimal Control for Reentry Flight

Axel J. Roenneke*   and   Klaus H. Well*

Abstract

We present a linear optimal control law to control an unmanned reentry vehicle to a reference. The control law is locally optimal minimizing a quadratic performance index at discrete points on the reference. Simulation results show that the controller eliminates trajectory errors resulting from off-nominal entry conditions and aerodynamics as well as atmospheric disturbances.

Nomenclature

$\mathbf{A}$	Reduced-order plant matrix
$\mathbf{B}$	Reduced-order control matrix
$\mathbf{K}(t_k)$	Feedback gain matrix
$\mathbf{P}(t_k)$	Solution of the algebraic Riccati eq.
$\mathbf{Q}$	Diagonal weighting matrix
$\delta\mathbf{x}(t)$	Reduced-order state error vector
C_D	Drag Coefficient
C_L	Lift Coefficient
D	Drag acceleration, normalized by G_o
G_o	Acceleration of gravity at sea level
g	Acceleration of gravity, normalized by G_o
L	Lift acceleration, normalized by G_o
m	Mass of vehicle
n_G	Aerodynamic load factor
q	Nondimensional dynamic pressure
R_o	Radius of the earth
r	Distance from the earth's center, normalized by R_o
S	Reference area
t	Flight time, normalized by $\sqrt{R_o/G_o}$
v	Earth-relative speed, normalized by $\sqrt{G_o R_o}$
γ	Flight-path angle
θ	Geodetic longitude
ρ	Atmospheric density, normalized by $m/(S\,R_o)$

*Universität Stuttgart, Institut für Flugmechanik und Flugregelung, D-70049 Stuttgart, Germany

σ Bank angle
ϕ Geodetic latitude
ψ Heading angle
ω Earth's angular velocity, normalized by $\sqrt{G_\circ / R_\circ}$

Reentry Trajectory Control

To serve space laboratories, orbit to ground transportation must be freely available at low operational cost. For this task, unmanned and re-usable reentry capsules are suggested for their payload capacity and structural efficiency[1-5]. A simple sphere or cone geometry, however, has the disadvantage of producing little aerodynamic lift. This restricts the vehicle's atmospheric maneuverability and thus its cross-range[2,6]. To get to European landing sites from low-earth orbit, multiple skip or orbital transfer maneuvers will be necessary[7]. The associated trajectories and the impact point are very sensitive to uncertain parameters, such as the encountered atmosphere and the aerodynamics of the vehicle[6,8]. For such missions, precise trajectory control is a pre-requisite, if expensive recoveries and long turn-around periods are to be avoided.

Reentry trajectory control closely depends on the guidance law and the available state information. Traditionally, perturbations around a fixed reference are used for both guidance up-dates and trajectory stabilization[9-12]. Feedback is provided by an inertial measurement unit aboard the spacecraft. For example, the Apollo guidance[12] tracks an acceleration reference, because this allows an analytical estimate of the target miss. Guidance commands are determined from target sensitivities to trajectory off-sets. The trajectory control system is designed to maintain this particular reference to keep analytical predictions within satisfactory error bounds.

Advanced guidance algorithms for future spacecraft will be able to determine new feasible trajectories aboard the spacecraft frequently during reentry[13-15]. Satellite navigation provides full information on current flight conditions[16]. This is an advantage also to the trajectory control system design. The control law can utilize feedback of original trajectory states, like altitude, velocity, and flight-path angle. For this purpose, we propose a linear control law which locally minimizes a quadratic performance index. This design approach has been presented in previous papers[17,18]. The results in this paper show that such a control system operates effectively while subject to off-nominal entry condition as well as aerodynamic and atmospheric variations.

Reentry Flight Model

The reentry trajectory is described by the translational motion of a rigid body. The equations of motion are derived for a rotating, spherical earth. The forces acting on the vehicle are gravity and the aerodynamic lift and drag. Wind is not considered.

The state of the vehicle is represented by standard earth-fixed variables[19,20]

$$[r \quad v \quad \gamma \quad \psi \quad \theta \quad \phi]^T$$

where r is the dimensionless distance from the center of earth and v is the velocity magnitude. Flight-path angle, γ, and heading angle, ψ, give the direction of the velocity vector. The angles θ and ϕ specify geodetic longitude and latitude, respectively. The control of the vehicle is represented by the air-path bank angle, σ, which is formed by the lift vector and the local vertical plane. The dynamics of the bank maneuver are disregarded in this analysis. In terms of these variables, the dynamic system is given by[20]:

$$\dot{r} = v \sin \gamma$$

$$\dot{v} = -D - g \sin \gamma$$
$$+ \omega^2 r \cos \phi \left[\cos \phi \sin \gamma - \sin \phi \cos \gamma \sin \psi \right]$$

$$\dot{\gamma} = \frac{1}{v} \{ L \cos \sigma - [g - \frac{v^2}{r}] \cos \gamma$$
$$+ 2 \omega v \cos \phi \cos \psi$$
$$+ \omega^2 r \cos \phi \left[\cos \phi \cos \gamma + \sin \phi \sin \gamma \sin \psi \right] \}$$

$$\dot{\psi} = \frac{1}{v} \{ L \frac{\sin \sigma}{\cos \gamma} - \frac{v^2}{r} \tan \phi \cos \gamma \cos \psi$$
$$\omega^2 r \frac{\sin \phi \cos \phi \cos \psi}{\cos \gamma}$$
$$+ 2 \omega v \left[\cos \phi \tan \gamma \sin \psi - \sin \phi \right] \}$$

$$\dot{\theta} = \frac{v \cos \gamma \cos \psi}{r \cos \phi}$$

$$\dot{\phi} = \frac{v \cos \gamma \sin \psi}{r}$$

The dot indicates differentiation with respect to nondimensional time, t. Normalized lift and drag accelerations, L and D, are related to the dynamic pressure, q, through aerodynamic constants

$$L = q \, C_L$$
$$D = q \, C_D$$
$$q = \rho \, \frac{v^2}{2}$$

where ρ is the normalized atmospheric density. The load factor is defined as the magnitude of the aerodynamic acceleration:

$$n_G = \sqrt{L^2 + D^2}$$

Vehicle data used in the following are taken from a European design study on a small reentry capsule[1]. The vehicle has a sphere-cone configuration with a hypersonic lift-to-drag ratio, C_L/C_D, of 0.52, a ballistic load, m/S, of 89 kg/m^2, and a resulting ballistic coefficient, $mG_\circ/(C_DS)$, of 2,500 Pa. The capsule's attitude can be changed by moving the center of gravity. In the simulations, this capability is modeled as variation of the air-path bank angle.

Linear Quadratic Optimal Control

The reentry dynamics can be separated into two different time-scales. The fast dynamics correspond to the motion in the vertical plane, represented by altitude, velocity, and flight-path angle. The slow eigenmodes correspond to the lateral motion of position and heading angles. Physically, the two motions are coupled through the rotation of the earth only. In reentry guidance, these two state manifolds are commonly treated separately[21]. This paper is concerned with the fast dynamics of the reentry equations. Directional and terminal control of the slow dynamics can be achieved independently by the guidance law.

The controller design is based on the trajectory error with respect to three reference states, $\delta\mathbf{x} = [r - r_R,\ v - v_R,\ \gamma - \gamma_R]^T$, and the control, $\delta\sigma = \sigma - \sigma_R$, the subscript R denoting the reference. For small deviations, a reduced-order linear system of state equations can be defined as:

$$\delta\dot{\mathbf{x}} = \mathbf{A}(\mathbf{x}_R)\,\delta\mathbf{x} + \mathbf{B}(\mathbf{x}_R, \sigma_R)\,\delta\sigma$$

System and control matrices, $\mathbf{A}$ and $\mathbf{B}$, are functions of the reference. For the controller design, they are approximated by the corresponding Jacobian submatrices of the state equations. This approximation is equivalent to assuming a stationary earth. Since the references, $\mathbf{x}_R(t)$ and $\sigma_R(t)$, are time-varying, $\mathbf{A}(\mathbf{x}_R)$ and $\mathbf{B}(\mathbf{x}_R, \sigma_R)$ are implicit functions of time. However, these functions are slowly varying[22] and the system can be treated as locally constant on the reference. Thus, at each time t_k, a linear control law can be suggested of the form

$$\delta\sigma = -\mathbf{K}(t_k)\,\delta\mathbf{x}$$

where the feedback gains, $\mathbf{K}(t_k)$, are constant. These gains are obtained by minimizing an infinite-horizon quadratic performance criterion

$$\lim_{t_f \to \infty} \int_0^{t_f} (q_1\,\delta r^2 + q_2\,\delta v^2 + q_3\,\delta\gamma^2 + q_\sigma\,\delta\sigma^2)\,dt$$

with non-negative penalties, q_i, on state and control errors. The locally optimal feedback gains are given by

$$\mathbf{K}(t_k) = \frac{1}{q_\sigma}\,\mathbf{B}^T(t_k)\mathbf{P}(t_k)$$

if $\mathbf{P}(t_k)$ is the solution of the following algebraic Riccati equation[23]

$$\mathbf{PB}\frac{1}{q_\sigma}\mathbf{B}^T\mathbf{P} - \mathbf{PA} - \mathbf{A}^T\mathbf{P} = \begin{bmatrix} q_1 & 0 & 0 \\ 0 & q_2 & 0 \\ 0 & 0 & q_3 \end{bmatrix}$$

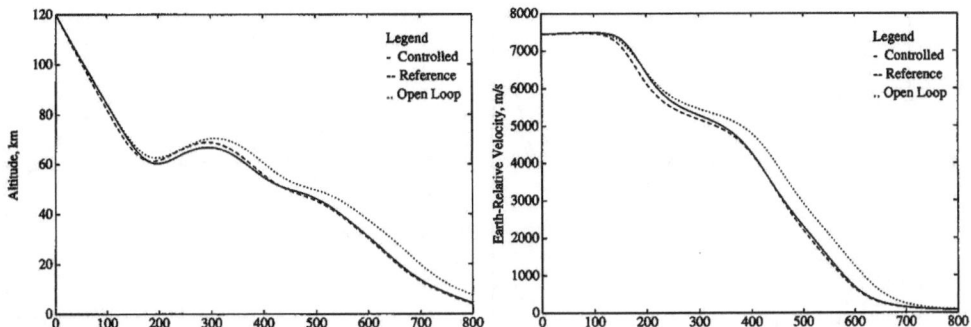

Figure 1: Time responses of the feedback system to entry conditions resulting from a 5% smaller de-orbit impulse.

These feedback gains are scheduled along the reference trajectory. The optimization weights are chosen with Bryson's rule[21]. Since the control input penalty, q_σ, directly scales the feedback gains, it is adjusted to keep the control input from reaching saturation limits.

Results

To evaluate the controller performance, several model parameters are changed simultaneously to an off-nominal value. To represent entry condition dispersions, the transfer orbit is modeled as a Hohmann[10] ellipse and the required de-orbit impulse is off-set by ±5 percent. Off-nominal atmosphere data is taken from measurements during Space Shuttle entry flights with north latitude[24]. This density profile is up to 50 percent lower than the U.S. Standard Atmosphere[25] between 120 km and 40 km altitude. To model design uncertainties, the nominal lift-to-drag ratio is varied by ±4 percent. This value compares with experience from the Apollo program[26]. Since the vehicle design[2] does not permit a lift-down command, the bank angle input is limited to ±90 degrees.

For the following simulations, reference trajectories are chosen with constant 30, 45, and 60 degree bank angles. At nominal entry and parameter conditions, the 45 degree trajectory is chosen as the design trajectory for the controller. Although these trajectories are chosen arbitrarily, they agree with the vehicle's design specifications[1].

Figure 1 shows the time histories of altitude and velocity due to a 5% smaller de-orbit impulse. Simultaneously, the lift-to-drag ratio of the controlled vehicle is increased by 4% to 0.54. The controller smoothly counteracts the higher initial speed and also approaches the desired altitude history. The controlled vehicle stays close to the desired trajectory after 400 seconds.

A second case with opposite parameter off-sets is presented in Fig. 2. The plots show the trajectories of altitude and flight-path angle versus velocity due to

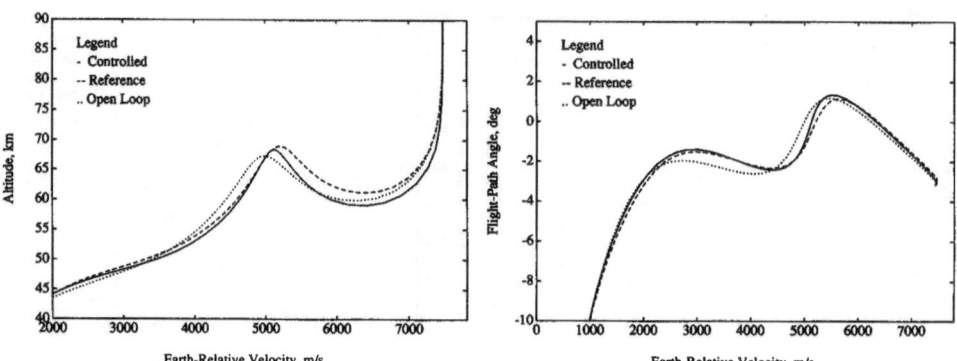

Figure 2: Altitude and flight-path angle versus velocity due to a 5% larger de-orbit impulse and a 4% lower lift-to-drag ratio.

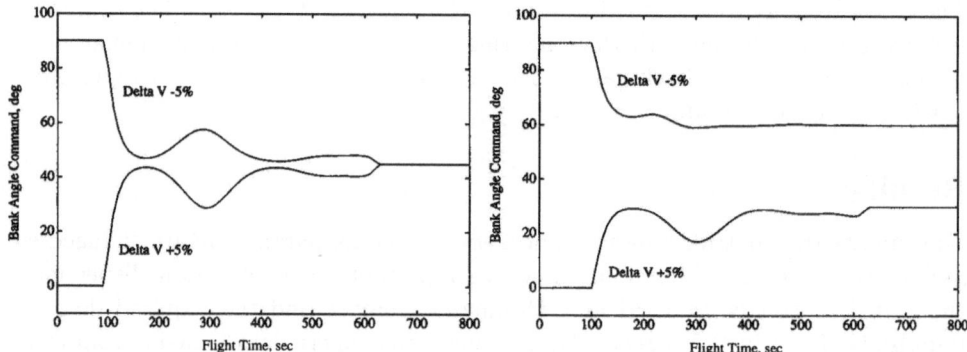

Figure 3: Bank angle input time responses as commanded by the controller for a 45 (left) and 30/60 degree reference trajectory (right).

a 5% higher de-orbit impulse, resulting in a lower entry speed and a more shallow flight-path angle. At the same time, the lift-to-drag ratio of the controlled vehicle is decreased by 4% to 0.50. Again, the controller is able to counteract these off-sets and control the vehicle to the desired trajectory during the skip phase. Altitude and flight-path angle histories approach the desired within satisfactory final error limits.

The bank angle histories for the above cases are plotted in Fig. 3 (left). It shows that the transients for opposite speed and angle off-sets are mirror images. For both cases, the bank angle histories are smooth functions with one oscillation. The required roll rate is less than one degree per second. This result shows that the presented control law is suitable for a vehicle with slow attitude dynamics.

Technically, one computed gain schedule is only valid for the particular reference trajectory for which it is designed. However, for practical applications, one would like to employ a single gain schedule on several neighboring trajectories.

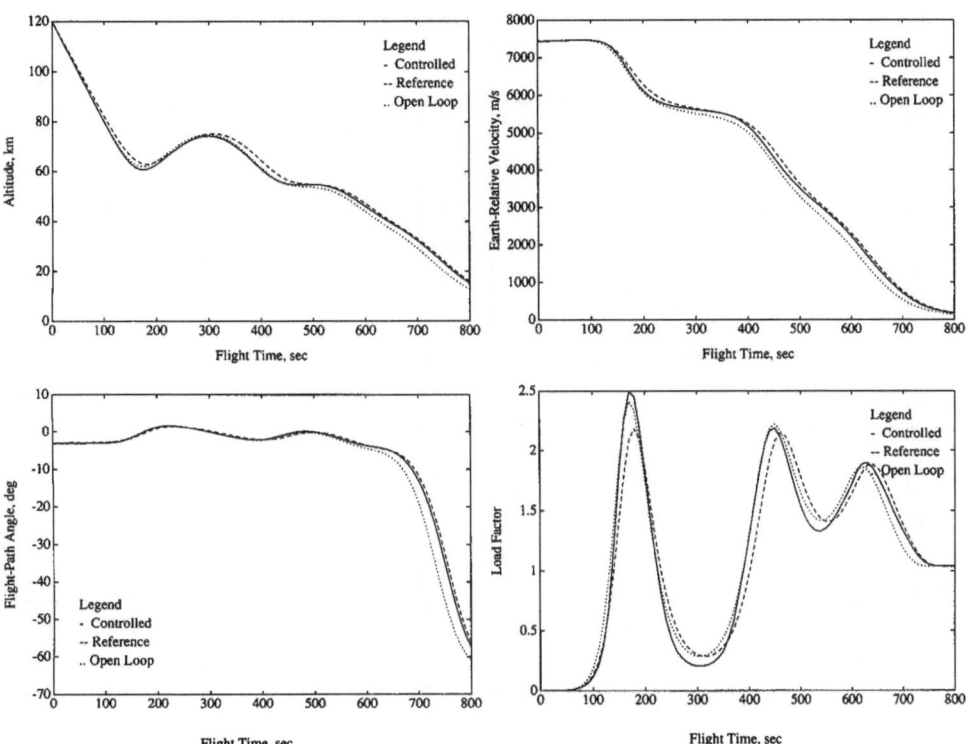

Figure 4: State trajectories of the feedback system due to entry conditions resulting from a 5% larger de-orbit impulse. The reference trajectory is computed for constant 30 degree bank angle.

Figure 4 shows the state trajectories when the vehicle is released with a 5 percent larger de-orbit impulse and controlled to a reference with constant 30 degree bank angle. Despite the sub-optimal controller gains, the controlled vehicle approaches the desired trajectory efficiently and within satisfactory steady state limits. The corresponding load factor (Fig. 4, bottom right) does not significantly exceed the nominal. Similar results are obtained for a 60 degree reference with a 5% smaller de-orbit impulse (Fig. 5). The corresponding bank angle inputs are shown in Fig. 3 (right).

These results demonstrate that the proposed state feedback law, which is designed for one particular reference trajectory, controls the vehicle to several reference trajectories. For practical applications, this implies that the number of gain schedules stored aboard the spacecraft may be limited without sacrificing performance. This, combined with the simplicity of the design, makes the described trajectory control system attractive for implementation into on-board prediction methods.

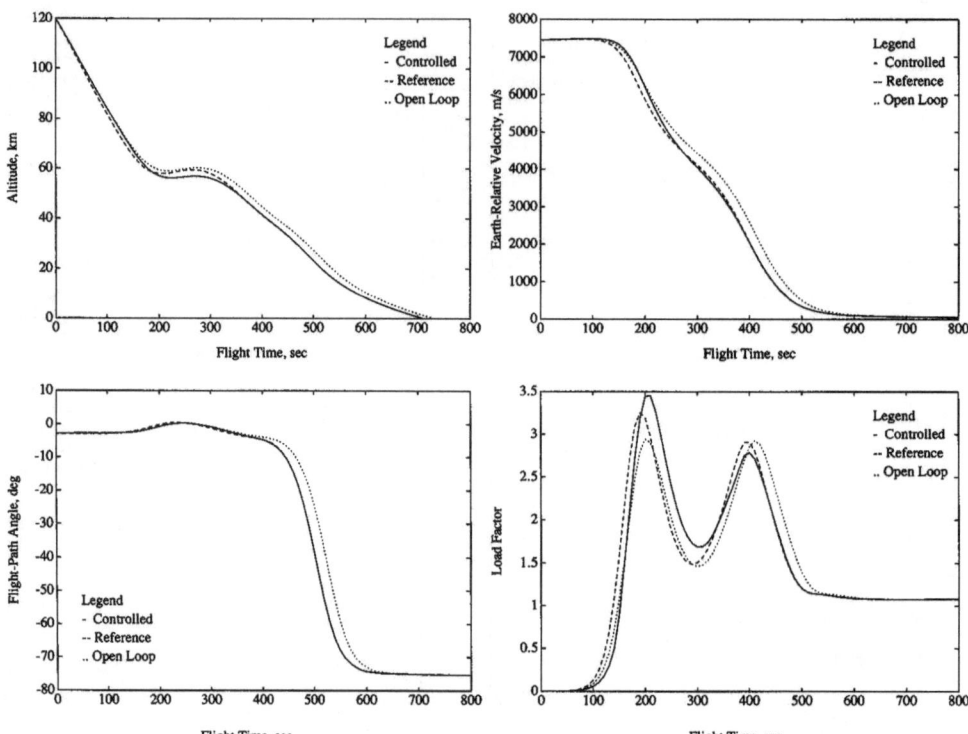

Figure 5: State trajectories of the feedback system due to entry conditions resulting from a 5% smaller de-orbit impulse. The reference trajectory is computed for constant 60 degree bank angle.

Summary and Conclusions

This paper presents a linear optimal control law to stabilize the reentry dynamics of a low-lifting reentry vehicle. A locally optimal control law restricts the vehicle to a reference defined by altitude, velocity, and flight-path angle. The controller design is based on reduced-order linear system models along a desired reference trajectory.

The results in this paper show that the proposed controller responds effectively to entry condition offsets resulting from de-orbit impulse perturbations of ±5%. Controller performance is indifferent to lift-to-drag ratio variations of ±4% as well as significant atmospheric disturbances. The presented results show that the controller works satisfactorily also on references in the neighborhood of the design trajectory.

We plan to combine this control concept with a predictive guidance law to achieve terminal control. The control system is to track accurately the predicted trajectory and reduce landing dispersions.

Acknowledgments. We gratefully acknowledge the helpful discussion with Dr. Ulrich M. Schöttle of the University of Stuttgart Institute of Space Systems. The first author would also like to thank Dr. Phillip J. Cornwell of Rose-Hulman Institute of Technology for his encouragement and support of the research carried out in Indiana.

References

[1]Aenishanslin, M. H., "Space Mail Feasibility Study," Sociétée Nationale Industrielle Aérospatiale, Les Mureaux, France, Columbus Preparatory Programme Report COL-TN-AS-0050, Nov. 1986.

[2]Schöttle, U., et al., "Conceptual Study of a Small Semiballistic Reentry Experiment Vehicle," 41st Congress of the International Astronautical Federation (Dresden, Germany), IAF-90-163, Oct. 1990.

[3]Mascey, A. C., et al., "The Reusable Reentry Satellite – Keeping it Up and Bringing it Down" *Proc. of the 1988 AIAA/AAS Astrodynamics Conf.* (Minneapolis, Minn.), AIAA, Washington, Aug. 1988, pp. 310–322. (AIAA 88-4257-CP)

[4]Albert, J. C., et al., "Online Guidance and Control of a Spacecraft for an Aeroassisted Orbit Transfer," *Proc. of the 12th IFAC Symp. on Automatic Control in Aerospace* (Ottobrunn, Germany), Intl. Federation of Automatic Control, Laxenburg, Austria, Sep. 1992, pp. 147–152.

[5]Gamble, J. D., et al., "Atmospheric Guidance Concept for an Aeroassist Flight Experiment," *Journal of the Astronautical Sciences*, Vol. 36, Jan.–June 1988, pp. 45–71.

[6]Morth, R., "Reentry Guidance for Apollo," *Proc. of the 2nd IFAC Conf. on Automatic Control in Space* (Vienna, Austria), Intl. Federation of Automatic Control, Laxenburg, Austria, Sep. 1967, pp. 735–759.

[7]ACRI and Laboratoire d'Automatique de Nantes, "Guidance and Control for Moderate Lift/Drag Re-Entry," ESA European Space and Technology Center, Noordwijk, The Netherlands, Contract Report 9359-91-NL-JG, May 1992.

[8]Seyler, T. A., and Florence, D. E., "Upper Atmospheric Disturbance Effects on Reentry Satellite Landing Accuracy," *Proc. of the AIAA/AAS Astrodynamcis Conf.*, AIAA, Washington, Aug. 1991, pp. 2371–2377. (AAS 91-495)

[9]Wingrove, R. C., "A Survey of Atmosphere Re-Entry Guidance and Control Methods," *AIAA Journal*, Vol. 1, Sep. 1963, pp. 2019–2029.

[10]Battin, R. H., *Astronautical Guidance*, McGraw-Hill, New York, 1964.

[11]National Aeronautics and Space Administration, *Guidance and Navigation for Entry Vehicles*, NASA Space Vehicle Design Criteria, NASA SP-8015, Nov. 1968.

[12]Graves, C. A., and Harpold, J. C., "Apollo Experience Report – Mission Plan-

ning for Apollo Entry," NASA TN D-6725, 1972.

[13] Stiles, J. A., "Predictive Entry Guidance for an Apollo-Type Vehicle," *Proc. of the 3rd IFAC Symp. on Automatic Control in Space, Toulouse, France*, Intl. Federation of Automatic Control, Laxenburg, Austria, Mar. 1970, pp. 732–740.

[14] Cramer, E. J., et al., "NLP Re-Entry Guidance: Developing a Strategy for Low L/D Vehicles," AIAA 88-4123-CP, 1988.

[15] Corban, J. E., et al., "Rapid Near-Optimal Aerospace Plane Trajectory Generation and Guidance," *Journal of Guidance, Control and Dynamics*, Vol. 14, No. 6, Nov.–Dec. 1991, pp. 1181–1190.

[16] Schänzer, G., and Kayser, D., "Precision Navigation – A New Approach for Re-Entry Vehicles," *Proc. of the 12th IFAC Symposiom on Automatic Control in Aerospace* (Ottobrunn, Germany), Intl. Federation of Automatic Control, Laxenburg, Austria, Sep. 1992, pp. 105–110.

[17] Roenneke, A. J., and Cornwell, P. J., "Trajectory Control for a Low-Lift Maneuverable Re-Entry Vehicle," AIAA 92-1146-CP, Feb. 1992.

[18] Roenneke, A. J., and Well, K. H., "Re-Entry Control of a Low-Lift Maneuverable Spacecraft," *Proc. of the 1992 AIAA Guidance, Navigation, and Control Conf.* (Hilton Head, S.C.), AIAA, Washington, Aug. 1992, pp. 641–652. (AIAA 92-4455-CP)

[19] Miele, A., *Flight Mechanics*, Addison-Wesley, Reading, Mass., 1962.

[20] Vinh, N. X., *Optimal Trajectories in Atmospheric Flight*, Elsevier, New York, 1981.

[21] Bryson, A. E., and Ho, Y., *Applied Optimal Control*, Hemisphere Publishing, New York, 1975.

[22] Roenneke, A. J., *Trajectory Control for a Low-Lift Maneuverable Re-Entry Vehicle Using State Feedback*, M.S. Thesis, Rose-Hulman Institute of Technology, Terre Haute, Ind., May 1991.

[23] Anderson, B. D., and Moore, J. B., *Optimal Control*, Prentice-Hall, Englewood Cliffs, N.J., 1990.

[24] Gamble, J. D., and Findlay, J. T., "Shuttle-Derived Densities in the Middle Atmosphere," AIAA 88-4352-CP, 1988.

[25] U.S. National Oceanic and Atmospheric Administration, *U.S. Standard Atmosphere 1976*, U.S. GPO, Washington, NOAA-S/T 76-1562, 1976.

[26] Crowder, R. S., and Moote, J. D., "Apollo Entry Aerodynamics," *Journal of Spacecraft*, Vol. 6, Mar. 1969, pp. 302–307.

International Series of Numerical Mathematics, Vol. 115, © 1994 Birkhäuser Verlag Basel

Steady-State Modelling of a Turbine Engine with Controllers

Zdeněk Schindler*

Introduction

Computer models of the behavior of a gas turbine engine coupled with a propeller are invaluable tools for engine design (Ping Zhu and Saravanamuttoo (1992), Rick and Muggli (1989)). Based on physical laws of mechanics and thermodynamics, and on experimentally given or computed characteristics of engine components, a mathematical description of the engine – propeller couple was proposed. It consist of a system of algebraic and differential equations. In Doležal et al. (1990, 1992) we described an implementation of the model of a moderately sized stand-alone gas turbine engine. The correspondence of computed results with measurements of real system was quite satisfactory.

Computer simulation of the behavior of the engine – propeller couple *connected with additional subsystems of controllers* is substantial as well. However, on a certain level of the engine development the dynamic properties of the controllers may not be known. We will propose a method for the computation of stationary states only with the limited knowledge of the controllers. The extension of the model brings new problems. We will show the way they could be solved using an optimization approach.

Basic Problem Formulation

The dynamics of the engine – propeller – controller system (Fig. 1) is described by the following system of differential algebraic equations (DAE)

$$x' = f_m(x, y, z, a) \tag{1}$$

$$\varepsilon \cdot y' = f_s(x, y, z, a) \tag{2}$$

$$\frac{1}{\mu} \cdot z' = f_\ell(x, y, z, a) \tag{3}$$

$$f_a(x, y, z, a) = 0 \tag{4}$$

where ε and μ express the time character of the process, $0 < \varepsilon \ll 1$, $0 < \mu \ll 1$. Equation (1) describes the motion of mechanical parts of the engine, i.e. the

---

*Institute of Information Theory and Automation, Academy of Sciences of the Czech Republic, Prague 8, Pod vodárenskou věží 4, Czech Republic, CS–182 08

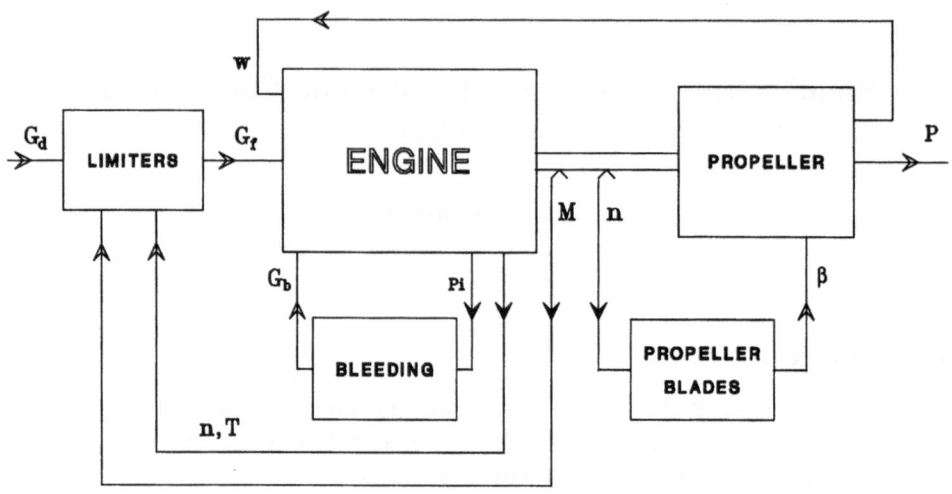

Figure 1: Scheme of the Engine – Propeller – Controller System

rotation of the shafts with the compressors and turbines. The state variable vector of the speeds of shafts is denoted by x. The rate of change of x can be classified as *medium*. Equation (2) stands for *fast* thermodynamic events (flow of working medium) with the state variable y. Equation (3) describes *slow* processes of heat transfer between the engine body and working gases (z is the temperature of engine parts). The solution of algebraic equations (4) resolves right-hand sides of (1) – (3). Equations (1) – (4) cover the description of control systems, too.

To use the described model in its complexity is practically impossible. More than a mathematical, it is a technical problem, as we are not able to determine the values of model parameters. For our purposes, the most important is the dynamics of mechanical parts. Therefore we simplify the system (1) – (4), assuming that

- thermodynamic processes (2) can be treated as infinitely fast. Instead they can be described by *algebraic* equations.

- dynamic processes of heat transfer (3) are neglected within the followed time horizon.

Then, the modified model contains only differential equations (1) extended with a system of algebraic equations (4). It is solved by an ODE solver while the right-hand sides are evaluated including a nonlinear equation solver for (4). We obtain time courses of more than fifty physical variables which converge to their steady-state values (Doležal et al. (1990)).

There is a substantial disadvantage of steady-state computation via dynamic simulation – the procedure is inefficient, the simulation consumes a considerable amount of time. Moreover, we lack the complete knowledge of dynamic properties

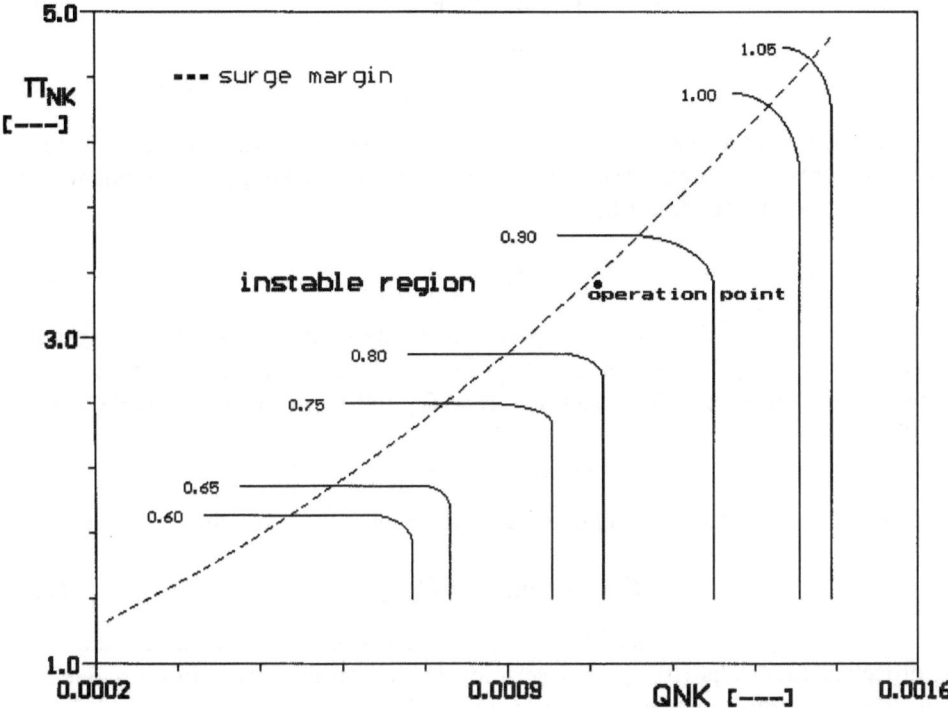

Figure 2: Compressor Pressure Characteristic and Region of Stable Operation Delimited by the Surge Margin $\pi_{\mathrm{crit}} = f(Q)$.

of the system parts. A question has risen, whether the correct steady-state can be computed using only asymptotic characteristics of individual elements.

Method of Solution

To describe the asymptotic behavior of the engine – propeller couple we set $x' = 0$ in (1). The DAE system (2), (4) then transcribes into a system of nonlinear algebraic equations with state dependent structure

$$F(x, a) = 0 \tag{5}$$

This problem was successfully solved despite many difficulties (cf. Doležal et al. (1990)).

Now, let us consider these two controller systems: limiters and a bleeding nozzle. The limiters guard the values of input low pressure turbine temperature T, shaft speeds n_L, n_H, and torque M

$$T \leq T_{\max} \tag{6}$$

$$n_L \leq n_{L,\max} \tag{7}$$
$$n_H \leq n_{H,\max}$$
$$M \leq M_{\max} \tag{8}$$

The bleeding nozzle keeps the compressors in the region of stable operation. In a compressor pressure characteristic it means, that the working point is below the curve of a surge margin (see Fig. 2)

$$\pi \leq \pi_{crit} \tag{9}$$

Thus the region of admissible operation points is determined by the system of inequalities (6)–(9). The control law of each controller defines a single point in this region. An actual fuel mass flow rate G_f must not cause the violation of (6)–(8)

$$G_f \longrightarrow \max, \qquad G_f \leq G_d \tag{10}$$

where G_f is a preset value of fuel flow rate. Bleeding should be minimal to prevent surging

$$G_b \longrightarrow \min, \qquad G_b \geq 0 \tag{11}$$

Combining the equalities describing the physical system (5) with the inequalities and objective functions (6)–(11) describing the controllers, we obtain a mathematical programming problem with multiple objectives.

The simplest solution of such problems transforms the vector objective into a scalar one. Frequently, such a transformation is done via the scalar product of an objective function vector with a suitable weighting vector, in our case

$$\alpha \cdot G_f + \beta \cdot G_b \longrightarrow \min, \quad \alpha < 0, \quad beta > 0 \tag{12}$$

The resulting problem is solved by common mathematical programming methods. Good results were obtained by the Recursive Quadratic Programming algorithm (Bartholomew-Biggs (1990)).

The proposed procedure — the transcription to a mathematical programming problem and its solution — may bring several complications. The principal difficulty of this approach lies in different principles which govern the iteration process of model compared to transition process of the real system. The trajectory of a stable engine starts in an initial point and terminates in a defined steady state point. Every intermediate point is a really existing operation point. On the other hand, the trajectory of the optimization process has no correspondence to real states and the termination point may be false.

The mathematical model of the controllers describes only the resulting terminal state, not their dynamic behavior. All control variables of the model change during the solution procedure *simultaneously*, being influenced by all violated constraints. The real controllers operate *independently*, they are activated only by respective input quantities. The bleeding nozzle of the engine stays closed as

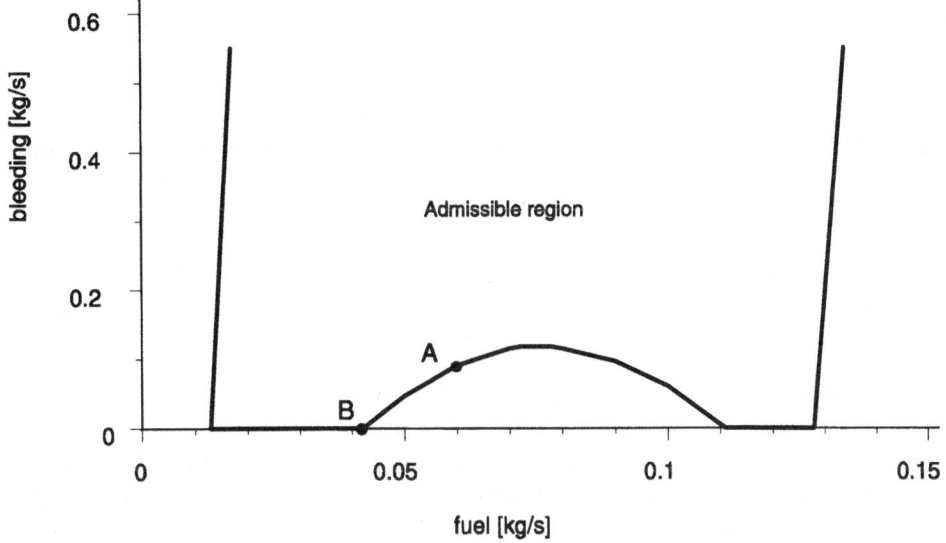

Figure 3: Maximal Admissible Region in $G_f - G_b$ Plane. Beyond the right part of the boundary, (6)–(8) are violated; below the central curve the compressor operation is unstable (surging). On the left, the region is bounded by idling of the engine. The vertical line of the inequality $G_f \leq G_d$ is not represented here.

long as the compressor operates in the stable region. It is actuated only by surging of the compressor. This means that

$$G_b = f(\pi)$$

However, the value of bleeding G_b influences variables T, n_L, n_H, M guarded by limiters, i.e. $T = f_1(G_b)$, $n_L = f_2(G_b)$, etc. The model attempts to fulfill all inequalities in (6)–(11) by all possible changes of variables. Therefore, the bleeding nozzle model may be activated also by violation of inequality constraints other than surge margin, that is G_b can be affected also by π, T, n_L, n_H, M. Thus, the computed G_f, G_b may differ from their actual steady-state values.

Let us specify all admissible operation points in a $G_f - G_b$ plane for one setting of the system parameters (Fig. 3). Using the transformation (12) e.g. with $\alpha = -2$, $\beta = 1$, we can see, that for $G_d = 0.06$ the actual operation point A differs from the result of computations B. To avoid a trap in the point B, we must change values of α and β. With increased absolute value of α, e.g. $\alpha = -10$, $\beta = 1$, we obtain correct results. Further increase of α may cause a failure on the right boundary of the region for high values of G_d.

It is hard to find a universal objective function capable to overcome these difficulties. Let us examine the following logical scheme for guiding the solution process:

Step 1. Solve the problem $(5), (6), (7), (8), (10)$ with $G_b = 0$ and the objective

$$G_f \longrightarrow \max,$$

Step 2. Now, if $\pi \leq \pi_{crit}$ then solve the problem $(5), (6), (7), (8), (10)$ with free G_b and with the new equation

$$\pi = \pi_{crit}$$

and the objective

$$G_f \longrightarrow \max.$$

Use the solution of Step 1. as an initial iteration.

We can follow the solution process in the plane $G_f - G_b$ in Fig. 3. At first we omit the bleeding nozzle and consider only one objective function. We stay on the x-axis. If the result of this simplified problem complies with (9), we terminate the solution. If not, we reduce the solution space and stay on the curve $\pi = \pi_{crit}$. It leads to a correct solution even in the worst case, when the right part of the boundary intersect its bottom (concave) part. Performed numerical experiments have proven that now the computed stationary states correspond to real operating points.

Conclusion

We suggested the computational scheme for the computation of stationary states of a gas turbine engine with a propeller and controllers based on an optimization algorithm. The proposed procedure removes incorrect solution points. It demonstrates how a detailed problem analysis may simplify and improve the efficiency of a mathematical model. However, in some models the structure of constraints may be so complicated that a similar analysis is impossible. Then, the only way how to find stationary states may be to perform full simulation of the dynamic transition. Then, unfortunately, we must know the dynamic properties of the control systems.

References

Bartholomew-Biggs M. C. (1990): A globally convergent version of REQP for constrained minimization. *IMA J. Numer. Anal.* 8, 253–271.

Doležal J., Schindler Z., Fidler J., Matoušek O. (1990): Modelling and simulation of turboprop engine behaviour. *Acta technica ČSAV* 35, No. 1, 1–27.

Doležal J., Schindler Z., Fidler J., Matoušek O. (1992): Optimization-based decision support system for turbine power units modelling and design. *Optimization-Based Computer-Aided Modelling and Design*, Beulens A. J. M., Sebastian H.-J. (eds.), Lecture Notes in Control and Computer Science 174, Springer-Verlag, Berlin, 28–37.

Ping Zhu, Saravanamuttoo H. I. H. (1992): Simulation of an advanced twin spool industrial gas turbine. *Transactions of the ASME, J. of Engineering for Gas Turbines and Power*, 114, No. 2, 180–186.

Rick H., Muggli W. (1989): Generalized digital simulation technique with variable engine parameter input for steady state and transient behaviour of aero gas turbines. *AGARD Conference Proceedings No. 324 on Engine Handling*, NATO, 26-1–26-20.

International Series of Numerical Mathematics, Vol. 115, © 1994 Birkhäuser Verlag Basel

Shortest Paths for Satellite Mounted Robot Manipulators

V. H. Schulz*   H. G. Bock*   and   R. W. Longman†

Abstract

In ground based robotics the most fundamental form of path planning uses linear interpolation, either in joint space or in cartesian space. It is the purpose of this paper to extend this fundamental path planning method to the space based robotics problem. Space based robotics will often require the manipulation of loads that have a mass that is not negligible compared to that of the satellite on which the robot is mounted. And the manipulation will often be performed with the attitude control system turned off in order to save attitude control fuel. It is nevertheless necessary that the satellite attitude be returned to its original attitude by choice of the robot path. Direct application of linear interpolation fails in this situation, and the analog developed here is to find minimum arc length solutions satisfying the attitude condition. The minimum is in the chosen space, i.e. in joint space, cartesian space for satellite fixed coordinates, or cartesian space for inertial coordinates. This paper presents methods of path planning, and examples of such robot paths, for the second of these alternatives.

Introduction

During on-orbit operations it will become increasingly common to have robots on satellites manipulating loads with masses that are not negligible compared to that of the satellite. Such operations present challenges to path planning (see for example [1-8]). Complications arise because the robot is not mounted on an inertially fixed base, but on a satellite which can translate and rotate in response to the robot motion. If the attitude control system is on during such operations, robot disturbances increase the use of attitude control fuel, a valuable expendable in space operation. This motivates the development of path planning techniques for use with the attitude control off. This requires that the paths chosen eliminate the attitude disturbances they produce by the end of the maneuver.

The required final load position and satellite attitude creates an implicit nonlinear boundary value problem in ordinary differential equations or differential-algebraic equations (see [8]), which has infinitely many different feasible solutions.

---

*Interdisciplinary Center for Scientific Computing, University of Heidelberg, D-69120 Heidelberg, Germany.

†Mechanical Engineering Department, Columbia University, New York, NY 10027

Unlike the ground based case where it is easy to develop many feasible paths, in the space based problem there is no simple method of such paths. Reference [4] gives an analytical method of constructing one such path, but it is complex and not desirable in practice. In [8] the authors extend their optimal path planning for ground based robotics [9-12] to address this problem, and develop methods to generate minimum length, minimum energy, minimum time, and minimum attitude disturbance solutions.

In ground based robot operation, the most common form of path planning is linear interpolation in joint or cartesian space. Linear interpolation does not produce a feasible trajectory for the space based problem, and the natural generalization to this problem is to find space based solutions that satisfy the same minimum arc length property that linear interpolation satisfies, and meet the required final satellite attitude condition. There are several choices to consider: 1) minimum arc length in joint space, 2) in the cartesian space of satellite fixed coordinates, and 3) in the cartesian space of inertially fixed coordinates. Unlike the ground based case, the mathematical complexity of all of these problems is similar, which makes the minimum arc length in joint space correspondingly less attractive than the other alternatives. In this paper we concentrate on the second of the options, considering that there is a natural appeal to minimization of the arc length as seen in robot base coordinates (i.e. satellite coordinates).

The problems considered here assume that the load carried by the robot can be modeled as a point mass, and that the robot links have negligible mass compared to the load and the satellite. In this case, it is possible in theory to generate robot trajectories that have the property that they produce no disturbance to the satellite attitude at all times throughout the maneuver, rather than simply return the satellite to its original attitude by the end of the maneuver. If the robot starts its maneuver by pulling the point mass load radially in to the system center of mass, and then completes the maneuver by moving the load radially outward to the desired final position, then there is no attitude disturbance. Of course, it is relatively unusual that such a maneuver can be performed without resulting in collision problems, but in some situations it can be practical. For example, in the case of the shuttle remote manipulator system, the center of mass point will likely be somewhere in the middle of the shuttle bay. Then it is possible to use such a trajectory when the initial and final load positions are both in a properly chosen region above the shuttle bay, the bay is empty and its doors are open.

The minimum arc length robot paths developed in this paper are compared to these zero attitude disturbance paths and seen to share some similarity. Note that the minimum arc length solutions are much more likely to avoid collision problems than the zero attitude disturbance paths, because they do not go all of the way to the system center of mass. When such a collision problem does arise, we suggest avoiding the difficulty by using additional teach points, or use a new code that is currently being generated which includes the collision constraints in the computation of the minimum arc length solution.

The Model

The basic assumptions used for the minimum arc length results are:

A1: Gravitational force effects, such as gravity gradient torques, are negligible.
A2: The satellite center of mass can be considered at rest in inertial space.
A3: The load mass is a significant fraction of the total mass of the satellite.
A4: The masses of the robot links can be considered negligible.
A5: The links are rigid and the controllers produce perfect tracking.
A6: The satellite attitude control system is turned off.

The definitions of variables used are: $\boldsymbol{R}$, a vector from the system center of mass to the center of mass of the satellite; r and $\boldsymbol{\rho}$ are vectors from the satellite center of mass to the point mass load and to the position of a differential element of mass dm, respectively; r^{ι} and ρ^{ι} are analogous vectors starting from the system center of mass; and $\boldsymbol{\omega}$ is the inertial angular velocity of the satellite. When any of these quantities appears in boldface it represents a vector quantity, and when the same variable appears without boldface, it represents the same vector written as a column matrix of components of the vector in the satellite fixed reference frame $\hat{x}_1, \hat{x}_2, \hat{x}_3$. We will choose this frame to be aligned with the principal axis directions. A tilde over a column vector such as $\tilde{\omega}$ represents a square antisymmetric matrix producing the matrix equivalent of a cross product operator (i.e., the (1,2), (1,3), and (2,3) elements are $-\omega_3, \omega_2$, and $-\omega_1$, respectively, with the other elements completed by the antisymmetric property). For purposes of the present computations, 1-2-3 type Euler angles, ϕ_1, ϕ_2, ϕ_3, going from inertial to satellite coordinates are used to represent the rotational motion. For routine application of the methods developed here, one would switch to quaternions or some other representation that avoids singularities. I is the inertia matrix in satellite coordinates for the satellite center of mass point, whereas E is used for the 3×3 identity matrix, and m_S and m_L represent the satellite and load masses, respectively.

In [8] we derived a system of ordinary differential equations governing the dynamics of the satellite-robot system using an elbow robot model. The robot structure was important in some of the computations, e.g. for minimum energy robot paths, and would be important here if we were to seek minimum arc length solutions in joint space instead of cartesian space. For the current minimum arc length problems with assumption A4 of massless robot links, the specific robot configuration has no influence on the shortest load paths. This fact allows us to derive the governing equations quickly. The system angular momentum about the satellite center of mass and its inertial derivative (indicated by presuperscript iota) are:

$$\boldsymbol{H} = \int_{Sat.} \boldsymbol{\rho} \times \frac{{}^{\iota}d\boldsymbol{\rho}}{dt} dm + m_L \boldsymbol{r} \times \frac{{}^{\iota}d\boldsymbol{r}}{dt}, \quad \frac{{}^{\iota}d}{dt}\boldsymbol{H} = (m_S + m_L)\frac{{}^{\iota}d}{dt}\left[\boldsymbol{R} \times \frac{{}^{\iota}d}{dt}\boldsymbol{R}\right]$$

Terms involving second inertial derivatives of ρ^{ι} and r^{ι}, which represent the sum of all external moments, have been set to zero according to assumptions A1, A6.

Other terms containing integrals of ρ^t, which represent the definition of the center of mass are also set to zero. The above expression can be integrated, and the constant of integration is a zero vector assuming that the system is at rest before every maneuver. H can also be calculated for each body: $H = H_{Sat.} + H_{Load} = I\omega + m_L \left(\tilde{r}\dot{r} + [r^T_r E - rr^T]\omega \right)$. Using the conservation of the system center of mass, and equating both expressions for the angular momentum, one obtains:

$$\omega = - \left((1+\nu)\hat{I} + [r^T_r E - rr^T] \right)^{-1} \tilde{r}\dot{r} \; ; \; \nu := \frac{m_S}{m_L} \; ; \; \hat{I} := \frac{1}{m_S}I \qquad (1)$$

$$\dot{\phi} = Q(\phi)\omega \; ; \; Q(\phi) := \begin{pmatrix} \cos\phi_2 \cos\phi_3 & \sin\phi_3 & 0 \\ -\cos\phi_2 \sin\phi_3 & \cos\phi_3 & 0 \\ \sin\phi_2 & 0 & 1 \end{pmatrix}^{-1} \qquad (2)$$

The first of these is the fundamental relationship between the velocity of the load mass with respect to the spacecraft, $\dot{r}$, and the angular velocity of the spacecraft, ω, at a given load position, r. The angular velocity expressed in terms of the 1-2-3 Euler angles and their derivatives is also given.

The Optimization Problem

The problem of finding a shortest path in the cartesian space of satellite coordinates, under the constraint that the initial and final satellite attitude should be equal, can be formulated as the following variational problem:

$$\min_r \int_0^T \|\dot{r}(t)\| dt$$

s.t. equations (1) and (2) are satisfied, and $\phi(0) = \phi_0$, $\phi(T) = \phi_T$

Since the integral and the constraints do not change when the path $(r(t), \phi(t))$ is reparametrized, e.g. using $(\bar{r}(t), \bar{\phi}(t)) = (r(g(t)), \phi(g(t)))$, with $g : \mathbb{R} \to \mathbb{R}$, the solution of this optimization problem is only unique up to reparameterizations. Stated in physical terms, one can pick any time history to use in following the shortest path, and the resulting history satisfies the variational problem. This implies that the minimizing solution is not an isolated solution. In order to facilitate the numerical solution of the variational problem, an additional requirement is imposed to insure an isolated minimum, i.e. we seek a solution which is parameterized by arc length using $\|\dot{r}\|^2 = 1$. From the resulting solution, all other optimal solutions for this minimizing path can be computed by choice of the reparametrization.

In the sequel we will solve this variational problem by use of a direct method as described in [8]. When the initial and final load positions lie within a principal plane of the satellite, we can state the following result which produces a simpler problem formulation for this case:

Theorem: Shortest path solutions to the above variational problem, which start and end in a principal plane, stay completely within this principal plane.

Sketch of Proof: Without loss of generality we may assume that the principal plane under consideration is the plane spanned by $\hat{x}_1$ and $\hat{x}_2$. In terms of the Eulerian angles used in equation (2), it is easily seen that any movement of the load within this plane changes ϕ_3 but not ϕ_2 or ϕ_1. Therefore there exist feasible paths from an arbitrary point within the principal plane to any other arbitrary point within the principal plane, which stay completely within the principal plane. Hence, there is a shortest path among all these feasible planar paths. In order to prove that this path is in fact a solution of the original variational problem, one imposes the constraint that the path must stay in this principal plane, and then shows that the constraint is inactive, i.e. using the Pontryagin Maximum Principle, one shows that the corresponding multiplier is zero. $\square$

Due to the simplicity of planar problems having the initial and final robot locations in a principal plane of the satellite, some of the numerical results presented were obtained using a code specialized for this purpose, which uses an indirect method. For ease of notation in such planar problems we introduce polar coordinates: s, distance of the load from the satellite center; θ, angle of the load vector with the $\hat{x}_1$ axis in satellite coordinates; ψ, orientation angle of the satellite in the principal plane; $\bar{I}$, normalized inertia of the satellite – now a scalar. The principal plane version of the variational problem is:

$$\min_{s,\theta} \int_0^T \sqrt{\dot{s}^2 + s^2\dot{\theta}^2}\,dt$$

$$\text{s.t.} \int_0^T F(s)\dot{\theta}\,dt = 0, \quad F(s) := \frac{s^2}{(1+\nu)\bar{I} + s^2} \tag{3}$$

$$s(0), s(T), \theta(0), \theta(T) \text{ given}$$

Again the solution of this optimization problem is only unique up to reparameterizations. In order to determine one of these solutions we look arbitrarily for that solution which is parameterized by arc length, i.e. $\dot{s}^2 + s^2\dot{\theta}^2 = 1$.

The corresponding optimal control problem is then:

$$\min_{(u_s,u_\theta)\in S} \int_0^T 1\,dt$$

$$S = \left\{ (u_s, u_\theta) : [0,T] \to \mathbb{R}^2 \,\middle|\, u_s^2(t) + s^2(t)u_\theta^2(t) = 1, \ t \in [0,T] \right\}$$

$$hboxs.t.\dot{s}(t) = u_s(t), \ \dot{\theta}(t) = u_\theta(t), \ \dot{\psi}(t) = -F(s(t))u_\theta(t) \tag{4}$$

$$s(0) = s_0, \ s(T) = s_T, \ \theta(0) = \theta_0, \ \theta(T) = \theta_T, \ \psi(0) = \psi_0, \ \psi(T) = \psi_T \tag{5}$$

Applying Pontryagin's Maximum Principle results in

$$u_s(t) = \left[\lambda^2(t) + p - qF(s(t))\right]^{-\frac{1}{2}} \lambda(t)$$
$$u_\theta(t) = \left[\lambda^2(t) + p - qF(s(t))\right]^{-\frac{1}{2}} [p - qF(s(t))]$$
$$\dot{\lambda}(t) = s(t)u_\theta^2(t) + qu_\theta(t)\partial F(s(t))/\partial s \tag{6}$$

where the constants p and q are determined by the boundary conditions. The system of equation (4), (5) and (6) form a well posed boundary value problem for s, θ, ψ, λ and the parameters p, q.

Numerical Results

In this section, we first compute paths having minimum arc length in satellite coordinates for planar problems according to the last section, and then study the way in which these solutions are altered when the initial and final robot locations do not lie in a principal plane of the satellite. The nonplanar solutions are computed using parameterization by arc length making use of u_1, u_2 in $\dot{r} = (\cos u_1 \cos u_2, \sin u_2, \sin u_1 \cos u_2)^T$ as the control variables. The numerical methods used are described in [8]. They are based on the following concepts: 1) discretization of the controls by a finite dimensional function space, 2) discretization of the ordinary differential equation by multiple shooting, and 3) a special sequential quadratic programming method for the solution of the resulting finite dimensional optimization problem. The planar solutions are computed as solutions of the boundary value problem derived in the last section, by using a multiple shooting technique for boundary value problems as described in [14].

We consider the satellite to be a rectangular solid of length 4ℓ, with both thickness and height equal to ℓ, where ℓ is an arbitrary length scale which is also used to produce the normalized radial position $\gamma = s/\ell$. The satellite to load mass ratio is $\nu = 2$. Then the normalized inertia matrix is $\hat{I} = \ell^2 diag(17,2,17)/12$ and for the planar problems $F(\gamma) = \ell^2\gamma^2/(3\ell^2 n^T \hat{I} n + \ell^2\gamma^2) = \gamma^2/(3n^T \hat{I} n + \gamma^2)$, where n denotes the normal vector to principal plane containing the motion.

Consider first, principal plane problems whose desired initial and final robot positions differ only by a rotation in this plane. The left half of Fig. 1 shows the shortest paths in satellite coordinates for robot rotations about the satellite $\hat{x}_3$ axis on the left, and on the right it shows the corresponding paths for the satellite $\hat{x}_2$ axis, for four cases $s_0 = s_T, \theta_0 = -\theta_T$ and $(s_T, \theta_T) \in \{2\ell, \ell/2\} \times \{\pi/16, \pi/2\}$. When the angle of rotation is small, whether at the larger or smaller radius, the minimum arc length solution is a smooth loop that can approach a circular shape. When the desired angle change is larger, the minimum arc length solution more closely follows the zero attitude disturbance solution. As discussed in the introduction, this solution is composed of the two radial lines going from the initial position to the satellite center of mass (which is then also the system center of mass) and then radially outward to the desired final position, as shown in the figure. By comparing the left half of the figure to the right half we can see that the effect of decreasing the inertia of the satellite is to decrease the size of the loop needed, and also to make the loops more circular. The decrease in loop size is reasonable, since the sole purpose of using a trajectory that is not a straight line connecting the endpoints, is to correct by the final time the attitude disturbances introduced by the maneuver; and these corrections are more easily made when the relative inertia of the load to that of the satellite is larger. One is tempted to look for some way in which the shape of the loops scale with the desired radius of the load position. However, the relationship is not a simple scaling, but instead any combination of radius, inertia, and mass ratio producing the same initial and final values of $F(s)$ in equation (3) will produce the same loop shape.

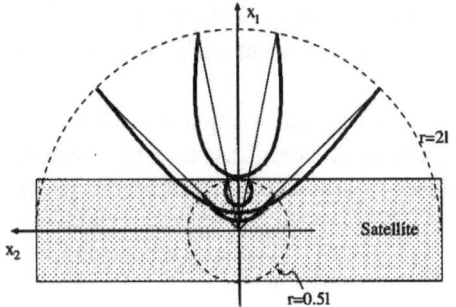 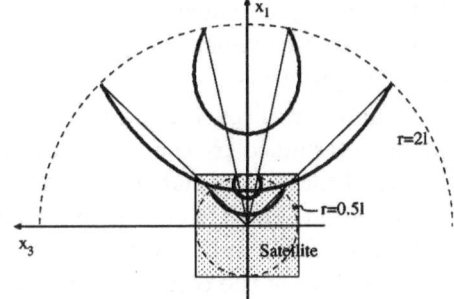

FIG. 1: Shortest Paths with Equal Initial and Final Distance

Consider now planar problems whose desired initial and final robot load positions differ only by a change in radial position. Of course, if the attitude boundary condition is ignored, then the minimum distance between these two points is a straight line. However, in this particular case, no attitude disturbance is introduced by using this straight line trajectory, and hence, it must be the minimum arc length solution. No figure is needed for this simple case.

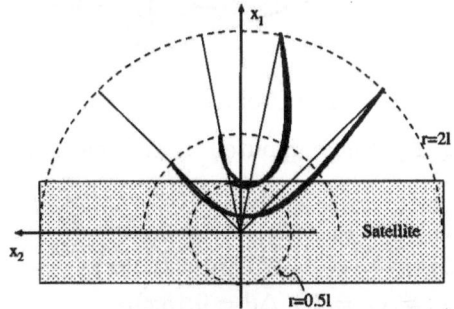 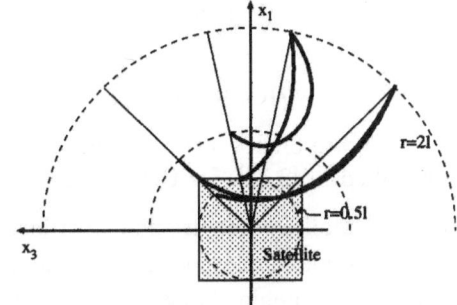

FIG. 2: Shortest Paths with Different Initial and Final Distance

The remaining planar maneuver case, when both the angle and the radial distance change, is studied in Fig. 2 which shows the shortest paths in satellite coordinates for $s_0 = 2\ell$, $\theta_0 = -\theta_T$ and $(s_T, \theta_T) \in \{\ell, \ell/2\} \times \{\pi/16, \pi/2\}$ for rotation about the $\hat{x}_3$ and the $\hat{x}_2$ axes, on the left and right of the figure, respectively. Note that there are four trajectories on the left of the figure. They are hard to distinguish without referring to the right half of the figure and to Fig. 1. Although these trajectories appear as if they may be a portion of a minimum arc length trajectory from Fig. 1 for properly chosen change in angle, this cannot be the case, since it is only at the end of the maneuver that the satellite attitude is returned to its initial value.

In the event that a minimum arc length trajectory produces a collision with the spacecraft, then it is easy to see that one can, in an ad hoc manner, insert intermediate taught points to avoid the difficulty. A more systematic and desirable approach is to use a new code, currently being generated which includes the collision inequality constraints in the minimum arc length computation. From the figures, it is the maneuvers involving large changes of angle, performed about the axis of largest moment of inertia of the satellite that are most likely to encounter collision problems.

In order to cover all possibilities, it remains to consider maneuvers where the plane determined by the initial and final desired load positions, and the system center of mass, does not happen to be a principal plane of the satellite. In order to accentuate the out of principal plane effects, minimum arc length solutions are calculated for the plane whose normal is $n = \frac{2}{3}(1, \frac{1}{2}, 1)^T$. This causes the plane to intersect two of the corners of the satellite.

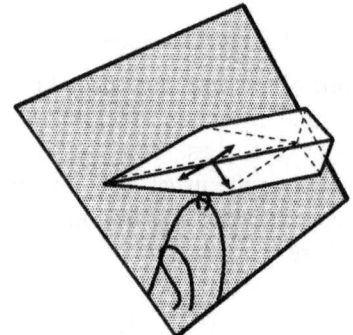 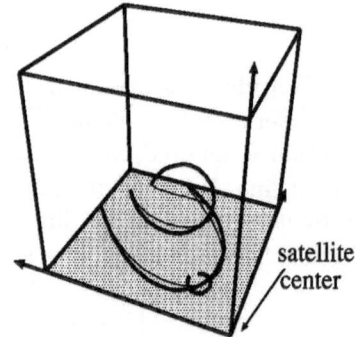

FIG. 3: Example Maneuvers for Non-Planar Problems

Figure 3 shows three minimum arc length paths, computed by the methods described in [8,13], for three cases:

(i)   a small change in angle at a large radius ($s_0 = s_T = 3\ell$, $\Delta\theta = 0.05\pi$),
(ii) a small change in angle at a small radius ($s_0 = s_T = 1\ell$, $\Delta\theta = 0.05\pi$),
(iii) a large angle change at a large radius ($s_0 = s_T = 3\ell$, $\Delta\theta = 0.15\pi$).

Here 2D-solutions calculated for $\bar{I} = n^T \hat{I} n$ serve as good initial guesses for these nonplanar problems. In the left half of the figure, the plane of interest through the satellite center of mass appears shaded, and only that part of the rectangular satellite above the plane is shown. The image of the three minimum arc length paths are shown, whether or not they are above or below the plane. In order to view the out of plane component of the paths, the right half of the figure shows both the minimizing paths and their projections onto the plane, as seen looking outward from somewhere above the satellite center of mass. Trajectories (ii) and (iii) stay quite close to the plane, and appear quite similar to these obtained above for the principal plane case. It is only the small angle change at a large radius that takes on a significant out of plane component and with a loop that appears somewhat twisted.

Conclusions

This paper develops an algorithm for space based robot path planning that is the natural extension of the most common form of path planning in ground based robots, i.e. linear interpolation between taught points. The algorithm produces paths joining the desired initial and final inertial load positions while returning the satellite attitude to its original undisturbed state by the end of the robot motion. And these paths have the property that they are the shortest possible paths as seen in robot base coordinates.

These minimum length paths are found to usually be close to planar, and to have a loop like shape. Instead of connecting the initial and final load positions with a straight line as in the ground based robot interpolation, the minimum length paths usually bring the robot load closer to the system center of mass than either the initial or the final robot position. Ground based interpolation in joint space is computationally simpler than interpolation in cartesian robot base coordinates, but in space based robotics the path planning is in fact simpler and more natural in base coordinates, as is done here. The resulting minimum length robot trajectories are independent of the robot configuration employed.

The path planning algorithm takes a matter of minutes to compute on a workstation level computer, so that it represents a practical path planning method in spite of the difficulty of the problem. The coding of the algorithm is specialized to optimal control problems, but has not been optimized for the application, so if needed one could produce substantially quicker computations. In addition, the algorithm is very naturally parallelizable, in case "real time" path planning is needed.

References

1. Longman R.W., Lindberg R.E., Zedd M.F., "Satellite-Mounted Robot Manipulators – New Kinematics and Reaction Moment Compensation", *International Journal of Robotics Research, Vol. 6, No. 3, pp. 87-103, 1987. Conference version appears in The Proceedings of the 1985 AIAA Guidance, Navigation and Control Conference, Snowmass, Colorado, pp. 278-290, 1985.*

2. Lindberg R.E., Longman R.W., Zedd M.F., "Kinematics and Reaction Moment Compensation for a Spaceborne Elbow Manipulator", *Paper No. AIAA-86-0250, AIAA Aerospace Sciences Meeting, Reno, Nevada, 1986.*

3. Lindberg, R.E., Longman, R.W., Zedd, M.F., "Kinematic and Dynamic Properties of an Elbow Manipulator Mounted on a Satellite", *The Journal of the Astronautical Sciences, Special Issue on Robotics in Space, Vol. 38, No. 4, pp. 397-421, 1990.*

4. Longman, R.W., "The Kinetics and Workspace of a Satellite-Mounted Robot", *The Journal of the Astronautical Sciences, Special Issue on Robotics in Space, Vol. 38, No. 4, pp. 423-440, 1990. Conference version appears in The*

   *Proceedings of the 1988 AIAA Guidance, Navigation and Control Conference,
   Minneapolis, Minnesota, pp. 374-381*, 1988.

5. Longman, R.W., "Attitude Tumbling Due to Flexibility in Satellite-Mounted
   Robots", *The Journal of the Astronautical Sciences, Special Issue on Robotics
   in Space, Vol. 38, No. 4, pp. 487-509, 1990. Conference version appears in The
   Proceedings of the 1988 AIAA Guidance, Navigation and Control Conference,
   Minneapolis, Minnesota, pp. 365-373*, 1988.

6. Longman R.W., Lindberg R.E., Guest Editors, Special Issue on Robotics in
   Space, *The Journal of the Astronautical Sciences, Vol. 38, No. 4*, 1990.

7. Xu, Yangsheng, Kanade, Takeo (Eds.), *Space Robotics: Dynamics and Control*,
   Kluwer Academic Publishers, 1993, (contains [3,4]).

8. Schulz V.H., Bock H.G., Longman, R.W., "Optimal Path Planning for Satellite
   Mounted Robot Manipulators", *Proceedings of the 1993 AAS/AIAA Spaceflight
   Mechanics Meeting, Advances in the Astronautical Sciences, Vol. 73, American
   Astronautical Society*, 1993.

9. Konzelmann, J., Bock, H.G., Longman, R.W., "Time Optimal Trajectories of
   Polar Robot Manipulators by Direct Methods", *Modelling and Simulation, Instrument Society of America, Vol. 20, Part 5, pp. 1933-1939*, 1989.

10. Steinbach, M., Bock, H.G., Longman, R.W., "Time Optimal Extension or Re-
    traction in Polar Coordinate Robots: A Numerical Analysis of the Switching
    Structure", *Proceedings of the 1989 AIAA Guidance, Navigation and Control
    Conference, Boston, Part 2, pp. 883-894*, 1989.

11. Konzelmann, J., Bock, H.G., Longman, R.W., "Time Optimal Trajectories of
    Elbow Robots by Direct Methods", *Proceedings of the 1989 AIAA Guidance,
    Navigation and Control Conference, Boston, Part 2, pp. 895-910*, 1989.

12. Steinbach, M., Bock, H.G., Longman, R.W., "Time Optimal Control of SCARA
    Robots", *AIAA Guidance, Navigation and Control Conference, A Collection of
    Technical Papers, Portland, Oregon, Part 1, pp. 707-716*, 1990.

13. Bock, H.G., Plitt, K.J., "A Multiple Shooting Method for Direct Solution of
    Constrained Optimal Control Problems", *Proceedings of the 9th IFAC World
    Congress on Automatic Control, Pergamon Press*, 1984.

14. Bock, H.G., *Boundary Value Problem Methods for Parameter Identification
    in Systems of Nonlinear Differential Equations*, (in German), Dissertation,
    Bonner Mathematische Schriften Nr. 183, University of Bonn, Bonn, Germany,
    1985.

International Series of Numerical Mathematics, Vol. 115, © 1994 Birkhäuser Verlag Basel

Optimal Control of the Industrial Robot Manutec r3

Oskar von Stryk*   and   Maximilian Schlemmer[†]

Abstract

Minimum time and minimum energy point-to-point trajectories for an industrial robot of the type Manutec r3 are computed subject to state constraints on the angular velocities. The numerical solutions of these optimal control problems are obtained in an efficient way by a combination of a direct collocation and an indirect multiple shooting method. This combination links the benefits of both approaches: A large domain of convergence and a highly accurate solution. The numerical results show that the constraints on the angular velocities become active during large parts of the time optimal motion. But the resulting stress on the links can be significantly reduced by a minimum energy trajectory that is only about ten percent slower than the minimum time trajectory. As a by-product, the reliability of the direct collocation method in estimating adjoint variables and the efficiency of the combination of direct collocation and multiple shooting is demonstrated. The highly accurate solutions reported in this paper may also serve as benchmark problems for other methods.

Introduction

With the increasing use of robotic manipulators the requirements of their abilities are also increasing. An essential part in design and application of robots is their dynamic behaviour. The discussion of optimal trajectories within the context of path planning and optimal design of parameters leads to the optimal control problems discussed in this paper.

Several methods for solving optimal point-to-point trajectory problems of robotic manipulators have been suggested and applied, e. g., in [7], [13], [14], [15], [23], to cite only a few of many papers.

As an extension to the previous cited work we investigate a non academic, highly nonlinear model of a commercially available robot, discuss several objectives for optimal trajectories and consider state constraints on the angular velocities that play an important role in the time optimal motion.

In our approach, we combine a direct collocation and an indirect multiple shooting method in an hybrid approach (cf. [28]) with a large domain of convergence and

---

*Department of Mathematics, Munich University of Technology, D-80290 München, Germany

†Institute for Robotics and System Dynamics, German Aerospace Research Establishment (DLR), P. O. Box 11 16, D-82230 Weßling, Germany

highly accurate solutions. The direct collocation method is easily capable to treat a wide variety of objectives and constraints on the state and control variables.

Problem Statement

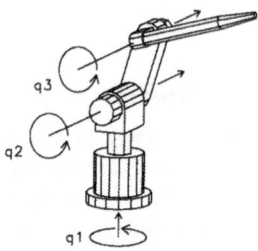

Figure 1: Three degrees of freedom in the DLR model 2 of the r3 robot.

We consider the Manutec r3 robot with 6 links. As the first 3 degrees of freedom (d.o.f.) are mainly responsible for the position and the last 3 d.o.f. for the orientation of the tool center point frame, we restrict ourself to the (first) 3 d.o.f. case. The DLR model 2 of the Manutec r3 robot was developed by Otter and Türk [19] and describes the motion of the links as a function of the control input signals of the robot drives

$$M(q(t)) \cdot \ddot{q}(t) = D \cdot u(t) + \chi^d(\dot{q}(t), q(t)) + \chi^g(q(t)), \quad t \in [0, t_f], \qquad (1)$$

where $q = (q_1(t), q_2(t), q_3(t))^T$ are the relative angles between the arms, the normalized torque controls are $u = (u_1(t), u_2(t), u_3(t))^T$, $D = diag(d_1, d_2, d_3)$ is a scaling matrix with $d_i = const[Nm/V]$, $M(q)$ is the positive definite and symmetric (3×3)-matrix of moments of inertia, $\chi^d(\dot{q}(t), q(t))$ are the moments caused by Coriolis and centrifugal forces, and $\chi^g(q(t))$ are the moments caused by gravitational forces. The final time t_f may be prescribed or free. The full data of the dynamic model can be found in [19]. Just to give an impression of the model we give the structure of the first element of the mass matrix M

$$
\begin{aligned}
M_{1,1}(q) = {} & c_1 \left(\sin(q_2 + q_3)\right)^2 + c_2 \sin(q_2 + q_3) \sin(q_2) \\
& + c_3 \left(\sin(q_2)\right)^2 + c_4 \left(\cos(q_2 + q_3)\right)^2 + c_5 \left(\cos(q_2)\right)^2 + c_6, \qquad (2)
\end{aligned}
$$

where $c_i = const$, $i = 1, \ldots, 6$, and of the driving forces $\chi^d(\dot{q}, q)$

$$\chi_i^d(\dot{q}, q) = \sum_{j=1}^{3} \left(\sum_{k=1}^{3} \Gamma_{i,jk}(q) \, \dot{q}_k \right) \dot{q}_j, \quad i = 1, 2, 3, \qquad (3)$$

$$\text{with} \quad \Gamma_{i,jk}(q) = -\frac{1}{2} \left(\frac{\partial M_{i,j}(q)}{\partial q_k} + \frac{\partial M_{i,k}(q)}{\partial q_j} - \frac{\partial M_{j,k}(q)}{\partial q_i} \right). \qquad (4)$$

The dynamic behaviour of the robot is now given either in an efficient implicit form of the right hand side of $\ddot{q} = M^{-1}(Du + \chi^d + \chi^g)$ by the subroutine R3M2SI [19] or explicitly by the output of a symbolic computation system given in the appendix of [19].

Point-to-point trajectories are to be considered, i. e.,

$$q(0) = q_0, \quad q(t_f) = q_f, \quad \dot{q}(0) = \dot{q}_0, \quad \dot{q}(t_f) = \dot{q}_f. \tag{5}$$

Here, we consider stationary boundary conditions, i. e., $\dot{q}_0 = \dot{q}_f = 0$. As objectives for optimal trajectories three criterions are investigated: The minimum time

$$J_1[u, t_f] := t_f \rightarrow \min! , \tag{6}$$

the minimum energy

$$J_2[u] := \int_0^{t_f} \sum_{i=1}^{3} (u_i(t))^2 \, dt \rightarrow \min! , \tag{7}$$

and the minimum power consumption (cf. [16], [21])

$$J_3[u] := \int_0^{t_f} \sum_{i=1}^{3} (\dot{q}_i(t) u_i(t))^2 \, dt \rightarrow \min! \tag{8}$$

The final time t_f has to be prescribed for J_2 and J_3 in order to obtain useful solutions. Otherwise, a free t_f will tend to become very large. Eighteen technical constraints have to be considered (cf. [19]): There are control constraints on the torque voltages

$$|u_i(t)| \leq u_{i,max} = 7.5[V], \quad i = 1, 2, 3, \tag{9}$$

state constraints on the angles

$$\begin{aligned} |q_1(t)| &\leq 2.97[rad], \\ |q_2(t)| &\leq 2.01[rad], \\ |q_3(t)| &\leq 2.86[rad], \end{aligned} \tag{10}$$

and state constraints on the angular velocities

$$\begin{aligned} |\dot{q}_1(t)| &\leq 3.0[rad/s], \\ |\dot{q}_2(t)| &\leq 1.5[rad/s], \\ |\dot{q}_3(t)| &\leq 5.2[rad/s]. \end{aligned} \tag{11}$$

The numerical results show that the latter constraints become often active during the time optimal motions. Thus they play an important role within the optimization.

Numerical Methods

In order to derive the necessary conditions of optimality and to apply the general numerical methods, a formal transformation of the second order system to a first order one has to be performed. First the notation

$$x = (x^1, x^2)^T = (x_1, \ldots, x_6)^T, \quad x^1 := (q_1, q_2, q_3)^T, \quad x^2 := (\dot{q}_1, \dot{q}_2, \dot{q}_3)^T \quad (12)$$

is introduced. The resulting system of first order differential equations is

$$
\begin{array}{rcl}
x_1 &=& q_1 \\
x_2 &=& q_2 \\
x_3 &=& q_3 \\
x_4 &=& \dot{q}_1 \\
x_5 &=& \dot{q}_2 \\
x_6 &=& \dot{q}_3
\end{array}
\Rightarrow
\quad
\begin{pmatrix} \dot{x}_1 \\ \dot{x}_2 \\ \dot{x}_3 \\ \dot{x}_4 \\ \dot{x}_5 \\ \dot{x}_6 \end{pmatrix}
=
\begin{cases}
\begin{pmatrix} x_4 \\ x_5 \\ x_6 \end{pmatrix} = x^2 \\[2em]
M^{-1}(x^1)\left(Du + \chi^d(x^1, x^2) + \chi^g(x^1)\right).
\end{cases}
\quad (13)
$$

For convenience, the functionals J_2 and J_3, resp., are transformed into a Mayer type functional by defining an additional state variable x_7

$$\dot{x}_7 = \begin{cases} \sum_{i=1}^{3} (u_i(t))^2, & J = J_2, \\ \sum_{i=1}^{3} (x_{i+3}(t)\, u_i(t))^2, & J = J_3, \end{cases} , \quad x_7(0) = 0, \quad \Rightarrow \quad J[u] = x_7(t_f) \to \min! \quad (14)$$

Necessary Conditions of Optimality

Necessary conditions of optimality are obtained via the minimum principle, cf., e. g., [3]. With the adjoint variables

$$\lambda = (\lambda^1, \lambda^2)^T, \quad \lambda^1 := (\lambda_1, \lambda_2, \lambda_3)^T, \quad \lambda^2 := (\lambda_4, \lambda_5, \lambda_6)^T \quad (15)$$

the Hamiltonian function of the unconstrained problem is

$$
\begin{aligned}
H^{free}(x, u, \lambda) \; :=& \; (\lambda^1)^T x^2 + (\lambda^2)^T M^{-1}(x^1)\left(Du + \chi^d(x^1, x^2) + \chi^g(x^1)\right) \\
&+ \begin{cases} 0, & J = J_1 \\ \lambda_7 \sum_{i=1}^{3} (u_i(t))^2, & J = J_2 \\ \lambda_7 \sum_{i=1}^{3} (x_{i+3}(t) u_i(t))^2, & J = J_3 \end{cases} .
\end{aligned}
\quad (16)
$$

To have a uniform treatment of active upper or lower state constraints we introduce the new state constraints

$$S_i := (q_i(t) - q_{i,max})^2, \quad S_{3+i} := (\dot{q}_i(t) - \dot{q}_{i,max})^2, \quad i = 1, 2, 3. \quad (17)$$

With the abbreviations for the total time derivatives of S

$$S_i^{(k)} := \frac{d^k}{dt^k} S_i, \quad i = 1, \ldots, 6, \quad k = 0, 1, 2, \ldots \quad (18)$$

we find

$$\frac{\partial}{\partial u_i} S_i^{(1)} = 0, \quad \frac{\partial}{\partial u_i} S_i^{(2)} \neq 0, \quad \frac{\partial}{\partial u_i} S_{3+i}^{(1)} \neq 0, \quad i = 1, 2, 3. \tag{19}$$

The functions S_i are second and the S_{3+i} are first order state constraints. Thus the Hamiltonian becomes (cf. Bryson, Denham, Dreyfus [2])

$$H(x, u, \lambda, \eta) = H^{free} + \sum_{i=1}^{3} \eta_i S_i^{(2)} + \sum_{i=1}^{3} \eta_{i+3} S_{3+i}^{(1)}, \tag{20}$$

where $\eta = \eta(t)$ is a multiplier. The necessary conditions from the minimum principle yield for the state and adjoint variables among others (cf. [20])

$$\dot{x}_i = \frac{\partial H}{\partial \lambda_i}, \quad \dot{\lambda}_i = -\frac{\partial H}{\partial x_i}, \quad \eta_i \geq 0, \quad \eta_i S_i = 0, \quad i = 1, \ldots, 6. \tag{21}$$

The optimal control is determined by the minimum principle

$$u^*(t) = arg \min_{v \in \mathcal{U}} H(x(t), v, \lambda(t), \eta(t)) \tag{22}$$

where $\mathcal{U}$ denotes the control space. If $J = J_2$ or J_3 then $\lambda_7 \equiv 1$ in $[0, t_f]$ because

$$\dot{\lambda}_7(t) = 0, \quad \lambda_7(t_f) = \frac{\partial J}{\partial x_7(t_f)}, \tag{23}$$

and $S_j^{(k)}$, $k = 0, 1, 2$, and $S_{3+j}^{(l)}$, $l = 0, 1$, $j = 1, 2, 3$, and the right hand sides of the first order differential equations (13), (14) do not depend on x_7.
Furthermore, it can be easily shown (for all three objectives) that $S_i(t) = 0$ and $S_{3+i}(t) = 0$, $i = 1, 2$, or 3, cannot occur at the same time (cf. [20], [22]).
In the time optimal motion ($J = J_1$) the controls appear linearly in H. Thus H is not regular. If no state constraint is active the i-th optimal control of bang-bang type is determined by the sign of the switching function W_i

$$W_i(t) := (\lambda^2)^T \left[M^{-1}(x^1) \right]_{i-th \ column}, \quad u_i(t) = \begin{cases} +u_{i,max}, & W_i < 0, \\ -u_{i,max}, & W_i > 0. \end{cases} \tag{24}$$

The case of a *singular* control, i. e., $W_i \equiv 0$ in a whole subinterval, did not occur in the point-to-point trajectories considered here, but might be possible, too. The minimum time t_f is determined by

$$H|_{t=t_f} = -\frac{\partial J_1}{\partial t_f} = -1. \tag{25}$$

As $dH/dt = 0$, $t \in [0, t_f]$, it follows that $H \equiv const = -1$ along the time optimal trajectory.

If the minimum power consumption $(J = J_2)$ or the minimum energy criterion $(J = J_3)$ are chosen the unbounded optimal control is determined by

$$\frac{\partial}{\partial u_i} H = 0, \quad i = 1, 2, 3, \tag{26}$$

if $S_i(t) \neq 0$ and $S_{3+i}(t) \neq 0$.

In addition, it can be shown that if $\dot{q}_i(0) = 0$ ($i = 1, 2$, or 3) then there exists $\epsilon > 0$ such that $|u_i(t)| = u_{i,max}$, for $t \in [0, \epsilon]$, and in the same way at t_f (cf. [20], [22]). If a state constraint is active, e. g., $S_i = 0$ or $S_{3+i} = 0$ then u_i is determined from $S_i^{(2)} = 0$ or $S_{3+i}^{(1)} = 0$, resp.

All in all, the necessary conditions can be stated as a well-defined multi-point boundary value problem if the optimal *switching structure* of state and control constraints is known. For more details we refer the interested reader to [20].

Multiple Shooting Method

The multiple shooting method has shown to be an effective tool in solving highly nonlinear multi-point boundary value problems. The method is described, e. g., by Bulirsch [4] and Stoer, Bulirsch [24]. Its application to a complicated state constrained optimal control problem is described by Bulirsch, Montrone, and Pesch [5]. Here, we used the code BNDSCO due to Oberle [18].

The main drawbacks when applying the multiple shooting method in the numerical solution of optimal control problems are **1.** the derivation of the necessary conditions (e. g., the adjoint differential equations), **2.** the estimation of the optimal switching structure, and **3.** the estimation of an appropriate initial guess of the unknown state and adjoint variables $x(t)$, $\lambda(t)$, $\eta(t)$ in order to start the iteration process. The great advantage of the multiple shooting method is the verification of the optimality conditions resulting in a highly accurate solution.

To overcome the first drawback a good knowledge of optimal control theory is required. Proper estimates of the switching structure and of the adjoint variables might then be provided by the use of a homotopy or continuation technique. This can be a very laborious task (cf. [5] for an example) that is especially cumbersome in our problem as none of the adjoint variables is given either at 0 or t_f. In this paper we will demonstrate how the drawbacks 2 and 3 can be overcome when a direct collocation method is used in a pre-computation to estimate the optimal switching structure, state and adjoint variables. In the derivation of the necessary conditions we used the symbolic computation method MAPLE due to Char et al. [6].

Direct Collocation Method

The basis of the direct collocation approach is a finite dimensional approximation of control *and* state variables, i. e., a discretization. Here, we choose a continuous, piecewise linear control approximation and a continuously differentiable, piecewise cubic state approximation, cf. Hargraves, Paris [11] and [25], [26], [28]. The differential equations, the state and control constraints are only pointwise fulfilled in this approach. The discretization results in a nonlinear optimization problem subject to nonlinear constraints. Convergence properties of the method and details of an efficient implementation are discussed in [26], [27]. Here, we used the code DIRCOL [27] where the resulting nonlinear programming problems are solved by the Sequential Quadratic Programming method NPSOL due to Gill, Murray, Saunders, and Wright [9]. The direct collocation method has a large domain of convergence and is easy to handle as the user has not to be concerned with adjoint variables or necessary conditions of optimality.

Combination of the Direct and the Indirect Method

Following [27] and [28] both methods are combined as follows: The direct collocation method is applied with a poor initial guess of the solution $x(t)$, $u(t)$, i. e., with an initial trajectory that interpolates the given values at initial and final time linearly. The obtained (suboptimal) solution provides reliable estimates of the state and adjoint variables and of the switching structure of state and control constraints (cf. [26], [27]). With this guess the multiple shooting method is applied to the multi-point boundary value problem resulting from the necessary conditions of optimality. For the derivation of the necessary conditions we used the symbolic computation method MAPLE [6]: The equations of motion are explicitly given in the appendix of [19] in the form of Eq. (1). First the explicit inverse of the mass matrix $M^{-1}(q)$ is computed. Then the partial derivatives of each component of the vector function $M^{-1}(Du + \chi^d + \chi^g)$ with respect to q_i, $\dot{q}_i$, $i = 1, 2, 3$, are derived. As output of the MAPLE program we obtain a FORTRAN code for M^{-1}, the adjoint differential equations, the Hamiltonian function and the formulae for the boundary controls from $S_i^{(2)} = 0$ or $S_{3+i}^{(1)} = 0$ in the state constrained case. The resulting FORTRAN codes for the inverse mass matrix are about 630 lines and for the adjoint differential equations are about 3350 lines long although in each step of the derivation several optimization strategies are applied in MAPLE to simplify the resulting formulae. For more than three degrees of freedom it might be more efficient to use automatic differentiation [10] and to make even more use of the special structure of the robotic dynamics in order to keep the number of the resulting arithmetic operations as small as possible.

Numerical Results

All computations have been performed on a SUN Sparc station 2 with 40 MHz. A special quarter rotation around the base of the robot is investigated with a load of 0 kg

$$q(0) = \begin{pmatrix} 0.00 \\ -1.50 \\ 0.00 \end{pmatrix}, \quad q(t_f) = \begin{pmatrix} 1.00 \\ -1.95 \\ 1.00 \end{pmatrix}, \quad \dot{q}(0) = 0, \quad \dot{q}(t_f) = 0. \tag{27}$$

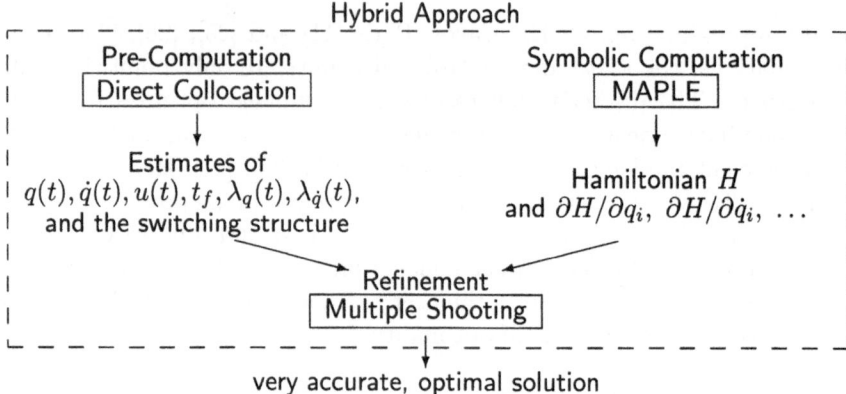

Figure 2: Combination of direct, indirect and symbolic methods in robot trajectory optimization.

The state constrained time optimal solution

The direct collocation method is at first applied to the state constrained minimum time problem with $\tilde{q}_i(t) = q(0) + (t/t_f)(q(t_f) - q(0))$, $\dot{\tilde{q}}_i(t) = 0$, $\tilde{u}_i(t) = 0$, $i = 1, 2, 3$, $t \in [0, t_f]$ and $\tilde{t}_f = 1[s]$ as initial estimates of the unknown solution. A first solution is obtained for 7 equidistant grid points. A refinement of the discretization yields a sequence of nonlinear programming problems with increasing dimensions ending up in an 81 grid point solution. The convergence history is shown in Table 1 where NGRID is the number of grid points of the direct collocation method DIRCOL, NY is the number of degrees of freedom of the nonlinear program, NLEQ is the number of nonlinear equality constraints of the nonlinear program, NITER is the number of iterations of the SQP-method NPSOL [9], NZJAC is the percentage of non zero Jacobian elements in the nonlinear program, CPU-Sec is the computing time in seconds, NDEQ is the number of differential equations of the multi-point boundary value problem of the necessary conditions, NKNOT is the number of multiple shooting nodes used in BNDSCO, NS is the number of switching points, NITER is the number of Newton iterations in BNDSCO.

Remark 1. The inequality constraints on state and control variables are treated as box constraints in the nonlinear program. Therefore no nonlinear inequality but only equality constraints appear in the discretized problem.

Remark 2. To compare the time optimal with the minimum energy solution a seventh state variable x_7 has been used in the computations. Therefore the computing time for the time optimal motion will in fact be less than the reported times if x_7 is not computed in the optimization.

Remark 3. It is a remarkable fact the the number of SQP-iterations is not increasing with the number of degrees of freedom in the nonlinear program. The increase in computing time is due to the fact that the sparsity patterns in the gradients are not yet used in the linear algebra of the quadratic programming subproblems.

NGRID	NY	NLEQ	NITER	NZJAC	CPU-Sec	t_f
7	58	42	162	33.3%	75	0.56901402
13	118	84	6	17.1%	43	0.49689417
21	198	140	9	10.4%	145	0.49574811
47	458	322	6	4.5%	960	0.49521483
81	798	560	6	2.6%	5248	0.49523283

NDEQ	NKNOT	NS	NITER		CPU-Sec	t_f
19	15	8	14	−	393	0.49518904

Table 1: Convergence history of the first time optimal motion.

sw. point	1	2	3	4	5	6	7	8
estimated	.0495	.1114	.1733	.2538	.330	.359	.390	.464
final	.04248	.10585	.16965	.25243	.3289	.3556	.3919	.4638

Table 2: Estimated and final switching points of the state constrained time optimal motion.

Much efficiency can still be gained if an appropriate sparse linear algebra is used (cf. Betts, Huffman [1] and Gill [8]).

From the solution for 81 grid points the switching structure is guessed, i. e., number and type of the switching points. With this switching structure and the state and the estimated adjoint variables from the direct collocation method the multiple shooting method is successfully applied to solve the multi-point boundary value problem of optimality conditions. The solution of the direct collocation method is shown with the solution of the multiple shooting method in Figure 3, the initial guess and the finally obtained switching points are listed in Table 2, and the qualitative behaviour of the switching structure is shown in Figure 4. The three dimensional motion of the robot is shown in Figure 6. **Remark 4.** The oscillating behaviour of the discretized controls results from the oscillating behaviour of the only pointwise fulfilled state constraints on $\dot{q}_i$. This is a common behaviour of direct collocation or direct shooting methods. Also, when entering the constraint on $\dot{q}_i$ the control u_i is not continuous and the angular velocity $\dot{q}_i$ is only continuous but not differentiable at the entry (and exit) point of the state constraint. Therefore a better approximation of state and control variables can be obtained either by increasing the number of grid points to a huge number or by taking the *switching structure* into account in a so-called *multi-stage* or *multi-phase* discretization, too ([26], [27]). I.e., the switching points are introduced as additional variables and the controls are allowed to be changed discontinuous and the states to be changed not differentiable at the switching points. This adopted

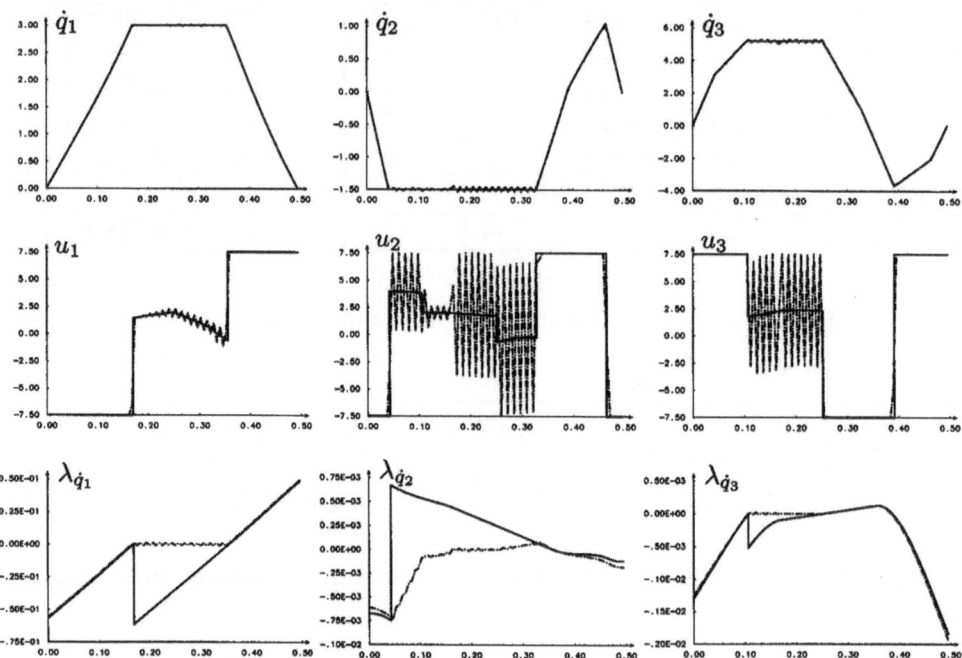

Figure 3: Solution curves for the state constrained time optimal motion of the direct collocation method $(-\cdot-\cdot-)$ compared with the multiple shooting method (———). In the curves of the angles $q_i(t)$ there are no visible differences between the two solutions. They are shown in Figure 5.

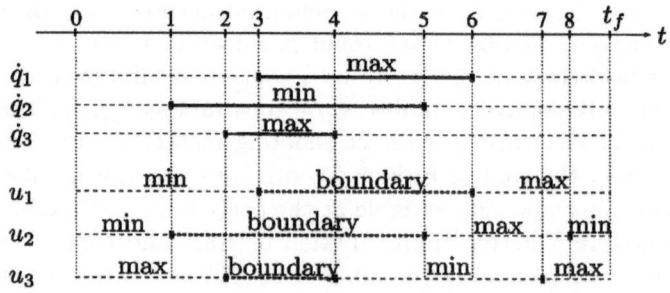

Figure 4: Switching structure of the state constrained time optimal motion.

trajectory\criterion	t_f	$\int_0^{t_f} \sum_i u_i^2(t)dt$	$\int_0^{t_f} \sum_i (q_i(t)u_i(t))^2 dt$
state cons. min. time	0.49518904*	51.351466	248.56509
uncons. min. time	0.44551780*	75.179701	906.30989
min. energy	0.53000000	20.404247*	43.089470
min. power consump.+	0.53000000	28.057499	35.911668*

Table 3: Comparison of results of the multiple shooting method for different objectives (correct figures of the direct collocation method underlined, * denotes the optimum value, + solution only computed by direct collocation).

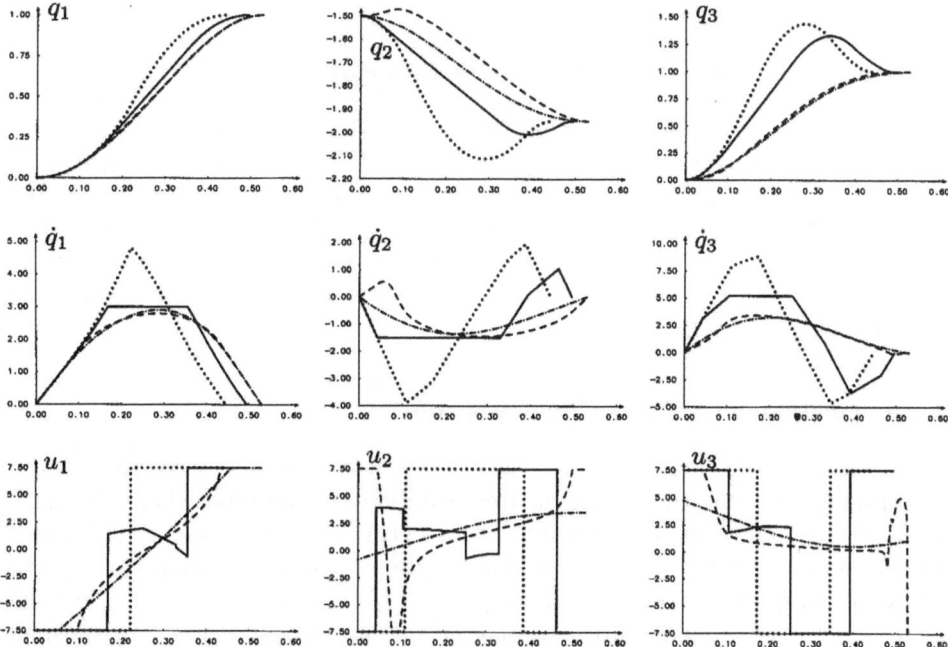

Figure 5: Solution curves of state and control variables for the state constrained time optimal motion (———), the minimum power consumption (– – –), the minimum energy motion (–·–·–), and the unconstrained time optimal motion (·····).

discretization results in a more accurate solution of the direct collocation method for the controls and the objective value with even less grid points (cf. [26] for an example and [27] for more details). The non adopted standard discretization is shown in Figure 3 to demonstrate the oscillating behaviour on state constraint subarcs.

Remark 5. The adjoint variables and their estimates seem to differ significantly on state constrained subarcs (cf. Figure 3). The estimates of adjoint variables obtained by the direct collocation method are related to the adjoint variables from the necessary conditions of Jacobson, Lele, and Speyer [12] where the state constraints themselves are coupled to the free Hamiltonian with a multiplier function $\bar{\eta}$

$$H(x, u, \lambda, \bar{\eta}) = H^{free} + \sum_{i=1}^{3} \bar{\eta}_i S_i + \sum_{i=1}^{3} \bar{\eta}_{i+3} S_{3+i}. \tag{28}$$

For the formulation of the multi-point boundary value problem the necessary conditions of Bryson, Denham, and Dreyfus (Eq. (20)) have been used. Both sets of adjoint variables differ only along the state constrained subarcs and can be transformed into each other [17], [27].

Solutions for the integral performance criterions

When the state constrained minimum time motion has been computed, solutions to the minimum power consumption and the minimum energy criterion can be computed to a prescribed final time that is only about 10-15 % slower than the minimum time and the constraints on the angular velocities do not become active. Here, we choose $t_f = 0.530[s]$ that is 7 % slower than the minimum time. The solution curves are shown in Figure 5 and compared with the minimum time solution.

The unconstrained time optimal solution

To analyze the impact of the constraints on the angular velocities on the time optimal motions, the minimum time solution is computed where the state constraints are not taken into account. The resulting minimum time is 10 % faster than the state constrained minimum time solution. But this solution violates the constraints on $\dot{q}_i(t)$, $i = 1, 2, 3$, and on $q_2(t)$ and is shown in Figure 5. Thus the constraints on the angular velocities play an important role in the time optimal motion.

Remark 6. All solutions shown in Figure 5 have been obtained by the combination of the direct and the indirect method besides the solution for the minimum power consumption criterion. The only visible significant differences between the solutions of both methods are in the state constrained time optimal motion and are shown in Figure 3. The differences in the objective values are listed in Table 3.

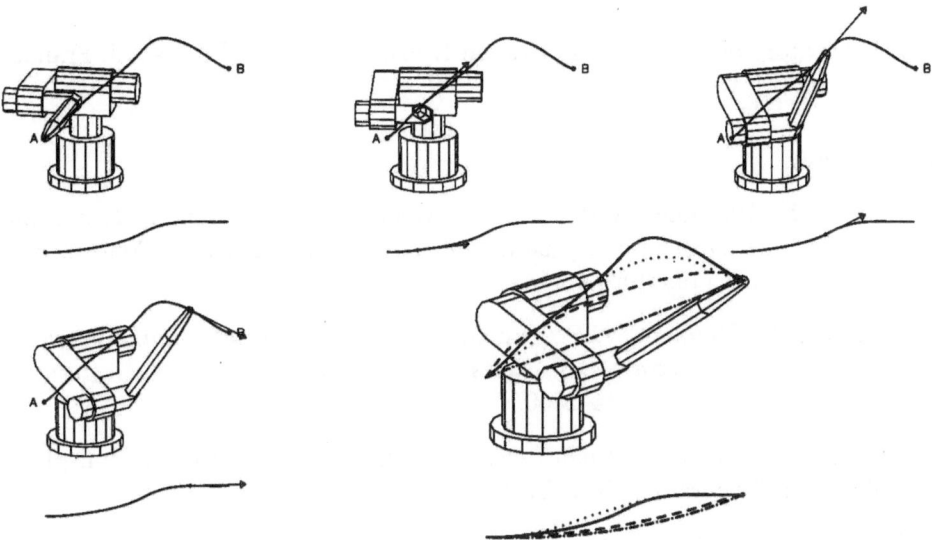

Figure 6: Five snapshots of the state constrained time optimal motion from A $(t = 0)$ to B $(t = t_f)$ at $t = i\,t_f/4$, $i = 0, 1, 2, 3, 4$, and a comparison of the four trajectories.

Conclusions

It has been demonstrated that even a simple manoeuvre can exhibit quite a difficult switching structure in the time optimal motion of an industrial robot. The state constraints on the angular velocities play an important role in the time optimal motion as they often become active. The knowledge of the fastest possible motion provides reliable bounds for fast minimum energy motions. Hereby, the stress on the links of the robot can be significantly decreased if an increase in time of about ten percent is accepted. Thus lifetime and reliability of the robot will increase.

The second link of the Manutec r3 robot is the weakest. This is indicated also by several other time optimal movements investigated by the authors where the constraints on $\dot{q}_2(t)$ become active during most parts of the motions. Thus a better design of robots might be possible if the investigation of optimal trajectories is included in the development phase.

The combination of direct and indirect methods, namely direct collocation and multiple shooting, is an efficient hybrid approach for solving highly complex, nonlinear, real life optimal control problems that amalgamates the benefits of both methods.

Acknowledgement. The authors acknowledge the helpful discussions with Priv.-Doz. Dr. H.J. Pesch and the colleagues from the Numerical Analysis and Optimal Control Group of Prof. R. Bulirsch at the Munich University of Technology and the valuable support by the colleagues from the Robotics Group of Prof. G. Grübel at the Laboratory of Robotics and System Dynamics of the DLR, esp. J. Franke, S. Lewald, and M. Otter.

References

[1] Betts, J.T., Huffman, W.P. *Path constrained trajectory optimization using sparse sequential quadratic programming.* AIAA J. Guidance, Control, and Dynamics **16**, 1 (1993) 59–68.

[2] Bryson, A.E., Denham, W.F., Dreyfus, S.E. *Optimal programming problems with inequality constraints. I: Necessary conditions for extremal solutions.* AIAA J. **1**, 11 (1963) 2544–2550.

[3] Bryson, A.E., Ho, Y.-C. *Applied Optimal Control.* Rev. Printing. (Hemisphere Publishing Corporation, New York, 1975).

[4] Bulirsch, R. *Die Mehrzielmethode zur numerischen Lösung von nichtlinearen Randwertproblemen und Aufgaben der optimalen Steuerung.* Report of the Carl-Cranz Gesellschaft, Oberpfaffenhofen, Germany (1971). Reprint: Department of Mathematics, Munich University of Technology, Germany (1993).

[5] Bulirsch, R., Montrone, F., Pesch, H.J. *Abort landing in the presence of windshear as a minimax optimal control problem.*
Part 1: Necessary conditions. JOTA **70** (1991) 1–23.
Part 2: Multiple shooting and homotopy. JOTA **70** (1991) 223–254.

[6] Char, B.W., Geddes, K.O, Gonnet, G.H., Leong, B.L., Monagan, M.B., Watt, S.M. *Maple V, language reference manual.* (New York/Berlin/Heidelberg: Springer, 1991).

[7] Chen, Y.-C. *Solving robot trajectory planning problems with uniform cubic B-splines.* Opt. Contr. Appl. and Meth. **12** (1991) 247–262.

[8] Gill, P.E. *Large-scale SQP methods and their application in trajectory optimization.* Proc. of the 9th IFAC Workshop on *Control Applications of Optimization*, Fachhochschule München, 1992, ed. by R. Bulirsch and D. Kraft.

[9] Gill, P.E., Murray, W., Saunders, M.A., Wright, M.H. *User's guide for NPSOL (Version 4.0).* Report SOL 86-2. Department of Operations Research, Stanford University, California, USA (1986).

[10] Griewank, A. *Automatic evaluation of discrete adjoints with logarithmic increase in storage.* Proc. of the 9th IFAC Workshop on *Control Applications of Optimization*, Fachhochschule München, 1992, ed. by R. Bulirsch and D. Kraft.

[11] Hargraves, C.R., Paris, S.W. *Direct trajectory optimization using nonlinear programming and collocation.* AIAA J. Guidance **10**, 4 (1987) 338–342.

[12] Jacobson, D.H., Lele, M.M., Speyer, J.L. *New necessary conditions of optimality for control problems with state-variable inequality constraints.* Journal of Mathematical Analysis and Applications **35** (1971) 255–284.

[13] Johanni, R. *Optimale Bahnplanung bei Industrierobotern.* Fortschrittsberichte VDI, Reihe 18, Nr. 51, 1988.

[14] Kraft, D. *TOMP – FORTRAN Modules for Optimal Control Calculations.* Fortschrittsberichte VDI, Reihe 8, Nr. 254 (1991).

[15] Lee, A.Y. *Solving constrained minimum-time robot problems using the sequential gradient restoration algorithm.* Opt. Contr. Appl. and Meth. **13** (1992) 145–154.

[16] Lewald, S.A. *Generierung von Robotertrajektorien für Industrieroboter mit 6 Freiheitsgraden.* Diploma thesis, Lehrstuhl für Steuerungs- und Regelungstechnik, Munich University of Technology, Germany (1990).

[17] Maurer, H. *Optimale Steuerprozesse mit Zustandsbeschränkungen.* Habilitationsschrift, University of Würzburg, Würzburg, Germany (1976).

[18] Oberle, H.J. *Numerische Berechnung optimaler Steuerungen von Heizung und Kühlung für ein realistisches Sonnenhausmodell.* Habilitationsschrift, Munich University of Technology, Germany (1982).

[19] Otter, M., Türk, S. *The DFVLR Models 1 and 2 of the Manutec r3 Robot.* DFVLR-Mitt. 88-13, Institut für Dynamik der Flugsysteme, Oberpfaffenhofen, Germany (1988).

[20] Pesch, H.J., Schlemmer, M., von Stryk, O. *Minimum-energy and minimum-time control of three-degrees-of-freedom robots. Part 1: Mathematical model and necessary conditions, Part 2: Numerical methods and results for the Manutec r3 robot.* In preparation.

[21] Pfeiffer, F., Reithmeier E. *Roboterdynamik.* (Teubner: Stuttgart, 1987).

[22] Schlemmer, M. *Zeit- und energieminimale Steuerung von Industrierobotern mit 3 Freiheitsgraden am Beispiel des Manutec r3.* Diploma thesis, Department of Mathematics, Munich University of Technology (1992).

[23] Steinbach, M., Bock, H.G. *Time-optimal extension or retraction in polar coordinate robots: A numerical analysis of the switching structure.* Proc. of the AIAA Guidance, Navigation and Control Conference, Boston, USA, AIAA Paper 89-3529-CP (1989) 883–894.

[24] Stoer, J., Bulirsch, R. *Introduction to Numerical Analysis.* 2nd ed. (Springer, 1993).

[25] von Stryk, O. *Ein direktes Verfahren zur Bahnoptimierung von Luft- und Raumfahrzeugen unter Berücksichtigung von Beschränkungen.* Z. angew. Math. Mech. **71**, 6 (1991) T705–T706.

[26] von Stryk, O. *Numerical solution of optimal control problems by direct collocation.* To appear in: R. Bulirsch, A. Miele, J. Stoer, K.-H. Well (eds.) *Optimal Control*, Proceedings of the conference on Optimal Control and Variational Calculus, Oberwolfach, 1991 (International Series of Numerical Mathematics, Birkhäuser).

[27] von Stryk, O. *Numerische Lösung optimaler Steuerungsprobleme: Diskretisierungen, Parameteroptimierung und Schätzung von adjungierten Variablen.* In preparation as Doctoral thesis.

[28] von Stryk, O., Bulirsch, R. *Direct and indirect methods for trajectory optimization.* Annals of Operations Research **37** (1992) 357–373.

ISNM

A series with a long-standing reputation

Since its foundation in 1963 more than 100 volumes have been published by Birkhäuser Verlag in the **International Series of Numerical Mathematics.**

John Todd's *Introduction to the Constructive Theory of Functions,* published as Volume 1, was a remarkable start. Proceedings volumes and further monographs such as Fenyö/Frey, *Moderne mathematische Methoden in der Technik,* Ghizzetti/Ossicini, *Quadrature Formulae,* Todd, *Basic Numerical Mathematics* (two volumes) and Heinrich, *Finite Difference Methods on Irregular Networks* followed, always presenting the state of the art in exposition and research.

Originally the Editorial Board consisted of Ch. Blanc, A. Ghizzetti, A. Ostrowski, J. Todd, H. Unger, A. van Wijngaarden. Despite a number of changes, it has shown long years of continuity; Prof. Ostrowski and Prof. Henrici, for instance, were members of the Board all their lives.

At present the series is being edited by
Karl-Heinz Hoffmann, München, **Hans D. Mittelmann**, Tempe, **John Todd**, Pasadena.

As in the past, we do not intend to restrict the series a piori to certain subjects. The series is open to all aspects of numerical mathematics with emphasis on mathematical content. At the same time, we wish to include practical applications in science and engineering, with emphasis on mathematical content.

Some of the topics of particular interest to the series are:
Free boundary value problems for differential equations, phase transitions, problems of optimal control, other nonlinear phenomena in analysis; nonlinear partial differential equations, efficient solution methods, bifurcation problems; approximation theory.

If possible, the topic of each volume should be discussed from three different angles, namely that of Mathematical Modelling, Mathematical Analysis, Numerical Case Studies.

The editors particularly welcome research monographs; furthermore, the series is to contain advanced graduate texts, dealing with areas of current research interest, as well as selected and carefully refereed proceedings of major conferences or workshops sponsored by various research centers. Historical material in these areas would also be considered.

We encourage preparation of manuscripts in LaTeX or AMSTeX for delivery in camera-ready copy which enables a rapid publication, or in electronic form for interfacing with laser printers of typesetters.

Titles previously published in the series

INTERNATIONAL SERIES OF NUMERICAL MATHEMATICS
BIRKHÄUSER VERLAG

ISNM 100   W. Desch, F. Kappel, K. Kunisch (Eds.): Estimation and Control of Distributed Parameter Systems, 1991 (3-7643-2676-X)

ISNM 101   G. Del Piero, F. Maceri (Eds.): Unilateral Problems in Structural Analysis IV, 1991 (3-7643-2487-2)

ISNM 102   U. Hornung, P. Kotelenez, G. Papanicolaou (Eds.): Random Partial Differential Equations, 1991 (3-7643-2688-3)

ISNM 103   W. Walter (Ed.): General Inequalities 6, 1992 (3-7643-2737-5)

ISNM 104   E. Allgower, K. Böhmer, M. Golubitsky (Eds.): Bifurcation and Symmetry, 1992 (3-7643-2739-1)

ISNM 105   D. Braess, L.L. Schumaker (Eds.): Numerical Methods in Approximation Theory, Vol. 9, 1992 (3-7643-2746-4)

ISNM 106   S.N. Antontsev, K.-H. Hoffmann, A.M. Khludnev (Eds.): Free Boundary Problems in Continuum Mechanics, 1992 (3-7643-2784-7)

ISNM 107   V. Barbu, F.J. Bonnans, D. Tiba (Eds.): Optimization, Optimal Control and Partial Differential Equations, 1992 (3-7643-2788-X)

ISNM 108   H. Antes, P.D. Panagiotopoulos: The Boundary Integral Approach to Static and Dynamic Contact Problems. Equality and Inequality Methods, 1992 (3-7643-2592-5)

ISNM 109   A.G. Kuz'min: Non-Classical Equations of Mixed Type and their Applications in Gas Dynamics, 1992 (3-7643-2573-9)

ISNM 110   H.R.E.M. Hörnlein, K. Schittkowski (Eds.): Software Systems for Structural Optimization, 1992 (3-7643-2836-3)

ISNM 111   R. Burlisch, A. Miele, J. Stoer, K.H. Well: Optimal Control, 1993 (3-7643-2887-8)

ISNM 112   H. Braess, G. Hämmerlin (Eds.): Numerical Integration IV. Proceedings of the Conference at the Mathematical Research Institute at Oberwolfach, November 8-14, 1992, 1993 (3-7643-2922-X)

ISNM 113   L. Quartapelle: Numerical Solution of the Incompressible Navier-Stokes Equations, 1993 (3-7643-2935-1)

ISNM 114   J. Douglas, U. Hornung (Eds.): Flow in Porous Media. Proceedings of the Oberwolfach Conference, June 21-27, 1992 (ISBN 3-7643-2949-1)

ISNM 115   R. Bulirsch, D. Kraft (Eds.): Computational Optimal Control, 1994 (ISBN 3-7643-5015-6)

Progress in Systems and Control Theory

Series Editor
Christopher I. Byrnes, Washington University, St. Louis, USA

Progress in Systems and Control Theory is designed for the publication of workshops and conference proceedings, sponsored by various research centers in all areas of systems and control theory, and lecture notes arising from ongoing research in theory and applications control.

We encourage preparation of manuscripts in such forms as LATEX or AMS TEX for delivery in camera-ready copy which leads to rapid publication, or in electronic form for interfacing with laser printers.

Systems and Control:
Foundations and Applications

A series of monographs and advanced graduate texts

Edited by
Christopher Byrnes, Washington University, St. Louis, MO, USA

Systems and Control is designed for the publication of research level monographs and advanced graduate textbooks in all areas of systems and control theory and its applications to a wide variety of scientific disciplines.

Asachenkov, A. / Marchuk, G. / Mohler, R. / Zuev, S.: Disease Dynamics (ISBN 3-7643-3692-7)

Aubin, J.-P. : Viability Theory (ISBN 3-7643-3571-8)

Aubin, J.-P. / Frankowska, H.: Set–Valued Analysis / SC 2 (ISBN 3-7643-3478-9)

Banks, H.T. / Kunisch, K.: Estimation Techniques for Distributed Parameter Systems / SC 1 (ISBN 3-7643-3433-9)

Basar, T. / Bernhard, P.: H^∞–Optimal Control and Related Minimax Design Problems (ISBN 3-7643-3554-8)

Bensoussan, A. / Da Prato, G. / Delfour, M.C. / Mitter, S.K.:
Representation and Control of Infinite Dimensional Systems
Volume I (ISBN 3-7643-3641-2)
Volume II (ISBN 3-7643-3642-0)

Chen, H.F. / Guo, L.: Identification and Stochastic Adaptive Control (ISBN 3-7643-3597-1)

Falb, P.: Methods of Algebraic Geometry in Control Theory I. Scalar Linear Systems
and Affine Algebraic Geometry / SC 4 (ISBN 3-7643-3454-1)

Keulen, B. van: H_∞-Infinity Control for Distributed Parameter Systems:
A State-Space-Approach (ISBN 3-7643-3709-5)

Kushner, H.-J.: Weak Convergence Methods and Singularly Perturbed Stochastic Control
and Filtering Problems / SC 3 (ISBN 3-7643-3437-1)

Zabczyk, J.: Mathematical Control Theory (ISBN 3-7643-3645-5)